ROYAL HORTICULTURAL SOCIETY

ENCYCLOPEDIA OF
GARDEN DESIGN

ROYAL HORTICULTURAL SOCIETY

ENCYCLOPEDIA OF
GARDEN
DESIGN

EDITOR-IN-CHIEF
CHRIS YOUNG

Contents

Foreword

Have you ever sat – just sat – in your garden, thinking, looking around, taking in the view? Not really looking at anything in particular, but thinking about anything and everything to do with your garden, asking yourself, "what if I planted a tree there?", or "if I moved those paving slabs, what would I put in their place?". Whether you were aware of doing this or not is, in a way, immaterial because what you have been doing is visually making this piece of land your own, and coming up with thoughts and ideas for improving your outside space. Welcome then – whether it be for the first or fiftieth time – to the world of garden design.

The concept of garden design is nothing new: when Man first cultivated land, and enclosed his arable crops and livestock, he was delineating usable space to its best advantage. This may not be design as we understand it now (obviously, aesthetics were of no practical value then), but he was making spatial relationships based on need. He was designing his environment to suit his individual daily, monthly, seasonal, and yearly requirements.

Since that time, the process of creating a garden has evolved according to style, fashion, prowess, skill, aptitude, wealth, travel, experimentation and history, but it can all be distilled down to that first need. In essence, garden making is all about a human being exerting some level of control over his or her own surroundings. And, really, that is all garden design is today.

As is set out by my fellow authors in this book, creating a garden can be an intricate and time-consuming process, but the fundamental starting point is to remember that garden design is about creating an outside space that you (or your client) want. Many discussions will ensue after that initial thought – from what style you want, to working out how sustainable your garden might be. But don't let the detail bog you down too

Welcome in
Successful garden design is about creating usable, attractive, and well-made spaces that suit the owner's personal needs.

much or too early in the process. Of course detail is essential for a successful garden, but holding on to that vision, that desire, is a key part of the process. This book will help you, not only with the nuts and bolts of garden making, but also to focus the vision and, I hope, help make it become a reality.

So why is there still a need for an encyclopedia such as this? In truth, because designing a garden can be something of a lonely experience. Even though we are constantly bombarded with images, suggestions, and information (books, internet, social media, and magazines), it is rare to be able to look in one place for everything – from plant selection to gravel colour, from fence posts to tree heights. The very nature of having so much choice can render the designer/gardener/client more than a little confused as to what they actually want from their garden. The activity of making a garden can also be influenced from so many quarters – by plants or hard materials – that a designer needs a refuge of sorts, where questions are answered and problems resolved. I hope this book will be that refuge in this ever-crowded, information-heavy world.

▽ **Plan your plan**
Putting your ideas onto paper, or computer, is an essential step when designing your garden.

▽▽ **Good form**
Successful designs use flower colour, leaf shape and tree stems to create a balance of colour and form.

Personal space
Good design should reflect the wishes, likes, and dislikes of the garden owner – regardless of the country or climate.

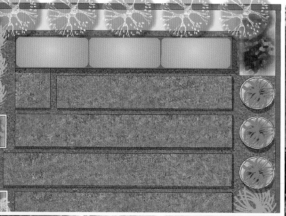

Often, coming up with an overarching vision for what you want your garden to be like is the easiest part of the process. It is translating that vision into a reality that takes the bulk of the time: working out how parts of a garden can sit together, how planting interest throughout the year can be sustained, deciding on hard landscaping materials that will work in all weather conditions, and so on. These are the stimulating – and at times frustrating – aspects of the process, but they make the difference between an unusable piece of land adjoining your property and a beautifully designed garden.

The chapters in this book take you through these very stages of garden design, helping to demystify the unknowns and clarify the unclear. I sincerely hope you enjoy it and, as a result, make the best garden you possibly can.

Chris Young
Editor-in-Chief

Considered style
Successful spaces are created when planting colours and combinations complement the hard landscaping materials.

▽ **Urban jungle**
Using foliage plants of different types and heights can help to provide privacy from neighbouring views and offers useful shelter.

▽▽ **Sense of scale**
When creating a plan, working to a scale allows you to be sure that all structures and details will work well on the ground.

Eye of the beholder
Sometimes, beautiful design expressions can be created by mirroring shapes, like this sculpture and round-flowered Allium.

What do you want to do in your garden?

Your garden is an extension of your home and it should provide a place for you to enjoy life to the full. When thinking about any changes that you plan to make to the garden, it is important to consider how you want to use the space, and not just at the current time but in the future as well. This can range from keeping very busy, to doing as little as possible. Ask yourself a series of questions about the garden's

GET INVOLVED

ENJOY THE PLANTS AND WILDLIFE

The active gardener
Digging, sowing and planting bring great rewards as plants grow and change throughout the seasons. Colours and textures evolve, and there is something new to see each week. Plants attractive to birds, bees and butterflies bring borders to life.

ENTERTAIN AND HAVE FUN

The room outside
Gardens are often described as "outdoor rooms" and can be planned as extensions of the house. Ensure continuity with features such as stylish furniture, screens, painted walls, canopies and planters. An open-air room can be used for entertaining and socializing, while also offering children space for energetic play.

many roles. Do you want a space for entertaining, a play area while the children are young, or do you simply want a peaceful but beautiful garden in which to relax? Bear in mind that your needs, and those of your family, are likely to change with time, and that it may be more difficult to make

significant changes to the garden in the future as it establishes and matures. Ideally, come up with flexible ideas that can be adapted. A range of different requirements might suggest the creation of separate and possibly hidden areas within the same garden.

APPRECIATE THE PICTURE

RELAX AND UNWIND

Simple solution
Gardens for busy people need to be easy to maintain, but they can still be lovely to look at. They require simple design solutions with a strong overall concept and a pleasing layout for long-term appeal, allowing owners to sit back and enjoy the view.

A peaceful space
One of the special joys of having a garden is that you can simply sit, doze, read or do nothing in the open air, surrounded by the sounds and scents of plants and wildlife. Gardens designed for this purpose can provide the perfect antidote to the stresses and strains of everyday life.

How do you want to feel?

Gardens stimulate emotions: upon entering a space people immediately respond to it. When planning a new design, you may choose to be bombarded with sensory stimulation, a riot of vibrant colour, textural diversity, or striking features to excite and energize. Or you might want a place for quiet reflection and contemplation, or perhaps even a space for therapy and healing, such as a calm, simple garden with

ENERGIZED

EXCITED AND UPBEAT

REJUVENATED

The dynamic garden
Exciting, stimulating sensations can be created using vibrant, hot colours, spiky plants, sharp lines, challenging artwork, varied textures, and bold use of lighting. But be warned: strident garden designs can be overpowering.

Refreshing space
The presence of water, creating sunlit reflections and offset by natural plantings, can help to evoke a feeling of energy, growth and rejuvenation. Soft colours and a complementary selection of natural materials enhance the mood. These are places for "recharging your batteries" after a long day at work.

evergreen trees and bushes, and a reflective pool. If you have enough outdoor space to play with, it may be possible to demarcate different areas for different moods by making effective use of screening or tall plants. Creating a new design for a garden provides an opportunity to change or enhance the atmosphere of each area through layout, distribution of paths and spaces, and light touches of detail and decoration. Colour, shape, fragrance, and foliage will also affect the tone, and by using these elements you can help to foster positive moods and emotions.

A SENSE OF WELL-BEING

PEACEFUL AND CALM

Restoring health
These gardens should be private, unchallenging spaces, and are often characterized by culinary, therapeutic and medicinal plants, such as herbs with their appealing scents, or healthy crops such as fruit trees. They provide a reassuring, relaxed and restorative environment.

Contemplative moods
Cool colours, simple flowing shapes, delicate scents, and restricted use of materials and planting will create a calm and peaceful mood in the garden. Simple focal elements, waterfalls, and carefully chosen lighting help to enhance these uncluttered spaces.

What will your garden look like?

Garden visits, shows and plant nurseries, as well as magazines, books, television programmes, social media, and websites, will provide anyone wishing to change their garden with a wealth of inspiration.

But remember, the key to successful design is not just collecting ideas and trying to combine them in your design. Rather, it is a process of reviewing and editing a range of ideas, with the aim of

TRADITIONAL

FILLED WITH FLOWERS

Grow your favourite flowers
Your garden can be a horticultural extravaganza, or a setting for favourite plants. These gardens are seasonal and offer change and continuous involvement. Try to work to a clear overall concept in terms of colour, texture, and structure.

A TROPICAL RETREAT

Sculpt with plants
Bold-leaved plants bring a sense of the exotic and can be used to create a lush, enclosed garden with a subtropical feel. Choose plants carefully to ensure that they will not get too big and are suited to your site's soil and climate.

A HINT OF HOLIDAYS

Recreate a summer break
Why limit your holiday to a fortnight, when you can pretend to be on a summer trip all year? Adapt ideas seen on your travels: for example, fragrant lavender beds and window boxes brimming with ivy-leaved geraniums for echoes of southern France.

developing a coherent overall appearance for your garden, whether you are revamping a mature plot or starting with a blank canvas at a new house. A good way of approaching this is to have a clear image of the look you are hoping to achieve and to carefully select elements, features, materials and plants that combine to produce a unified composition. Make notes, collect pictures, sketch ideas. Some starting points are given below, from the traditional to the modern, to the imaginative and quirky. Use them as a prompt to see which style suits you best.

A SPACE TO REFLECT

Make a sanctuary
A tranquil setting, characterized by straight lines, simple shapes, subtle lighting and a coherent layout, provides a comfortable space for retreat from modern-day life. Avoid clashing materials and keep planting simple.

CHIC AND MINIMAL

Cut out the clutter
Restrict yourself to no more than three complementary materials and a muted colour palette, but combine them beautifully. A large, dramatic water feature or sculpture adds a dynamic quality to a pared-down design.

FUN AND FUNKY

Show your creative side
Perhaps better suited to show gardens or temporary installations, these quirky gardens are attention-grabbing but require artistic flair and confidence to be successful. Not for the shy or retiring, but they can be great fun while they last.

How much do you want to do?

The amount of time you have to devote to your garden on a daily, weekly, or monthly basis should be a major consideration when thinking about an overall design and its future maintenance. Unless you have a very simple, easy-care garden, with hard landscaping and evergreen planting, the list of tasks normally changes seasonally, with less to do in the cooler winter months. In a high-maintenance garden with

GET INVOLVED

EVERY DAY

Regular upkeep
Most small gardens will not need attention more than two or three times a week at most, although a plot filled with lots of containers will require daily watering in hot, dry spells. Generally, larger gardens with lawns, mixed borders, a diverse range of plants and productive growing areas will take up more of your time.

ONCE A WEEK

The weekend gardener
This is the most common category, especially for people who only have spare time at weekends. Formal lawns require weekly mowing and edge-trimming in summer, and weeds need to be kept in check throughout the garden.

mixed flower borders, lawns, fruit trees and a vegetable plot, spring and summer are very busy seasons. Lawn-mowing, hedge-trimming, pruning and feeding fruit trees, sowing and transplanting vegetables, plant propagation and ongoing cultivation, all take time. This may be the garden you want, but be realistic about how much time you can spare to keep it looking good. Working in your garden, watching it mature and admiring the results, is immensely pleasurable, but do plan for maintenance in advance, and budget to bring in help if necessary.

TWICE A MONTH

SIX TIMES A YEAR

Keep it practical
Most shrubs, climbers, and perennial plants require attention at intervals. Seasonal pruning may be required in spring and autumn, borders need weeding and feeding, and flowering plants such as roses should be dead-headed regularly. Lawns are impractical in this category, although meadows are an option.

Minimal maintenance
Gardens requiring infrequent attention will exclude lawns and hedges. Plan for "low" rather than "no" maintenance. Many trees and shrubs need only an annual tidy-up, and hard landscaping just occasional attention, such as sweeping or cleaning.

First principles

Designing your garden is all about finding solutions. It can seem daunting at first, but if you start with a clear idea of your aspirations and practical needs, your basic design will soon begin to take shape.

Begin by pulling together all your inspirations, using magazines, photographs, and online sources to create a book or folder of ideas. Your images may include plants and landscapes you love, and perhaps furniture or art you admire. To help clarify your thoughts, you could then draw a simple bubble diagram that identifies areas for different activities, such as eating, seating, or play space for the children.

The routes of paths, shapes of structures, and the spaces between elements all have an impact on the look and feel of a design, and need to be considered before you draw up a finished plan. For example, sinuous paths and organic shapes combine to create relaxed and informal designs, whereas straight paths and symmetrical layouts convey a formal look.

A strong pattern unifies different materials.

Every site will have its own particular challenges, whether your garden is on a steep slope and needs terracing, or if it is tiny or an awkward shape.

Whatever the problem, an understanding of how to use lines, shapes, height, structure and perspectives will help. You can also employ a range of techniques to lead or deceive the eye, creating an illusion of space in a small garden, or diverting attention to focus on specific features.

When it comes to creating atmosphere and mood, the colours, patterns, and textures that you choose have a powerful impact. Colour also affects the impression of size and space in the garden – cool blues and whites tend to make an area feel bigger; warm reds and yellows make spaces appear lively and more compact. Pale colours and white reflect light into gloomy plots. Texture can be used to great effect, too, creating exciting contrasts by combining rough with smooth, or shiny with matt.

There are no rights or wrongs in the world of garden design, so have fun and experiment.

Plans help you to organize design ideas.

Understanding plans

A plan is a two-dimensional representation of a three-dimensional garden and provides a useful thinking tool. It allows you to develop and share ideas easily with others about how your space can be organized and where various elements should be located. You can produce a simple sketch or a more detailed, scale plan to illustrate your design; the plans shown here explain the different types and how to use them.

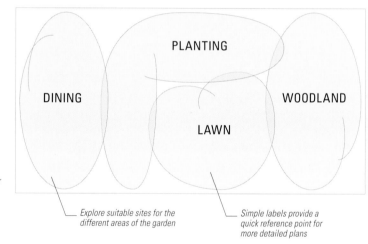

The finished garden
Sara Jane Rothwell, owner of the design practice London Garden Designer, produced both an overhead and a planting plan (opposite, top, and middle) to show clients the new design.

Working plans

These plans don't need to be accurate or drawn to scale, but they can be used to experiment with ideas, especially the relationship of horizontal surfaces (built and planted) with the locations of walls, screens, trees, and other main features. They can also include connecting elements such as paths and views.

Explore how best to create perspective by siting elements such as trees

Think about whether you want to replace existing elements, like this fence

Consider whether vertical features, such as a wall and steps, will work well

Overlaid photos
Perspective drawings are difficult to master, so cover a photo of your garden with tracing paper and sketch ideas on top to give a three-dimensional view of the changes.

PLANTING

DINING

WOODLAND

LAWN

Explore suitable sites for the different areas of the garden

Simple labels provide a quick reference point for more detailed plans

Bubble diagram
A basic bubble diagram helps you explore relationships between areas within the garden. It is an ideal way to experiment quickly before drawing a more detailed plan.

Garden plan symbols

These common symbols for plans form a visual design language that enables builders and other professionals working in your garden to read the plan quickly and understand what is being proposed. The symbols illustrated here are those that are most often used and most widely understood, and can be reproduced in black and white or colour.

WATER

Still water

Fountain

Water around rocks

PLANTING

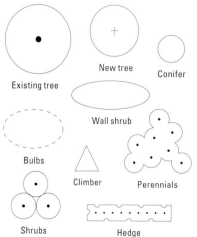

Existing tree

New tree

Conifer

Wall shrub

Bulbs

Climber

Perennials

Shrubs

Hedge

LANDSCAPING

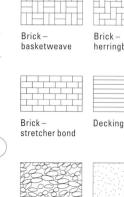

Brick – basketweave

Brick – herringbone

Uniform paving

Square-cut stone

Brick – stretcher bond

Decking

Granite setts

Random-cut stone

Cobbles or pebbles

Gravel

Rough grass

Mown grass

Finished plans

Plans that have been drawn to scale and show accurate arrangements, locations, and dimensions of proposed structural elements, planting, and features are known as finished plans (*see pp.114–121 for detailed advice on how to draw a plan*). These plans are intended mainly for construction purposes and will need to be read and understood by builders or contractors who use them to measure areas and lengths (for costing purposes), and to identify exact locations on the ground. Changing ground levels are shown as separate cross-sections, or by annotating the change of level on the overhead plan.

Overhead plan

An overhead plan should show the correct sizes and locations of all proposed elements, such as horizontal surfaces, areas of planting (topsoil), locations and alignments of linear elements (walls, fences, screens, hedges), and singular components (trees, specimen shrubs, pools, stepping stones, steps, lights, drainage points, and so on).

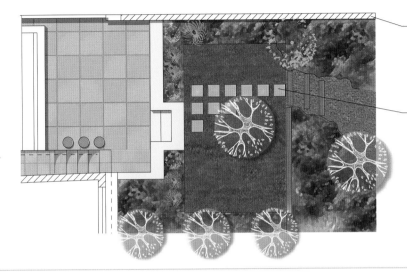

Include the site boundaries and any relevant buildings, doors, and windows on your plan

An overhead plan needs to include the correct materials and measurements of all hard landscaping features

Adding the details
In small-scale overhead plans, the individual materials can be shown; larger scale plans usually illustrate these materials more symbolically (see also p.118).

Planting plan

A planting plan is important for calculating the correct number of plants in the garden and identifying their exact locations. It also shows the position of larger specimens, as well as groups or drifts of the same species. This plan is most useful, and needs to be most accurate, when planting is being carried out by a contractor without the designer present. If you are doing the planting, a plan can help you accurately calculate the number of plants you'll need and show how to set them out prior to planting (*see pp.122–129 for more on creating a planting plan*).

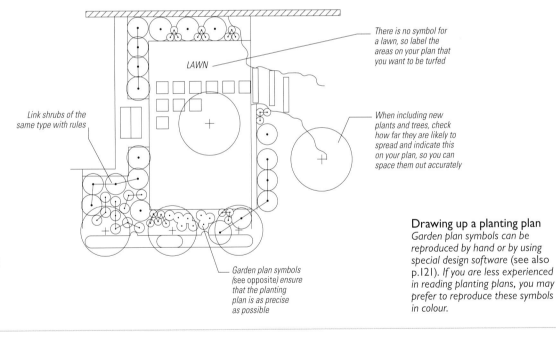

LAWN

Link shrubs of the same type with rules

There is no symbol for a lawn, so label the areas on your plan that you want to be turfed

When including new plants and trees, check how far they are likely to spread and indicate this on your plan, so you can space them out accurately

Garden plan symbols (see opposite) ensure that the planting plan is as precise as possible

Drawing up a planting plan
Garden plan symbols can be reproduced by hand or by using special design software (see also p.121). If you are less experienced in reading planting plans, you may prefer to reproduce these symbols in colour.

Cross-section

If you have a sloping garden and want to make changes to it, you may need a plan to show the impact of these alterations. For steeply sloping gardens, employ a land surveyor to draw a cross-section, or elevation plan. This will show the significant levels before and after any changes. More complex slopes may need additional plans.

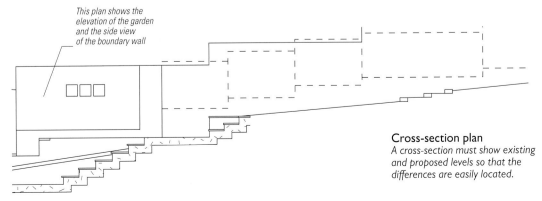

This plan shows the elevation of the garden and the side view of the boundary wall

Cross-section plan
A cross-section must show existing and proposed levels so that the differences are easily located.

Gathering inspiration

How do we find ideas for our outside spaces? For most of us, inspiration may initially come from other gardens, whether they are our friends' or pictures we have found online or in books, magazines, or newspapers. While this is a good starting point, and probably the best stimulus for anyone who is still developing their confidence in making design decisions, it can ultimately constrain the creative process. Most successful designers look outside their own discipline for other influences to help develop their concepts and push the boundaries, so seek inspiration from a variety of sources or select a theme. You can then create a "mood board" of appealing ideas to help you develop your own unique design.

Finding inspiration

By focusing on aspects of experiences that you like – for example, places you have visited on holiday, natural landscapes that you love, the work of favourite artists or architects, interior designs, or ideas you have seen on websites, such as Facebook, Pinterest or Houzz, or TV programmes – you can build up a picture of a garden you will enjoy. Also scroll through nurseries' websites for images of plants that you favour, and make a note of these too.

Bright colours and sculpture – mosaics?

Mediterranean fishing boat – blues and greens

You can collate your images and ideas by printing out pictures and sticking them into a notebook or onto an A3 sheet of paper to create a mood board. Alternatively, source a website that allows you to upload your images to make a mood board online, which you can then easily refer to on your phone, tablet, or computer. Whichever method you choose, continue to build up your portfolio of images until you are ready to start the garden design process.

Remember that you do not need to include all of your design influences in your final plan. In fact, professional designers often start with the bare bones of an idea and build on that, rather than cramming in everything on their, or their clients' wishlist from the start.

Also narrow down your plant list to about 20 key varieties (you can always introduce more at a later stage), and look through your images for colours that appeal, again keeping to a simple palette – see the information on Introducing colour and the colour wheel on pp.46–7 for guidance.

Beach-themed garden – props?

Coastal wild plants

Pebble pathway idea

Yellow flowers for an accent colour

Using a mood board
Collate photographs, images from websites, and pictures from magazines to create a mood board of creative and planting ideas. You can then use these as the inspiration for a totally new garden design or a starting point for the renovation of an existing plan.

Beach hut style – storage?

Case study: a seaside theme

A coastal theme is a natural choice for anyone who has been inspired by a holiday by the seaside. Study scenes, plants, and other features while you are away, and start compiling a sourcebook of ideas, photographs, and even pressed flowers that capture the essence of the garden you want to create at home.

Also look at colours, shapes, and materials that reflect the location. These may include the turquoise water, local costumes, or landscaping materials used for houses or walls. However, remember that developing a design is not about copying exactly what you have seen elsewhere, nor is it combining all your ideas into one busy area. Good design evolves when a theme is carefully adapted to suit a planned space. So consider all the elements that inspire you and see whether they work together well before you draw up your final plan.

You may also find it useful to sketch a bubble plan (*see p.22*), marking the different areas and functions you are planning for your new garden. Then file your inspirations under those headings, as shown here.

△ **Main inspiration**
An inspiring holiday by the sea will provide a wealth of ideas. Here, the light through the trees adds a romantic ambience.

◁ **Seaside planting sources**
Recreate coastal shallow soils and drought conditions – for example, with gravel borders – to mimic the environment in which these plants would naturally grow.

▽ **Seaside furniture**
Furniture that is in keeping with the overall mood, such as these casual deckchairs, helps to create a coherent look, as well as providing a welcome area of relaxation.

Devising play areas

Sand and water continue the seaside theme, and are obvious magnets for children. A micro-environment that includes these elements not only makes a great play area that will provide children with hours of fun, it also looks attractive when not in use. If you have very young children, you may prefer to avoid the potential danger of open water and just install a sand pit. If you are wary of vast quantities of sand ending up in the pool (or in your house), substitute small rounded pebbles to make your "beach".

▷ **Sun and sand**
A practical play area combined with an organic layout and seaside plants makes a delightful feature.

▷▷ **Swinging idea**
If you have room in your garden, allocate a space for a swing. Use recycled, hardwearing rope and driftwood for the seat, and cover the ground beneath with bark chips.

Shapes and spaces

Choosing the basic ground shapes for your plot is a good starting point for a design: one simple shape is best for small gardens, but larger areas can accommodate a variety. How you fill the spaces between the shapes also determines the final look.

How to use shapes

When choosing squares, rectangles, or circles for a design, also consider the size, shape, and location of the surrounding buildings and boundaries. Experiment with different options: try layouts based on existing features, the structure of the house, and the way the garden will be viewed and used. In general, shapes with straight sides are easier and cheaper to build than circles and ovals.

Right-angled shapes

A variety of these straight-sided shapes easily divide the garden into separate areas, provide a strong sense of direction and exploit both long and short views. A long axis running straight down the garden will lengthen it visually; a diagonal layout creates more interest; blocks laid across the plot foreshorten the garden and take the eyes to the sides, making it feel wider.

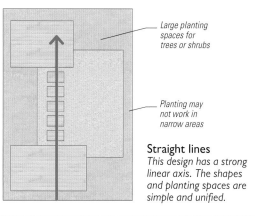

Large planting spaces for trees or shrubs

Planting may not work in narrow areas

Straight lines
This design has a strong linear axis. The shapes and planting spaces are simple and unified.

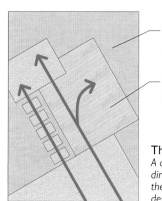

Several large interesting triangular spaces for planting

The full width of the garden is defined by the tilted shapes

The long view
A diagonal layout directs the eye towards the corners. The overall design evokes energy.

Circular shapes

Circles are unifying shapes, and while combinations can create pleasing effects, they do leave awkward pointed junctions that can be difficult to plant or designate. Work with geometric principles: for example, a path should lead you into the centre of the circle; if set to the side, the design will appear unbalanced. Ovals have a long axis, providing direction and orientation.

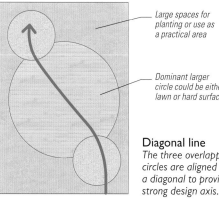

Large spaces for planting or use as a practical area

Dominant larger circle could be either lawn or hard surface

Diagonal line
The three overlapping circles are aligned along a diagonal to provide a strong design axis.

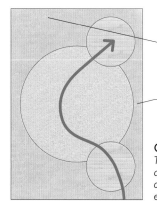

Moving circles to one side allows for a larger planting area

Awkward area needs to be taken into account

C-shaped curve
The restricted access and pleasing asymmetry of this design create an enticing space to explore.

Mixing shapes

Combining various shapes creates more interest, but throws up problems when a curve and a rectangle meet, or different materials connect. Generally, keep the layout simple, experimenting with scale and proportion to work out how many opposing shapes can be employed. Planting can be used to "glue" the shapes together, and to blur the joins between awkward junctions.

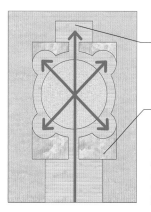

A focal point here will draw the eye down the central axis

Planting separates the different shapes

Classic match
A traditional symmetrical layout, mirrored along a central axis, is the basis for a formal design.

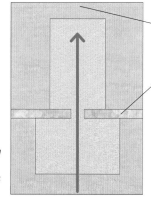

Use planting or a focal point to provide a visual full stop

Planting partly obscures the different areas

Simple approach
Changing the size and orientation of a shape delivers a dramatic and imposing layout.

Clean lines
Interlocking, steel-edged rectangular "trays" are the basis for this simple design. The metal cladding on the building creates a focal point and an effective visual full stop.

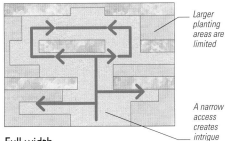

Larger planting areas are limited

A narrow access creates intrigue

Full width
A series of parallel divisions, with offset gaps for planting or practical structures, forces movement and views around the garden. The design draws you in.

Large planting pockets

Long axis directs the eye

Smooth flow
Using ovals instead of circles adds a smoother flow to the layout because the eye is taken along their lengths, rather than in all directions as in a circle.

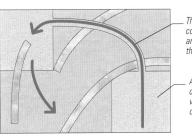

The path connects and unifies the spaces

Access could be via a patio or terrace

Secret corners
In this mixture of rectangles and curved hedges, only one part of the garden can be seen at any time. This allows the hidden areas to have different themes.

Using spaces

Densely planted spaces, using height and filling the garden's width, will create an enclosed space, while sparse, airy planting hugging the boundaries gives an open, spacious feel. Spaces can also be used to disguise the size and shape of a garden. For instance, a jungle effect in a small garden can imply the existence of more space by blurring the edges, but exposed boundaries may make it appear smaller. Conversely, in a large country garden, open spaces can blend seamlessly with the surrounding landscape, making the plot appear even bigger. Consider, too, existing planting and structures and work with the spaces they create.

Mixed moods
This garden is densely planted by the house, allowing close inspection of the flowers and plants, and then opens up on to a spacious lawn, creating two moods.

Open aspect
A narrow space between tall boundaries will be claustrophobic and oppressive. Here, in a design dominated by a lawn or hard landscaping, low vegetation creates an area exposed to more light, longer views, and with a connection to the sky above. It will feel open, but intimate areas may be lost.

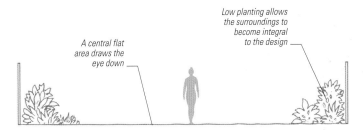

A central flat area draws the eye down

Low planting allows the surroundings to become integral to the design

Enclosed feeling
The same space filled with vegetation of different heights will be darker, much more enclosed, and with no views to the sides. The path will appear as a corridor through the centre and can lead to different parts of the garden, divided by the planting into separately designated areas.

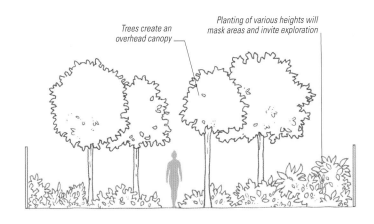

Trees create an overhead canopy

Planting of various heights will mask areas and invite exploration

Balanced approach
The same path now moved to the side also creates a corridor-like effect, but this time views are allowed under the canopy to the right, across a narrower strip of planting into the brighter space beyond. To the left, secret, intimate places can be created with a pergola or arbour amongst the mixture of high and low planting.

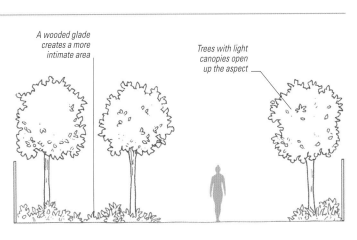

A wooded glade creates a more intimate area

Trees with light canopies open up the aspect

Routes and navigation

The location, width, pattern, and choice of materials of your path network will affect the way the garden is used. The routes determine how the area is navigated, as well as revealing views and framing spaces. Not all paths have the same role: some, the primary routes, will dominate the vista and dictate the garden plan. The secondary routes are used occasionally, guiding you off the main thoroughfare to access areas hidden from sight, whether for practical or design purposes.

Primary routes

The main route or pathway through the garden not only links together the different areas, but also determines the basic design. For example, a main path laid straight down the centre suggests formality, while a curved route snaking through the garden creates the template for an informal plan. A wide path offers an open, inviting entrance, welcoming in visitors, and a narrow winding path, flanked by tall planting that obscures the view, adds mystery. To punctuate the end of the route, use a focal point, such as a bench, statue, or container, to create a visual full stop. By its nature, a primary route will be heavily used, so materials need to be durable as well as complementary to the overall garden style. Consider, too, how the shape and appearance of path edges fit into the design.

Central paths

Paths converge in the centre — A container provides focus

Classic layout
A formal design is often built around a series of geometric and symmetrical paths. They are used to frame planted areas and meet at a specific focal point. There is usually no opportunity to deviate.

Winding paths

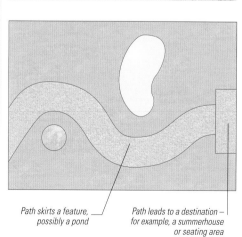

Path skirts a feature, possibly a pond — Path leads to a destination — for example, a summerhouse or seating area

Enticing curves
Routes that snake through the plot add a flowing sense of movement and an air of intrigue. They can be used to move around or join up key elements, as well as provide a few unexpected surprises.

Diagonal paths

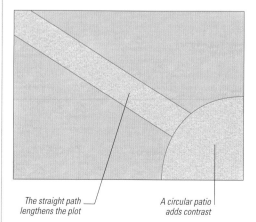

The straight path lengthens the plot — A circular patio adds contrast

Illusion of size
Setting a path on a diagonal allows the garden to be viewed along its longest axis, thereby creating the illusion of greater space and depth in small spaces, drawing the eye away from the back boundaries.

ROAM FREE

Random paving with planted crevices creates a slightly erratic, informal design. With no defined route, the eye — and body — can move in several directions across the whole area.

Circular paths

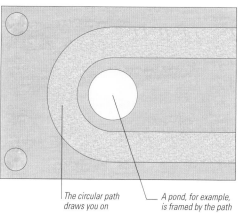

The circular path draws you on

A pond, for example, is framed by the path

Continuous flow

A circular path takes you on a journey around the garden. It can be planned to provide alternative views of key features and different elements, depending on the direction in which you travel.

Secondary routes

While primary routes can determine the style of a garden, secondary routes should be less intrusive and subtly incorporated into the design. They can be both practical and ornamental, providing occasional access to a seating area, shed or compost heap, or leading you off the main path on an intimate journey to view a concealed corner. They can even cut through large flowerbeds, allowing you to experience colours and scents up close. Access routes need not be as durable as main paths, and can be created from softer, organic materials, or mown through an area of grass or wild flowers.

Access paths

While helpful in offering access to other areas, plan secondary routes carefully and use sparingly to avoid a maze-like confusion of paths that make the design look muddled. They can be obvious (as right), or hidden in some way, either deliberately behind planting (see below left), or concealed within the design (see below right).

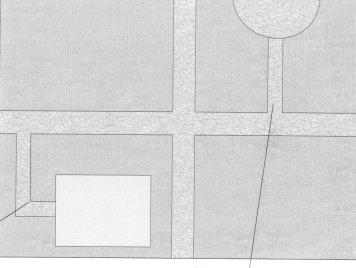

Path to shed

Path to patio

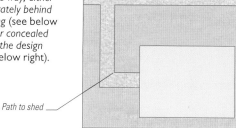

Practical solution

A path tucked away at the back of this formal design is not obvious, but it provides a practical, hard-surfaced route to the shed and compost bins.

Hidden approach

The gravel to right and left of the path, while part of the design, also provides a direct, hard-wearing pathway to the garden's seating and play areas.

Secret way

Visually, it appears as if the main pathway stops at the lawn, but concealed behind low hedging, a side path takes you off to a secluded area of the garden.

Subtle link

A path laid in the same paving material as the main circular route links the off-set dining area without impinging on the cleanness of the design.

Scenic route

The journey through this urban garden has been lengthened with a sinuous timber pathway that snakes through the centre of the plot and traverses a rill. A curved path helps create an illusion of greater space and presents the garden from different angles by obliging visitors to look one way and then another.

DESIGNER Adam Frost

Creating views and vistas

Your garden may look out over countryside or towards a block of flats, but either way, the views within your space can be enhanced with careful planning. A combination of framing and screening, using barriers, archways, and pergolas, can create a memorable experience as you move through your plot, glimpsing the next view as you go.

Planning your route

One ingeniously planned vista is gratifying, but a sequence of changing views is even more rewarding. Different views can be devised by varying the size of open spaces, using screens to mask change of use, and adding focal points. Creating viewing positions by placing a seat or orientating a path along a vista will also direct attention. Remember to consider the view looking back from the end of the plot, as well as the main view from the house. Follow the blue walking route through the plan (above right) of this long, thin family garden, by Fran Coulter; the numbered viewpoints correspond to the surrounding images and help demonstrate how these ideas work in practice.

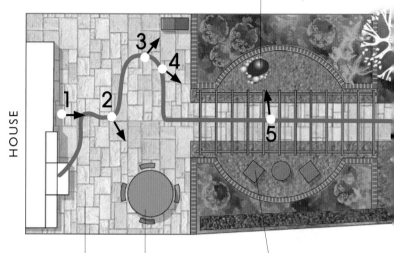

The bubble pool draws the attention to the side of the main path

HOUSE

The paving here is also used for the path through the garden, providing continuity

Circular table and chairs for outdoor relaxation and entertaining

From a second, more secluded seating area, the eye is drawn towards the bubble pool

2 Eating outside
The table and chairs are near the house, and are set against a simple green hedge, which creates a comforting sense of seclusion.

4 Looking through planting
From this angle, looking across the planting to the seats beyond, the pergola looks quite different and the garden takes on a more organic, less formal appearance.

1 View from house
This is the most important view in the garden and dictates the layout. The pergola reinforces and frames the view, and the inclusion of a flower-filled container as a focal point in the middle distance draws the eye forward.

3 The tool shed
The slim shed on the patio is both decorative and functional, adding a focal feature to this area of the garden.

5 Water feature
A glance to the side reveals another eye-catching feature. Hostas and grasses frame a discreet, low bubble pool.

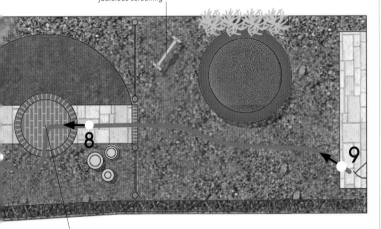

The play area is almost hidden from the house by judicious screening

8

9

A container of white-flowering roses stands on this brick circle, drawing the eye down the garden from the house

KEY

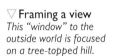
route through the garden

direction of viewpoint

Borrowing beautiful views

If you can see the surrounding landscape from your house, try connecting it visually to your own garden. Consider framing a key view, or opening up your garden, using a discreet barrier, such as a low hedge or picket fence, to link it to the wider landscape. Think about the view in different seasons and consider what it will look like in winter when trees and hedges are more open. You may also need to adapt your own garden planting to blend it into the landscape.

▷ **Blending in**
Here, there is no clear boundary between the garden and the land beyond. One becomes the other, and the garden seems to stretch as far as the horizon.

▽ **Framing a view**
This "window" to the outside world is focused on a tree-topped hill.

6 Shady corner
Beyond the pergola, the garden is more open and has a different character. This area is hidden from the house, and quite shady, providing the owner with an opportunity to use a different range of plants, such as leafy hostas.

8 Focal point
Circular features break up and soften long, straight lines. The large pot is a focus for this circular space and can be viewed from all sides.

Disguising unattractive views

Not all views are good. Within a garden, especially a small one, there will be areas of utilitarian clutter, such as sheds or household bins, which are not especially attractive and may need screening. Neighbouring houses may overlook the property, spoil the view, and compromise privacy. Tall planting or screens can help to hide eyesores, but if these are not an option, try adding an attractive focal point elsewhere in the garden to distract and lead the eye away.

△ **Covering an old shed**
Garden sheds are often unwelcome focal points. This rambling climber is a good summer disguise, less effective in winter.

▷ **Screening neighbours**
The tall bamboo screen blocks the view to the neighbouring property and provides an attractive backdrop to the planters.

7 Relaxing family area
This swing seat is tucked around the corner, just beyond the pergola and faces towards the brick circle and the shade garden.

9 Play area
The play area is hidden behind a semi-transparent screen, which separates it, both physically and visually, from the rest of the garden.

Geometric designs

Small, symmetrical, rectangular-shaped plots, often found in towns and cities, are ideal for geometric layouts, although some large rural gardens are also highly geometric. Most are based on simple combinations of rectangles and squares, with linear elements, such as walls, screens, hedges, and steps used to reinforce the formality of the design.

Descending planes
A progression of levels, low block walls, rectangular beds, strip lighting and matching recliners produces a series of parallel lines, giving this contemporary garden a dynamic feel. The planting is simple, so it does not detract from the strength of the overall design.

Layering shapes

By adding a variety of layers above ground level to offer different views and experiences, gardens can be made more visually exciting and functional. These layers can be set directly above the ground pattern, or angled so that the shapes above eye level have a different, but complementary geometry. Pergolas, clipped-tree canopies, and roof-like structures all offer opportunities to layer your design.

Overlapping layers
The arrangement of elements in this small garden breaks up a dull rectangular plot, and creates different spatial effects.

Canopies provide shade and create a layering effect

Raised decks are quick and easy to build

Screens and hedges provide height

Hard-wearing paving is best at ground level

Level changes
To create visual interest, introduce subtle changes of level using a range of different materials, including water.

Circular designs

Layouts based on circles, arcs, and radiating patterns help to create spaces that are full of movement. However, they are difficult to build from hard landscape materials, and getting the geometry wrong will look unattractive. Organic layouts (see pp.38–39) should be considered as an alternative, if this is likely to be a problem.

Dominant shapes can be softened by planting

Circular shapes draw the eye to the centre of the garden

The converging lines of the patio connect the house to the lawn

Formal approach
A central lawn surrounded by a radiating pattern of low beds and clipped hedges combines a sense of order with rhythm and movement.

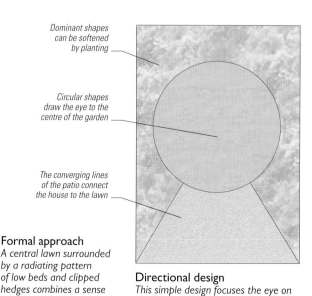

Directional design
This simple design focuses the eye on the centre of the garden. A container or sculpture could be used as a focal point.

Shapes on a diagonal

A classic design trick for long, linear, and narrow plots, is to rotate a rectilinear geometric pattern so that it is orientated along diagonal lines. These layouts on a bias draw your eye down the garden and encourage views to the sides.

Dynamic angles
The diagonal lines of staggered beds, patchwork wooden decking, and a raised pool make a bold statement, and direct visitors through the space.

Twists and turns
A diagonal path with steps traces a zig-zag line through the garden, providing areas to linger and enjoy the wide beds and colourful planting.

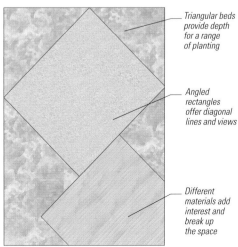

Triangular beds provide depth for a range of planting

Angled rectangles offer diagonal lines and views

Different materials add interest and break up the space

Defining shapes
Here, rectangles of hard landscaping, set side-by-side and edged with planting, make the garden appear wider than it is.

Symmetrical layouts

Throughout the world (except in the Far East), from the middle ages to the early 18th century, gardens were not only geometric, but also symmetrical. Inspired by Islamic and classical designs, they transformed the landscape into a controlled work of art. These formal layouts complemented classical architecture and reinforced the belief that beauty derives from order and simplicity.

Contemporary symmetry

Contemporary layouts can adapt classical symmetry to meet the requirements of modern living, such as creating space for outdoor entertaining or for growing herbs and vegetables. Good design also involves an understanding of a wide range of hard landscape materials and the way in which they can be combined to make a simple and elegant framework for the planting.

Cool control
A chequerboard of white paving and emerald grass against a dark hedge offers a modern interpretation of a traditional format.

▷ **Perfect harmony**
This sophisticated garden illustrates classical symmetry and demonstrates the importance of proportion and scale.

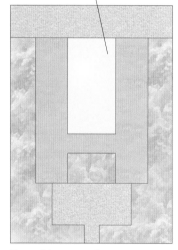

Create a striking central feature to accentuate design symmetry

Formal framework
A combination of rectangles with block planting gives a strong structure that works well in a contemporary setting.

Informal planting

Symmetrical layouts are often less obvious when viewed from eye level, especially when taller plants are used. A variety of forms, textures and colours will also soften hard lines and sharp edges. The combination of formal design and more relaxed, informal planting is a tried-and-tested formula, but requires skill and discipline if it is to work well. The balancing effect of a restricted colour palette and repeated plants, perhaps mirrored along a path, help to develop and reinforce the symmetrical theme.

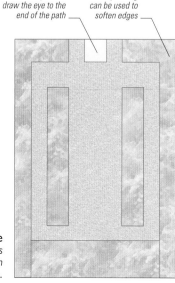

Use a focal point to draw the eye to the end of the path

Lush planting can be used to soften edges

Mirror image
In a symmetrical garden, dominant shapes are repeated and guide you through a sequence of harmonious spaces.

△ **Softened lines**
The subtle haze of herbaceous planting spills out from flower beds onto the path and contrasts with the formal garden layout.

▷ **Repeated planting**
Leading the eye through the garden, this long, airy avenue of grass demonstrates the compositional power of symmetrical planting.

Traditional and formal

Traditionally, it was the symmetrical pattern on the ground, such as a parterre of low hedging laid out around a central axis, that dominated garden layouts. These geometric designs are still popular in vegetable and herb gardens today, where they allow easy access to tend the beds. In the classical gardens of large estates, a sequence of focal points, such as ornamental pools and fountains, dramatic sculptures or large urns, were added to enhance key points and to make the pattern more interesting from eye level. Nowadays, when many planting styles are used, the geometric approach works best when the overall design can be viewed from a terrace or house above.

Visual journey
Well-positioned focal points, such as this nautilus sculpture, create a strong sense of direction. The domes of box and clipped yew lining the path accentuate this effect.

Planting can be changed seasonally for different effects

Crossing paths lend themselves to Islamic-style gardens

Planting edged with dwarf box hedging reinforces the formal pattern

Circles and squares
Reminiscent of a celtic cross, this layout divides the garden into quadrants with a central focal area, ideal for an ornament.

Permanent patterns
This formal layout of box-edged beds is infilled with spring flowers, which will be replaced as summer approaches.

Organic shapes

Organic shapes and layouts work well in large gardens and are especially suited to rural or semi-rural locations, but they can also work in small spaces. They are characterized by flowing lines, soft curves, sympathetic use of landscaping materials, and relaxed planting schemes. These naturalistic gardens evolve over time as planting matures, blurring the original layout.

Simple curves
Generous curves, wide beds, and the addition of a pinch-point draw the eye around the garden.

Interlocking circles

Developing two areas of the garden, separated by a pinch-point, leads the eye from one space to another, and offers both open and enclosed areas. The organic layout provides a setting where some shrubs and trees can be allowed to grow to their natural size, creating a backdrop for lower plants at the front of the beds. The narrow space between the circular forms can also be used to bring colour and interest into the centre of the design (*right*). This figure-of-eight layout makes the garden appear larger, as all areas are not visible from a single vantage point.

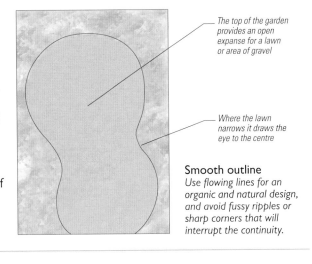

The top of the garden provides an open expanse for a lawn or area of gravel

Where the lawn narrows it draws the eye to the centre

Smooth outline
Use flowing lines for an organic and natural design, and avoid fussy ripples or sharp corners that will interrupt the continuity.

Fluid lines

A simple device to draw the eye along the garden, and to give the illusion of movement and space, is to adopt an S-shaped design. Two circular areas are connected by a single fluid line, which can be developed into a snaking path or a flowing lawn. If used as a path, the spaces at the top and bottom are ideal for planting, a seating area, or an ornamental feature, such as a pool. If these two areas are different in size, the path may be tightly coiled at one point and then more relaxed, providing contrasting experiences.

◁ **Serpentine path**
A coiling stone path leads through robust planting to a cave-like chamber in this children's play garden.

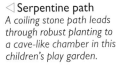

▽ **Curved decking**
The sinuous lines of the deck and lawn complement the subtle shades of the surrounding foliage.

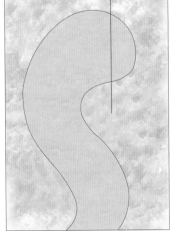

An ideal spot for a pool or feature to be viewed from a winding path

Meandering route
This curvaceous shape provides many different views and vistas as you move through the garden.

Sweeping curves

Curved lines may evolve to avoid an obstacle, such as a tree, pond or building, or added to make a path that leads to a particular destination. These are the fluid lines found in the natural world and lend an organic character to shapes and forms. They are frequently used to create calm, relaxing, and unchallenging garden designs.

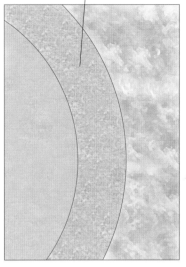

Use gravel or bark for a soft organic look

Bold statement
Curving round a bench, this dynamic raised bed adds colour and momentum to a paved circular terrace.

Gentle arc
Wide curvilinear paths create generous space on either side for deep planting beds or expansive water features.

Continuous journey
This C-shaped gravel path guides the visitor between still water and soft planting. The view around the curve is partly obscured, which adds a sense of mystery.

Multi-level layouts

Sloping sites provide an opportunity to create beautiful spaces full of movement and drama. Working a plan around the site's natural slope will create a more natural effect, while terraces offer structure and shape for formal and contemporary designs. Drainage is an important consideration, as any changes to slopes will affect the movement of water (see *pp.104–105*).

Terraced slopes

Terracing makes a dynamic statement and can be used to extend the architecture of buildings into a sloping landscape. Retaining walls and steps are solid, permanent additions and a long-term investment. Measuring and building them are skilled jobs at both the design and construction stages. Wooden decking is a cheaper solution; materials are lighter, but not as long-lasting.

▷ **Steep terrace**
Tiered wooden sleepers behind a low wall provide perfect conditions for sun-loving plants.

▽ **Tree platform**
Decked platforms are easier and less costly to build than terraces, which involve major earthworks.

Gentle slopes

Gentle changes of level in a garden offer visual interest and depth to the design. For practical purposes, gardens with only a slight incline can be treated as a flat site. However, if completely level areas are needed, for example, to accommodate a table and chairs, it will be necessary to level the ground and carefully consider the route between changing elevations. A combination of walls, steps, ramps, and terraces can be introduced as required, to suit any design.

Gradual progress
Shallow steps, with space for decorative pots, bridge a small pond and provide an easy route up to the seating area beyond.

Designing with steps

When building steps, the proportions of the tread (horizontal) and riser (vertical) are both important. Generally, they are more generous outdoors than inside a building, with treads 300–500mm deep (12–20in) and risers 150–200mm high (6–8in). Materials should complement those used elsewhere in the garden, especially adjacent walls.

Steep steps
These are a good option if space is limited, or when more drama is required, but they hinder fast movement and can be dangerous, so install a handrail too.

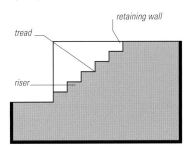

Shallow steps
Although they take up more space, shallow steps allow a relaxed progress through the garden. The depth of the treads also provides space for decorative pots.

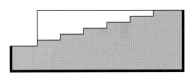

Stepped ramp
A stepped ramp is easy to negotiate and, if shallow enough, can accommodate wheeled transport. It can be useful where there is not enough room for a ramp.

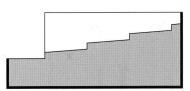

Continuous ramp
Invaluable for wheelchairs, bikes, etc, ramps also provide a useful route for wheelbarrows. They need seven times more horizontal space than steps.

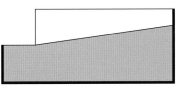

Natural hillside

The best advice when dealing with a hillside garden is to change a natural slope as little as possible. The soil is likely to be shallow and held together by the existing vegetation. Drainage will be complex and removing the native plant material may result in soil erosion and landslides, as the soil-binding roots are lost. Try to work with the unique contours of the landscape and make small, thoughtful interventions over time rather than significant alterations all at once.

Nature's way
Uneven, weathered stone steps meander romantically up through a secluded and naturalistic woodland setting.

Adding a landing
A landing is desirable at the top of a flight of steps, and to provide a resting place every ten or eleven steps within a long flight. It is also required when there is a change of direction.

Using height and structure

The plants or features that give height and structure to a scheme greatly enhance the way a garden is perceived and used. This is especially true of a straight-sided, horizontal plot, where introducing different heights will create movement and dynamism. There are certain principles to bear in mind, such as the rules of perspective, and it is useful to remember that the closer you are to a structure, the larger it will appear. Use hard landscaping and planting to create the effects you want.

Height levels

It is practical to think about height levels in terms of how they relate to the adult human body, which affects how they are viewed and experienced. Anything below knee height is viewed from above. Waist-high elements are seen at an angle and form a screen, partly blocking views to anything immediately behind them. At shoulder and head height, dense or opaque elements (such as closely planted tall shrubs, hedging or high screens) will completely block a view. Structures above head height, for example a tree canopy, can create a sense of seclusion as the sky and nearby buildings are obscured. Hard landscaping provides fixed elements but all further interest comes from planting. Indeed, combining plants of different heights is one of the key aspects of a successful garden. Few built elements can compete with a mature tree for interest and drama.

A see-through trellis distracts the eye from a shed

The tree lifts the gaze upwards

A painted, rendered wall forms the boundary

Low walls double as seating

The lowest plane is lawn

Planting is repeated at intervals to provide rhythm

An outer wall gives a sense of enclosure

Stones add a change of texture

▷ **Varying heights**
This multi-level design shows the clever relationship between the fixed height of the parallel low walls, and the natural variations achieved with perennials, grasses, shrubs, and trees.

▽ **Height levels explained**
This diagram shows the relationship between the human form and height levels within the garden. Planting, hard landscaping, and screens have all been planned to vary viewing angles throughout. The three low walls interrupt the planting but do not obscure the view beyond.

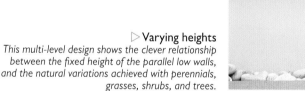

Planting at waist height is seen at an angle

Low walling around knee height punctuates the space

An area laid to lawn creates open space in the scheme

A see-through screen stands above head height

Paving adds a different texture at ground level

The highest element is the rendered wall, creating a backdrop

Planting breaks up the flat expanse of wall

Above head height

Head height

Waist height

Knee height

Ankle height

Introducing height

A range of height levels gives variety and interest to a garden, whatever its scale. Elements that create instant height include barriers (walls, fences, screens, or trellis), overhead structures (pergolas, arbours, or canopies), and play equipment, such as a child's swing. Planting options are varied and include trees, many shrubs, bamboos, climbers, hedges, and perennials for seasonal variation. Bear in mind that young trees and shrubs need not be expensive, but take time to gain height. Built structures cost more, but are quickly realised and make permanent features.

△ Contrasts of height
The stature of these elegant olive trees is given greater emphasis by the low planting below.

▷ Shielding neighbours
A combination of trees and shrubs behind trellis screens provides partial screening and privacy from neighbours. The painted frame adds height and structure to what would otherwise feel like a small space.

Temporary screens

While pergolas and other built structures provide height and solid overhead planes, they need support and can fill small gardens with posts. If uprights would be a problem in your garden, consider suspending temporary canopy screens to create shade and make the garden feel more intimate. Sail-like screens are a good solution and they can be taken down when not required. They need to be attached securely, but can be an excellent way of creating privacy in a small garden.

Nautical screen
A lightweight and elegant sail canopy provides shade, does not clutter the garden with posts, and conveys a feeling of intimacy to small urban gardens.

Using perspective

There are two important principles to consider when using perspective (the way in which objects appear to the eye). The first is that parallel lines in the viewer's sight appear to converge at a point in the distance, known as the "vanishing point". The second is that objects nearer to the viewer appear larger than those further away. A large tree or work of art, for example, may look too dominant placed in the foreground, but in proportion sited farther away. By carefully positioning elements of different heights in the garden, the rules of perspective can be exploited. It is even possible to produce slight optical illusions, for example, by repeating motifs at intervals to make a garden look longer.

Repetition of this broad, shallow curve makes the garden seem longer and wider than it actually is

The sculpture at the far end makes an appealing focal point in the distance

Tricking the eye
The use and orientation of parallel lines, and the repetition of shapes, draw the eye forward to the sculpture, creating a sense of depth.

Transparent screens

Trellis, glass, and other transparent and semi-transparent screens help to separate garden spaces without diminishing light. They are useful in smaller plots, where they allow visual connections to be made, while breaking up the space into different areas, and adding a change of mood. Transparent screens also make attractive features in their own right.

Versatile trellis
The open latticework of trellis associates well with plants and climbers and may be left open or screened with evergreens.

Glass panels
This patterned glass panel allows light through but slightly obscures the visual connection to the next area of the garden.

Choosing structural elements

Boundaries are the frame within which your garden sits and form the backdrop to the space, especially in a newly planted garden. Screens allow you to divide the garden into smaller areas, and come in a variety of forms and materials, while some garden structures may even be works of art in themselves.

Boundary options

The main boundary choices are walls, fences or hedges. Walls are an investment, making a permanent addition to the property, and can connect garden and house visually. Fences are cheaper but shorter-lived, so bear in mind that they will need replacing in time. Hedges take time to grow, and need clipping, but form a soft, natural boundary.

▷ **Wooden screen**
A trellis clad in clematis makes an inexpensive decorative screen.

▷▷ **Mixed materials**
Panels of concrete, painted timber, and a planted living wall create striking textural contrasts.

◁◁ **Bright squares**
The mix of brightly coloured opaque and transparent screens makes a bold statement.

◁ **Green colonnade**
An interesting alternative to a traditional continuous hedge, these tall clipped conifers form a strong background feature.

Internal screens

Adding screens and panels within the garden divides it into smaller, more intimate spaces. They are especially useful in predictable rectilinear plots where they can add interest and heighten mystery. Panels below waist height allow views across the garden, taller screens separate different areas, and gaps allow tempting glimpses of the garden beyond. Consider the effect of opaque and transparent screens and introduce colours and textures to add visual contrasts. Supports and other frameworks should form an important part of the design and, if well planned, will help to reinforce the overall composition.

Using natural forms

Structural elements can be introduced using planting alone. A range of trees and shrubs can be trained to form hedges and screens with great results. Patience is needed while slower-growing plants mature, but this is a rewarding process. Natural forms suit traditional gardens, but are not out of place in a modern design, where clipped shapes, such as "lollipop" trees and sculptural plants like bamboos, add spheres or lines to a design. Accentuate the vertical lines of small trees by placing low-growing plants at the base.

◁ **Bamboo screen**
This bold planting of tall Phyllostachys sulphurea f. viridis *is reflected in the pool in front.*

◁◁ **Clipped trees**
Here clipped "lollipop" bay trees emerge from box-framed lavender beds, demarcating the dining area. The slate terrace lends textural contrast.

Sculptural structures

Screens and garden dividers of all kinds can be decorative in their own right and, equally, a work of art can play a dual role and have a structural function in a garden. By introducing a strikingly different material, such as glass or metal, into a design filled with plants, you can add exciting accents and heighten the drama. Glass may be frosted or clear, printed with patterns or moulded in different ways, although even toughened glass may not suit a family garden. Metal adds gleam and reflection to an otherwise matt series of surfaces. Site sculptural structures where they can be fully appreciated.

▷ **The path ahead**
This unusual elliptical, wire mesh tunnel, a work of art in itself, invites use and functions as both a screen and a walkway.

▷▷ **Frosty looks**
The image printed on the transparent and frosted screen acts as additional "planting". Both the screen and the seat appear to float within the garden.

Introducing colour

Colour is a powerful tool in garden design, influencing our senses and the way in which we respond to the environment around us. Colours can also convey an atmosphere, mood or message: warm, vibrant colours generate a feeling of immediacy, liveliness, and excitement, while cool colours create a calm, spacious, often tranquil atmosphere.

Colour wheel

The language of colour is best understood using a colour wheel – a device employed by many artists and designers to explore the visual relationships between colours and the effects different ones can create when placed together. In particular, it helps us to see why some combinations work better than others, and why one colour can dramatically influence another to produce a startling contrast or confer harmonious continuity.

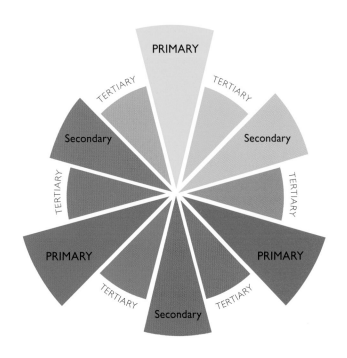

Primary colours
Red, blue, and yellow, the largest slices of colour on the wheel above, are primary colours, from which all other colours derive. These three hues cannot be mixed or formed by combining other colours.

Secondary colours
Two adjacent primaries will create a secondary colour when mixed together. These secondary hues are green, orange, and purple.

Tertiary colours
These are made by mixing adjacent primary and secondary colours in different quantities, until the wheel becomes a circular rainbow.

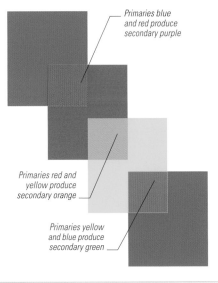

Primaries blue and red produce secondary purple

Primaries red and yellow produce secondary orange

Primaries yellow and blue produce secondary green

Hues, tints, shades, and tones
The true colours or "hues" are in the third ring of this wheel. The two central rings are light "tints", which are mixed with white. The outer rings show how adding black makes darker "shades". If grey were added, it would make a "tone".

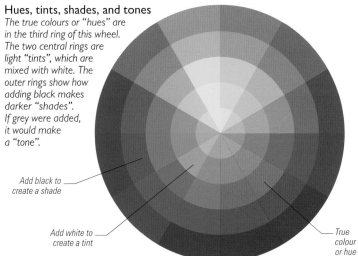

Add black to create a shade

Add white to create a tint

True colour or hue

Introducing colour in the garden

Planting combinations
Creating a variety of colour combinations with plants and flowers is exciting. You can alter the palette to produce changing colours for each season.

Hard landscaping
When nothing is in flower, hard landscaping can provide colour and interest. The effect is consistent, although weather conditions may affect the colours.

Paint
Earthy tones, derived from natural pigments, work well in more natural contexts, while bright, bold colours create a feeling of energy, excitement and optimism.

Combining colours successfully

The opportunity to combine different tints and shades of various colours makes garden design an exciting challenge; using a colour wheel can help our understanding of which combinations create the best effects. The key concept involves working with harmony and contrast to develop a visual experience to engage the viewer. Those colours allocated the most space in your design will become dominant.

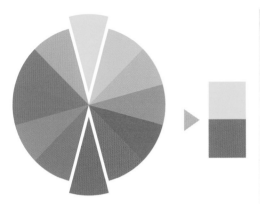

Opposite colours
Two colours from opposite sides of the wheel are considered to be complementary, for example, yellow and purple, and red and green. The high contrast of these colours creates a vibrant look, but they can cause eye strain, too, and should be used sparingly.

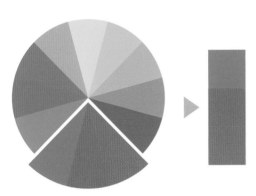

Adjoining colours
Harmonious colours, selected from adjoining hues (also called analogous colours) match well, are pleasing to the eye and create a sense of order. Choose one colour to dominate, and others to support it. Adjoining colour groups create a "warming" or "cooling" effect.

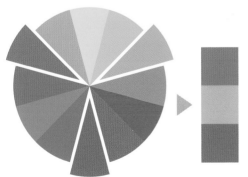

Triadic colours
Selecting three colours that are evenly spaced around the wheel can instil a sense of vibrancy. This works best with flower and foliage colour rather than with hard landscaping materials, where triadic combinations can be overdone and appear chaotic.

Colour effects

In a garden, colour is never perceived in isolation and should always be considered as part of an overall design composition that includes form, line, texture, and scale. Other elements, such as the intensity of sunlight and shadow, can also influence how colours are seen in an outdoor space. It is important to understand how and where to use different colours in your design to achieve the best effects.

Colour influence

You can use colour to attract attention to a particular feature or area; the more an object contrasts with its surroundings, the more visible it becomes. Hues (saturated colours) are dominant and offer the highest level of contrast when placed together. Darker shades or lighter tints contrast less, although small areas of light against dark, or vice versa, can create an accent. Recessive colours, like cool blue or green, give the illusion of distance.

▷ **Shorten a view**
A dominant colour (red) placed behind a recessive colour (green) will bring the background forward. This is particularly effective if they are complementary.

▷ **Lengthen a view**
If a dominant colour (purple) is in the foreground with a harmonious recessive backdrop (green), the garden appears longer.

Outline
Without colour, the outline of this tree doesn't stand out from the background.

Red on green
When red is placed on its complementary, green, the colours appear to "vibrate".

Green on red
The intensity is the same, but as red is dominant, the green tree is less clear.

△ **Warm contrasts**
This group of yellow flowers is highlighted against the dull red brick wall. The drift of mauve flowers in the distance contrasts with the dark woods behind and the lighter green field.

▷ **Bright white**
While purple and green are closely related on the colour wheel, adding white creates a stronger composition. As pure white reflects the most light, these pots stand out against the purple wall.

Creating highlights

You can achieve some bold effects in a garden using colour highlights. Try contrasting one hue against another, or combine adjoining hues in close proximity (see p.47). For example, plants with complementary colours (red and green, purple and yellow) will intensify the brightness of each other when placed together, while plants with hues that are close to each other on the colour wheel (see p.46) (purple, red, and pink) blend to form a harmonious effect. The introduction of a single, intensely coloured plant against a recessive background (such as green or blue) will make the bright plant stand out. Combinations of warm and cool colours can also result in eye-catching compositions that highlight the more dominant colour. (Note that white may appear recessive or dominant depending on the light.)

△ Colour boosting sunlight
The strong sunlight has a brightening effect on the yellow wall, and on the sizzling intensity of the red flowers in pots and on the hedge in the background.

△ Nature's neutral colours
Beautiful effects can be achieved by combining a variety of soothing greys, blues and greens with light-catching whites and yellows, which brighten up a shaded area.

Light and shade

Responding to colour is a sensory reaction, like smell and taste, and the way in which our eyes read a colour is dependent upon the amount, and intensity, of light that is reflected from that colour. Sunny areas make colours appear bolder and more concentrated, while shaded areas reflect more muted hues. This means that flat areas of colour – for example, a painted wall – may look quite different depending upon their aspect and orientation. Similarly, the hues of flowers and leaves will change depending on their location, the degree of shade cast on them, and the time of day.

THE PROPERTIES OF COLOUR

Warm colours (reds, yellows, and oranges) can make spaces appear smaller and intimate. Cool colours (blues, whites) make areas look larger and more open. Green is a neutral colour.

REDS
Reds and oranges suggest excitement, warmth, passion, energy, and vitality. They stand out against neutral greens, and work best in sunny sites but, if over-used, can be oppressive.

YELLOWS
Yellows are sunny and cheerful. Most are warm and associate well with reds and oranges. Greenish-yellows are cooler and suit more delicate combinations.

BLUES
Deep blues can appear very intense, lighter blues more airy. Blues suggest peace, serenity, and coolness. Purples carry some of the characteristics of both reds and blues.

GREENS
The most common colour in the plant kingdom, green comes in many variations, ranging from cool blue-green to warm yellow-green. Greens suggest calm, fertility, and freshness.

WHITES
White is common in nature. It is a combination of all other reflected colours, and suggests purity and harmony. White spaces seem spacious; the downside is they can feel stark.

BLACKS/GREYS
Blacks and greys are the absence of colour, when light rays are absorbed and none are reflected back. Black is glamorous when used sparingly, but depressing when extended over large areas.

Tints, shades, and tones

A general guideline to remember is that pure hues or saturated colours are more intense, while colours that have been mixed together are less vibrant. Black and grey are rare in nature, but they do exist in the form of shadows. A tinted colour, which has been "diluted" with white, will be lightened and appear more airy and farther away. A shaded colour, which has been "diluted" with black, will appear to be nearer. Tones mainly occur when a colour is cast into shade. However, the quality of light in a garden, such as on a bright sunny terrace or in a shady border at twilight, will affect the way that colours are perceived.

Tints
Hue + white = tint. The more white added, the lighter the colour. Tints recede, but pure white may advance.

Shades
Hue + black = shade. Darker shades advance. They are warmer and appear closer than pale tints.

Tones
Hue + grey = tone. Seen mainly in shadows, tones are less intense and appear muted.

Theory in practice

The planting scheme in this garden demonstrates the colour wheel in action. The palette is dominated by oranges and blues, which, as opposite colours, produce a brightly contrasting effect. Neutral whites and greens help to soften the impact, while a steely grey backdrop tempers the heat and picks up blue tones in the planting.

DESIGNER Catherine MacDonald

Applying colour

We tend to be more adventurous with colour in the garden than we are in our homes, perhaps because the outdoor environment feels brighter and less confined. The neutral greens of foliage and blues and greys of the sky also have a softening effect on more strident or clashing colours.

Vibrant colours

Strong colours can be used to dramatic effect in the garden: as bright pinpoints that energize more subtle plantings, or surprise pockets of colour separated by greenery. In a flower border you can build up from quieter blues and purples to crescendos of fiery reds and oranges. These hot colours will stand out all the more by combining them with a scattering of lime green, dark bronze and purple foliage.

▷ **Radiant hues**
Use glowing flower shades for hot, sunny aspects where the colours will really sizzle in the light.

▷▷ **Hot seats**
The colours used in this seating area create an upbeat atmosphere – the ideal setting for stimulating lively conversation.

Relaxing colours

The muted greys, purples and blue-greens typical of Mediterranean herb gardens create a restrained atmosphere, perfect for a contemplative retreat. Plantings that pick up the heathery colours of distant hills make a space appear larger. However, a calming palette doesn't have to be muted; it can also include fresh greens and pastels, which will work well in most settings.

◁ **Refreshment**
Fresh white, lemon, and green combine with a brighter pink to create an uplifting but essentially restful planting. Perfect for an intimate seating area tucked somewhere away from the house.

◁◁ **Country calm**
The lavender and purple sage add to the serene colour palette of this formal garden with a Lutyens-style seat.

Neutral colours

Earthy browns and biscuit tones are reminiscent of harvest time and appear warm and nurturing, contributing to a calm, relaxed atmosphere. Weathered wood elements are perfect for gardens with a country look. In urban locations, you can feel closer to nature by utilizing reclaimed timbers, wicker and bamboo for screens, raised beds, and furniture. For flooring, consider sandstone paving, decking, or a shingle beach effect with pebbles.

◁ **Muted tones**
As they die back, perennials and grasses continue to inspire, creating winter interest and a harmonious palette of browns.

△◁ **Rustic simplicity**
Basket-weave stools and a table made from a tree trunk blend seamlessly with a rustic-style garden.

◁◁ **Nature room**
Blocks of wood provide a muted backdrop for birches and the intermingling greens of the grasses and foliage plants.

Monochrome colours

Hard and soft landscaping in a restrained palette of black, grey and white, with the addition of green foliage, produce refined, elegant designs. The approach is perfect for smart period gardens with a formal layout. White blooms and silver foliage also work well with metallics in a chic city courtyard. Use cream or white flowers to enliven shade, and combine with variegated and lime-green leaves.

▽ **Spring whites**
This elegant scheme comprises white forget-me-nots, tulips, daisies and honesty with hostas and silver astelia foliage.

△ **Black diamonds**
Flanked by crisp green woodruff and a low clipped box hedge, this stylish grey and cream gravel pathway with a black pebble mosaic, makes an eye-catching focus for the small front garden of a town house.

ARTIFICIAL COLOUR

Colours that are rarely seen in nature tend to be the most attention grabbing. Contemporary designers use Day-Glo coloured materials and lighting to give a space a more futuristic or avant-garde look. You can include these colours with furnishing fabrics, Perspex screens and LED lights.

Day-Glo colours
Bold, cartoonish colours, such as bubblegum pink, lime green, orange and turquoise are so vivid they seem to glow. Attention grabbing but use sparingly.

Painting with light
LED lighting is available in any colour and can also be programmed to create a sequence of changing hues to produce spectacular effects in the garden.

Integrating texture into a design

It is easy to be seduced by colour when selecting plants and materials for the garden, but form and texture are equally important. Whether the design is a success or not depends on how well you combine the various shapes and textures, not only on a large scale but also at a more detailed level. To emphasize the contrasts, try to visualize in monochrome the hard and soft landscaping elements you are considering using. Also pay particular attention to how light affects different forms.

Types of texture

Experiencing different textures in the garden is a crucial part of our sensual enjoyment of the space. You can often tell what something is going to feel like just by looking at it, but there may be more surprises in store as you explore. Certain forms and surfaces invite touch and the visual and physical effect is heightened when there is great textural contrast. There are a number of basic categories describing texture, some of which relate to how something feels and others to how light affects a material's appearance.

Rough
For rough textures choose stone chippings, dry stone walls, wattle hurdles, peeling tree bark, or prickly plants.

Smooth
Choose flat or rounded surfaces like concrete cubes and spheres, plain pots, smooth bark and water-worn cobbles.

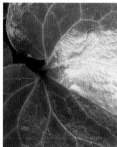
Gloss
Shiny, mirrored surfaces include many evergreens, polished granite, stainless steel, chrome, still water and glazed ceramic.

Matt
Ideal for combining with glossy elements, matt surfaces include cut timbers, galvanized metal planters and sandstone.

Soft
Impossible to ignore, soft, felted, furry-leaved plants are irresistible to the touch, as are fluffy seedheads and grass-like stems.

Hard
Non-pliable solid surfaces can be matt or gloss: cast metal, stone and concrete walling, flint, granite setts and terrazzo pots.

Combining textures

To introduce a variety of textures, combine plain with patterned surfaces, shiny with matt, smooth with rough, and so on, but don't overdo the number of materials or the garden could end up looking too busy. Accentuate the contrast between two elements by making the difference marked. Pair strongly vertical plants with horizontal decking, for example, or a glittering, stainless steel water feature with matt-textured ferns and hostas.

▷ **Textural contrasts**
Combinations of textures create the visual excitement in this harmonious design. Horizontal lines on the planter echo the lines of irregular stones bedded in concrete, while the rill provides a glittering contrast.

Rough with smooth
This walled courtyard marries gravel and rough-cut stone with smooth spheres to dramatic effect. The dry stone water feature cuts the sheer rendered wall in half.

Gloss with matt
Shiny glass and metal doors echo the visual qualities of the swimming pool. These elements are separated by the smooth paved terrace and matt rendered wall.

Soft with hard
The wooden walkway, circular terrace and snaking wall are perfectly opposed by luxuriant "soft" plantings of hostas, irises, grasses and marginals.

Choosing materials

It is not just planting that defines a garden. The texture and shape of the hard materials you select, whether for surfaces, boundaries or structures, are an integral part of the design. Different materials add shape, colour and movement, to lure you in and to determine where the eye is drawn, while materials sympathetic to the house or the local environment produce a more pleasing aspect.

When making your selection, consider the view from the house. Do you want to soften large areas of hard landscaping by incorporating a mixture of materials – slate with gravel, or wood with crushed shells, perhaps? Paths that are heavily used need to be solid, but a secondary walkway can be constructed from gravel, bark, or stepping stones. Using the same material for a path and a terrace creates continuity; a change further along will suggest a different area of the garden.

Laying materials lengthways or widthways draws the eye onwards or to the side, and obscuring paths invites exploration. Walls and solid screens shut out the vista, while open screens and apertures provide teasing glimpses of what lies beyond.

Furniture should be in keeping with the style of the garden. Ensure any timber pieces carry the Forest Stewardship Council (FSC) logo to show that the wood comes from sustainable forests. Also consider the siting: if you want a large dining table and chairs, you may have to build a terrace big enough to accommodate them.

Tall metal containers form a divide in a gravel garden.

Many gardens will have a spot for a water feature, as well as a piece of art. If you plan to include lighting, the electricity supply and cables must be installed by a qualified electrician; photovoltaic lighting has to be accessible to good light levels. Outdoor heating is popular, too, but consideration should be given to its environmental impact.

Permeable materials provide environmentally friendly parking.

Materials for surfaces

Large areas of paving or decking are visually dominant features, and have a significant impact on the appearance of a garden. Select materials that reinforce your style, complement the colours and textures used, and mix different types to develop patterns and lead the eye around the garden. (See also pp.352–363 for more on materials.)

Paving and decking

A strong design statement, or simply a block of uniform colour, can be achieved with large paved spaces. Bear in mind that when using slabs, pavers, or bricks, the joints will form a pattern, too; the smaller the unit, the more complex the pattern will be. Rectilinear paving can be combined to form larger rectangles or grid layouts, or use fluid materials, such as gravel and poured concrete, for curved edges to make organic shapes. All paving must be constructed on a solid base, and should slope to allow drainage (see *opposite*).

Decking with a twist
Decking is easy to cut and a good option for both geometric and organic layouts, and intricate designs such as this, with its inlay of blue tiles.

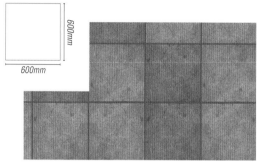

Large pavers may need cutting
When planning an area to be paved, try to avoid cutting by making the overall area an exact multiple of units. If it is not, larger slabs may require more cuts to fit.

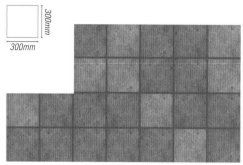

Small pavers fit tighter spaces
Smaller units provide greater flexibility, and are more likely to fit exactly the dimension of your patio. They are also easier to cut, when required.

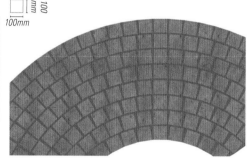

Small units best for intricate designs
Using small units or even mosaic tiles allows you to create intricate shapes and patterns more easily, but these designs are often very time-consuming to build.

△ **Horizontal paving**
Bricks are used to frame the edge of this formal path, and stone slabs laid horizontally shift the focus to the planting.

◁ **Bricks following direction**
The cottage planting is complemented by a traditional brick path which leads the eye to the gate.

◁ **Textured surface**
This random paving pattern is framed by a strip made from the same material, giving a clean, sharp edge. Although difficult to construct, the light-coloured textured path works well against the still water of the pond.

Paths and walkways

Paths are the arteries of the garden. Materials should be selected to enhance the journey along the path, and to complement the planting on either side. Pavers, and the joints between them, can run lengthways to give a sense of motion, or laid perpendicular to the direction of travel to slow walking pace, and attract attention to the surroundings. Choose paving that matches the garden style: bricks or gravel are good for a cottage-style garden, and more up-to-date materials, such as concrete and composites – or traditional materials used with a contemporary twist – suit a modern space.

Mixing materials

Assorted materials, as well as different textures and levels, can be used to dramatic effect in paving and decking designs. Use different materials to highlight key features, or to define and separate areas, such as a raised wooden deck over a stone-tiled floor. Colours may be complementary or strongly contrasting, but it is best to select pre-sized, coordinating materials to avoid extra work and higher costs; more complex construction techniques may be required when working with materials of varying thicknesses and where a different foundation may be needed.

△△ Wood and slate
This mix of hard and soft materials, with contrasting colours but similar tones, has been combined on four levels to great effect.

△ Stones and mosaic
Set on a concrete foundation, these small stone blocks and mosaic tiles create a decorative pattern around the trees and a foil for the gravel.

◁ Complementary textures
Four materials combine here – pebbles, granite, slate, and gravel – to give interest and texture to a threshold between two paths.

Edging ideas

Most paving materials, except *in situ* (poured) concrete, or those set on a concrete slab, will require an edge to contain the material. The edge can be detailed or functional depending on the style of your garden, and also connect or separate different materials, or areas of planting. However, you may not need an edge if you intend to allow planting to invade your gravel pathway.

▷△ Pebbles
Loose pebbles make an informal edge between the deck boards and the rill.

▷▷ Slate and setts
This bold design is created by slate paving butting up to stone granite units.

▷ Gravel and paving
Make a design statement with a clear, decorative edging pattern.

Planting opportunities

Plants add colour and texture when squeezed into joints and crevices; take care to choose those that tolerate trampling, are relatively drought-resistant, and ideally produce a scent when crushed. Think carefully about joints when combining paving and plants – a solid foundation, while necessary for most paving, will also contaminate the soil.

Plants between paving
Contrasting colours and textures are combined in this beautifully executed pavement, where mind-your-own-business (Soleirolia soleirolii) frames the paving.

Drainage issues

All surfaces should slope to allow water to drain or be collected, and even gravel surfaces may need extra drainage if laid on clay-rich soil. Ensure that rainwater runs away from buildings into collection points, such as gullies; water from small areas of paving can be directed into planting beds.

gully for runoff

patio sloping away from house

Slightly sloping patio
Create a slope away from buildings towards a collection point. Patios made from rougher materials will need to slope more steeply than smooth ones.

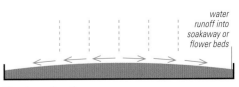

water runoff into soakaway or flower beds

Cambered path
Paths can be profiled to allow water to run off on both sides, where it can be collected in channels, or allowed to drain into planting beds.

Materials for screens and boundaries

Walls and boundary features, and the materials they are made from, have a major impact on the look of a garden. Traditionally, boundaries were constructed from local materials, such as stone, brick, timber, or hedging, but today your options are much broader, and modern gardens may make use of smooth rendering, metal screens, or reinforced concrete. If you share a boundary your choice may be limited, but if not, you can make it as subtle or as dominant as you wish, and add a personal touch with your choice of material, colour, shape, and texture.

Walls and solid screens

Brick, stone, or rendered walls enclose spaces and form a framework around the garden. Solid foundations and specialist construction skills may be required, and these boundaries can demand a large proportion of your building budget.

The colour of stone and brick walls is best left unaltered, so take this into account when making your choice. Consider the size and shape of the units, too, which can range from random rubble to expensive dressed stone blocks. Man-made materials, such as concrete, offer almost endless possibilities in terms of both colour and shape, providing clean lines or fluid structures.

▷ **Stone**
Well-constructed stone walls should last for ever, but require an expensive initial investment.

▷ **Brick**
Brick has been used for centuries and is durable and useful for creating patterned designs.

▷▷ **Rendered**
For flexibility and quick and easy construction, consider using rendered concrete walls.

Enhancing walls

Once you've decided on a material, think about any details you could add, whether for aesthetic or practical purposes. You could consider adding colour to all or some of the wall, depending on the material. Masonry walls, especially those made with mortar,

render, or clay bricks, benefit from capping or coping to frame the top of the wall and allow water to run off. However, ensure that it is in proportion to the size of the structure. Planting in crevices is another possibility, but select species carefully.

UNUSUAL MATERIALS

As long as walls are stable and shed water, most materials that are suitable for outdoor use can be used. Visit websites, look at books, or visit trade shows, but remember that specialist construction techniques may be required.

Planting pockets
Plants will soon establish in pockets of soil at the top or on the face of a wall. Limited water will be available to them, however, so choose species that can survive and flourish in dry conditions.

Rendered coping
Coping keeps the body of the wall dry and protects it from frost damage. It also forms an important visual element and can make a useful horizontal surface for a decorative effect, or for seating.

Textured wall
The walls of this small urban garden have been covered with old billboard vinyl, for a dramatically individual, textured look.

Fencing and trellis

Timber and metal fences do not require strong strip foundations or heavy building materials, and so are usually cheap and easy to build. Most are made from strips of material, and you should think about a design based on a combination of these "lines". To unify the design of an existing garden, it may be best to simply repeat or copy the original fencing styles. However, for new designs you can create patterns using different lengths, widths, and shapes of timber. In exposed areas, leave gaps in the fencing to allow some wind to pass through (see diagrams below).

Effective windbreaks
Solid screens do not allow any wind to pass through them and create turbulence on the leeward side. Use a perforated screen, such as a trellis, to solve this problem.

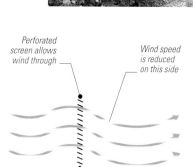

Wind forced up and over solid screen

Turbulence created on this side of fence

Perforated screen allows wind through

Wind speed is reduced on this side

△ **Solid fence**
This tall, close-boarded fence creates privacy, and has been stained grey to enhance the overall composition.

▷ **Perforated fence**
The strong pattern of this fence complements the garden, and acts as a decorative windbreak.

Gates and apertures

While screens and boundaries enclose space, they also create barriers that restrict movement and views. Punctuating these with doorways, gates, windows, and other apertures allows access or visual links to other parts of the garden. Importantly, these features provide further opportunities for attractive details, and should not be dismissed as utilitarian access points. Choose complementary materials and consider how apertures can frame vistas and views. Also, design doors and gates that look attractive when both open and closed.

△ **Picket fence**
When closed, this picket gate blends in with the rest of the fence; the only breaks in continuity are the posts and braces required for structural stability.

◁ **Classic doorway**
A traditional ledge-and-brace door makes a beautiful contribution to this old brick wall, as well as providing access. When left ajar, it gives an enticing glimpse through to another part of the garden.

△ **Modern aperture**
This perforated, reinforced concrete screen would be difficult to construct, but the beautiful results link the contemporary structure to the natural planting beyond.

Windows on the world

Dividing an outdoor space into different "rooms" helps to make it look larger, but solid screens can be imposing and create unwanted shadows, especially in a small garden. Sliding glass panels are used to separate the spaces in this ingenious design, bringing the architecture of the house out into the garden with a deft lightness of touch.

DESIGNER Pip Probert

Materials for slopes and structures

Raised beds, retaining walls, and similar structures that hold soil need to be constructed from water-, frost-, and stain-resistant materials. Natural materials, such as stone and some metals, are obvious choices, but rendered concrete and even sheet metal could be used for a more contemporary look. For garden structures such as pergolas and sheds, choose materials that are lightweight and easy to fit together, and that provide an opportunity to combine colours, textures, and patterns.

Retaining walls

Heavy or strong materials, such as stone, concrete blocks, bricks, timber, sheet metal, or reinforced concrete, are necessary for a retaining wall. Your wall needs to hold water as well as soil, and will require a drain to relieve the build-up of water, unless you have used a permeable material such as dry stone. You should consult a structural engineer for advice on any impermeable retaining wall above 1m (3ft) in height. Consider coordinating your wall with the house, a water feature, or screen to help unify your garden style.

◁◁ **Dry stone walls**
A dry stone wall works well in rural gardens. Place landscape fabric behind the wall to trap soil but allow water to pass through the gaps in the stones.

◁ **Wooden walls**
Timber walls are reasonably simple to construct: the individual sections will need to be screwed together for added strength and stability.

Raised beds

Essentially low retaining walls, raised beds do not need to be as strong or as heavy as larger structures. They can also be more elegantly designed, rather than serving a purely functional purpose. Line beds with heavy-duty plastic (with drainage holes punched in the bottom) to retain soil moisture and avoid leakage and staining. Also choose materials that complement the plants you plan to use, as well as the composition of your garden.

◁ **Contemporary beds**
Although susceptible to knocks and dents, metal lends a contemporary note to raised beds. Lighter coloured and galvanised metals do not conduct heat as well as darker metals, and plants are therefore less likely to suffer from scorched roots.

▽ **Country charm**
For vegetables and native planting, consider woven beds to complement your scheme. They are comparatively short-lived and will need replacing after a few years, but add rustic charm to a kitchen or cottage garden.

△ **Elegant containers**
Beautifully detailed and finished timber beds can add to the quality of a crisp, modern design. The addition of a gravel margin will keep the timber pristine.

Garden structures

Many suppliers produce pre-fabricated garden structures, or you may prefer a bespoke design if you have something specific in mind, and your budget allows. If you have a small garden, a structure can dominate the space, so plan carefully to ensure that it makes a positive contribution to your design. The materials you choose for the structure can reinforce a particular style. For a sharp, modern look, combine clean-sawn timber with glass and stainless steel, or consider rough-sawn timber for a rustic shed in a woodland-style garden. Hardwood is expensive but durable and does not require treating, but ensure that you use only FSC-certified woods from sustainable forests. A cheaper option is softwood, pressure treated for durability and stained with a coloured preservative, or recycled timber. Metal structures can be light, elegant, and contemporary, and galvanised steel, painted if desired, is a popular choice. Self-oxidising metals such as Cor-Ten steel and copper (ideal for roofs), which develops a green patina as it ages, should last indefinitely.

△ **Open structure**
This pergola is constructed using powder-coated aluminium combined with a wood trim (see pp.272–273 for more information on constructing a pergola).

▷ **Blending in**
The choice of dark stain allows this large garden office to recede into the background, while the stainless steel staircase gives a modern touch.

Step style

To prevent timber and metal steps rotting or rusting, they need to be supported on a solid framework above soil level. Stone slabs can also be constructed in the same way. Alternatively, solid blocks of stone, concrete, or timber can sit directly on the ground on a slope, or smaller units, such as paving slabs, can be used with a retaining edge. Consider the surrounding planting – you can allow it to "intrude" on to, or grow through your steps – and the material used for areas around the steps.

Bound chippings
These stylish steps are made from galvanized metal risers and bound crushed CDs (an alternative to gravel).

Metal steps
Strong and durable, these stainless steel grid steps allow planting to creep between them.

Wooden stairs
Timber steps supported on posts and bearers, like these, can be built to any height.

Materials for water features

When choosing and planning your water feature, make sure that it fits in with the composition of your garden, perhaps using materials that feature elsewhere in the design. Water features can be complex, so consult an expert or research water gardening in detail before planning one. Remember that you will need to ask a qualified electrician to bring an electricity supply into the garden, and some specialist water feature mechanisms and materials may also require expert installation.

Containing water

Waterproof masonry, such as concrete, will seal in the water in your feature, whether it is a raised or sunken pool. Any material with joints, such as bricks, will leak, so add a specialized render to the inside of your pond, which can then be coloured or clad with tiles; alternatively, line it with a waterproof membrane, such as butyl. Take care not to add any decoration that could puncture the waterproof layer or liner, and ensure that any joints where pipes enter the pool are fully watertight.

◁◁ **Raised pool**
A pond like this can be created with a pre-formed fibreglass liner, and enclosed with brick walls that match other garden features or the house.

◁ **Wildlife pond**
Covering the edge of a butyl liner with flat stones will protect it, but ensure that they are smooth-edged to prevent punctures.

Edging and lining streams

Natural-looking water features, such as artificial streams or wildlife ponds, are usually irregularly shaped, and lined with flexible butyl (see p.276). Ensure that the pond is deep enough in places to allow the required rooting depth for your chosen aquatic plants (see p.98). Streams require a "header pool" or reservoir at the top of the slope, into which water is pumped from the lowest pool. Cover the edges of your pool or stream with planting or flat stones to conceal the waterproof membrane.

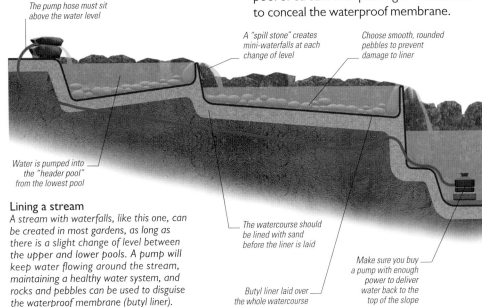

The pump hose must sit above the water level

A "spill stone" creates mini-waterfalls at each change of level

Choose smooth, rounded pebbles to prevent damage to liner

Water is pumped into the "header pool" from the lowest pool

Lining a stream
A stream with waterfalls, like this one, can be created in most gardens, as long as there is a slight change of level between the upper and lower pools. A pump will keep water flowing around the stream, maintaining a healthy water system, and rocks and pebbles can be used to disguise the waterproof membrane (butyl liner).

The watercourse should be lined with sand before the liner is laid

Butyl liner laid over the whole watercourse

Make sure you buy a pump with enough power to deliver water back to the top of the slope

Naturalistic waterfall
This artificial pond is on two levels and has been lined with a membrane covered with flat stones; large stones overhang the edge of each level to protect the liner from damage and to create mini-waterfalls.

Design materials checklist

The following table will allow you to quickly compare various materials, and their general suitability for the garden design and features you have in mind. This is intended as a guide, and you should consult other sources (especially product websites) for more comprehensive information when making your choice of materials.

MATERIAL	USE	DURABILITY	IMPACT ON ENVIRONMENT	COST	CONSTRUCTION
POURED CONCRETE	Foundations, walls, pools, surfaces, steps	♦♦♦	High	££	Simple construction easy; can be highly specialist
PRE-CAST CONCRETE	Paving units, blocks, building units, reconstituted stone	♦♦♦	High	££	Easy, but requires skill to achieve high-quality finish
RENDERING	Joints, surface finishes	♦♦	Medium–high	£–££	DIY possible, but skill required to achieve high-quality finish
AGGREGATE	Paving, foundations, drainage, decorative finishes	♦♦♦	Depends on source	£–££	Easy, except wall finishes
BRICK	Paths, surfaces, walls, retaining walls	♦♦♦	Medium	£–£££	DIY possible, but skill required to achieve high-quality finish
EARTH CONSTRUCTION	Walls, retaining walls	♦♦♦	Low	££	DIY possible, but skill required to achieve high-quality finish
LOCAL STONE	Paving, walls, structures	♦♦♦	Medium	£–£££	Variable: irregular stone needs skill for all but basic walling
IMPORTED STONE	Paving, walls, structures	♦♦♦	High	£–£££	Variable: irregular stone needs skill for all but basic walling
CERAMIC TILES	Decorative finishes	Mostly ♦♦♦	High	£–£££	DIY possible, but skill required to achieve high-quality finish
SOFTWOOD TIMBER	Construction timber, fences, gates, decks, paving, structures, furniture	♦–♦♦	Low–medium	£	Easy, but requires skill to achieve high-quality finish
HARDWOOD TIMBER	Decorative details, fences, gates, decks, paving, structures, furniture	♦♦♦	High if from unsustainable source	££	DIY possible, but skill required to achieve high-quality finish
NATURAL WOVEN TIMBER	Fences, hurdles, planters	♦	Low	£	Quite easy, but requires skill to achieve high-quality finish
MILD STEEL	Fences, railings, fixings, structures	♦♦ if not protected	Medium	££	Difficult – requires specialist skills
STAINLESS STEEL	Fences, railings, fixings, structures	♦♦♦	High	£££	Very difficult – requires specialist skills
SPECIAL STEEL ALLOYS	Fences, railings, fixings, structures	Mostly ♦♦♦	Variable	£££	Very difficult – requires specialist skills
ALUMINIUM	Lightweight structures, greenhouses	♦♦♦	Medium	££	DIY possible, but skill required to achieve high-quality finish
COPPER	Pipework, decorative cladding	♦♦♦	High	££	Difficult – requires specialist skills
ZINC	Planters, decorative cladding	♦♦♦	Medium	££	Difficult – requires specialist skills
GLASS	Screens, barriers, windows, surfaces, glasshouses	♦♦	High	£££	Very difficult – requires specialist skills
PLASTICS	Pipes, furniture, fixings, decorative facings	♦♦	High	£	Variable – DIY possible
PERSPEX/ PLEXIGLAS	Screens, structures, windows	♦♦	High	££	Difficult – requires specialist skills

Designing with furniture

A well-placed bench, lounger, or chair is an invitation to spend time relaxing in the garden. Whether permanent or temporary, garden furniture can have a marked effect on the look and feel of an outdoor space. The sculptural qualities of a particularly eye-catching or stylish piece of furniture could even be viewed as garden art. Of course, looks aren't everything, so do ensure that your chairs and tables are comfortable and practical.

Matching your garden style

Furniture has the potential to strengthen a design and create focal points within it. When the style of a plot is distinctive, such as in a Japanese garden, it's best to choose elements that follow the theme faithfully or that have a strong visual relationship to it. For example, cottage garden seating is likely to have a softer, more rustic and homespun feel. You might use wicker or Lloyd Loom chairs or reclaimed farmhouse kitchen furniture. In contrast, seating for contemporary settings works best if it has sleek, minimalist lines and is made from modern materials and fabrics, such as aluminium, plastic or synthetic rattan. The architecture of the house often influences garden style and in the grounds of a period property, pieces from the wrong era can stand out like a sore thumb. You don't have to source originals however: many companies offer quality reproductions.

Integrating furniture into a design

The size and shape of the available space will influence the type of furniture you choose; intimate corners surrounded by planting may, for example, only have room for a couple of foldaway seats. For outdoor dining, carefully calculate the size of table and chairs you can accommodate, to ensure a comfortable fit, and select furniture that mirrors the shape of your terrace or patio – a round table on a circular patio not only fits perfectly, but also accentuates the curved layout. A decorative seat can make an excellent focal point.

△ **Secret corner**
Simple foldaway furniture, light enough to carry around, is ideal for making use of different areas of the garden. Consider painting it to create highlights.

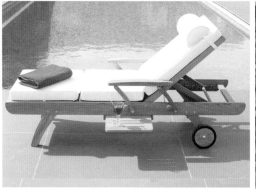

△▷ **Minimalist lines**
Large pieces of furniture, like this modern sunlounger, need space and a simple backdrop to allow their sculptural qualities to be fully appreciated.

▷ **Integrated design**
Try to match furniture to your design. This quirky, rustic site is enhanced by the bespoke wooden bench seat constructed from reclaimed materials.

▷▷ **Space to lounge**
Consider the size of the available space before buying furniture, or design your garden around chosen pieces. This sofa fits perfectly beneath its modern canopy.

Practical considerations

If you want to keep your furniture outside all year, check before you buy that it is resistant to rain and UV damage. Modern synthetic rattan furniture and plastic or resin pieces often come with guarantees, but while sofas and chairs with all-weather cushions will dry quickly after a shower, it is a good idea to cover them when they are not in regular use – an outdoor storage locker could prove useful for this. To retain the original patina on wooden furniture – which weathers and may change colour if left outside all year – clean, oil, or varnish it regularly, and, if possible, cover it during the winter.

△ **Outdoor sitting room**
Buy plush, upholstered furniture with fade-resistant, shower-proof covers, and ensure that the frames are sufficiently durable for outdoor use. Protect heavy pieces in situ.

◁ **Dining in style**
Hardwood, aluminium, and synthetic woven mesh combine in this stylish yet durable dining table and chairs for a contemporary urban garden by Wilson McWilliam Studio.

Environmental factors

Tropical hardwoods like teak have long been used to manufacture garden furniture because of their natural durability. However, this type of timber is not always obtained from a sustainable source, and uncontrolled logging is having a devastating effect on the environment. Always check the source before you buy; temperate hardwoods such as oak or more durable softwoods are likely to be "greener". Also look for furniture that has been manufactured from reclaimed wood, which can add a rustic quality to a design.

Greener options
Check for Forest Stewardship Council (FSC) certification on wooden furniture, to ensure that forests have been managed in an environmentally responsible way.

STORAGE IDEAS

In small urban gardens in particular, the lack of space available outdoors to store items such as gardening equipment, furniture cushions, and children's toys can present a real problem. One option is to choose garden seating that also provides storage, such as benches with hinged lids for access. Use a liner inside your storage to create a waterproof area to keep more delicate items safe. Alternatively, buy garden cupboards and boxes specifically designed to store cushions over winter from specialist furniture suppliers.

Storage lockers double as garden seating.

Furniture styles

Larger garden centres and DIY stores stock outdoor furniture through spring and summer, and you can often pick up bargains by waiting until later in the season to buy at sale time. However, if you cannot find what you are looking for locally, check out magazines and newspaper adverts, or search the internet for furniture specialists. Once you start looking for furniture you'll realize that the choice is vast, so persevere to find pieces that fit your garden style perfectly.

Traditional rustic

In more relaxed country- and cottage-style gardens, sleek furniture could well look out of place, though modern pieces with organic forms based on natural shapes may be appropriate. Quirky, reclaimed furniture is worth seeking out, as well as woven and wicker sets. The latter will weather rapidly, so you'll need a convenient storage place, such as a summerhouse or shed. Don't be afraid to mix and match country styles with classic pieces: lightweight, portable foldaway tables and chairs made from wood and metal can work well in period gardens with authentic-looking reproductions, such as Victorian fern seats or Lutyens-style benches.

Simple style
Traditional, hard-working or utilitarian designs add to the relaxed atmosphere of a cottage- or country-style garden.

Willow weave
Though not as durable as wood, wicker furniture, like this circular tree seat, adds romantic charm to an old-fashioned plot.

Chic modernist

A seating area dressed with designer furniture makes a strong statement, particularly in urban courtyards and on roof terraces, where the garden often functions as an extension of the house. Modern, minimalist items made of steel and synthetic mesh fabric or man-made rattan can add style and comfort to a contemporary design, while all-weather beanbags add colourful highlights. This look is about bringing interior style outdoors, so cushions and matching light fittings and containers play an important linking role.

◁ **Sixties model**
This up-to-the-minute design echoes the styling of the 1960s. The doughnut form contrasts well with the sparse backdrop.

△ **Inside out**
Only glass doors separate the house and garden, while the deck, with its stylish furniture, blurs the boundary further.

Contemporary looks

It's hard to put your finger on why certain furniture styles have an up-to-date feel, but, in general, clean lines and plain, neutral-coloured fabrics coupled with man-made elements like steel, glass, and chrome appear modern. Sometimes a traditional item or seating shape is updated for the 21st century using hi-tech materials; sometimes designs from previous decades experience a revival. Today's designers are increasingly developing the architectural role of furniture, as well as working on integrated or site-specific designs.

△ Spiral appeal
This curving, raised walkway culminating in a seat that "floats" on transparent glass is a piece of sculpture in its own right.

◁ Samurai seats
The Japanese influence in this modern set, with its minimalist lines, helps create an atmosphere of calm in a green oasis.

Furniture as art

There's no doubt that the sculptural qualities of certain furniture items, typically in wood, metal, ceramic, or resin, puts them into a different category from everyday functional seating. You can order sculptural furniture online and find artists via their websites, but it is also worth visiting the studios of local craftspeople, as well as gardening shows and galleries, to commission bespoke items. If possible, allow the artist to see the garden and the site for the piece, or provide as many photographs as possible, as this can really affect the success of the design.

△ Modern abstract
The organic form of snail shells has been the inspiration for this original bench with a carved wood seat.

▽ Sleek in steel
These boldly sculptural chairs are constructed from a perforated steel that softens their impact in the overall design.

Integral seating

You can create impromptu seating simply by utilizing steps, sunken areas, and the walls of raised beds: just add a few cushions, and you can accommodate a large group of people with ease. Elsewhere, a seat or table could follow the contours of a landscape feature, such as a serpentine wall.

Built-in beauty
Integrated seating can have an intimate feel. A cosy nook for relaxation could be created in a wall alcove, as here, or perhaps carved into a tall hedge.

Temporary seating

As your garden changes through the year, different areas will become more or less attractive or accessible. A portable seat, such as a director's chair, allows you to take advantage of particular settings, or to follow the sun around the garden.

Deckchair classic
The wonderful thing about collapsible furniture is that you can easily move it to where it's needed, and view the garden from different angles.

Integrating sculpture into a design

When choosing sculpture, you don't need to be limited by what's on offer in your local garden centre. Many objects take on sculptural qualities when placed in a garden, including beautifully shaped ceramic vases, driftwood, rounded boulders, or even pieces of disused machinery, so be as imaginative as possible. Think carefully about the relationship of your sculpture to the rest of your garden, where you will position it for best effect, and how its appearance will change over time.

Choosing sculpture

The appeal of a sculpture depends largely on your emotional response to it. You may prefer abstract shapes for the garden, especially if the style of your plot is sleek and modern, but wildflower gardens or woodland can also provide an exciting setting for a contemporary piece. Equally, classical statuary can add an element of surprise in a modern rectilinear layout, and will enhance an urban space. In cottage gardens, try figures of domestic animals, beehives, or rustic farm equipment.

▷ **Plant form**
This rusting iron sculpture, reminiscent of a flowering plant, works well in the Mediterranean-style setting. As the surface weathers, the patina will subtly change.

△△ **Figurative**
With one toe dipping into the water, this figure adds a relaxed and humorous touch to this contemporary landscape.

△ **Topiary**
Clipped greenery, a type of living sculpture, has many forms and includes Japanese cloud pruning.

◁ **Abstract**
The rectangular leaping salmon wall art is perfectly balanced here by the tall, narrow sculpture set amongst the planting.

Positioning sculpture

Take time to find the right spot for garden art and to integrate it into your design. Some pieces work best surrounded by reflective water, or by plants in a border. Contrast simple, solid shapes with diaphanous grass heads, for example, or view them through a haze of lavender. Intricately detailed sculptures look best with a plain backdrop, such as a rendered wall or clipped yew hedge. Matt surfaces like natural stone or weathered timber create a foil for highly polished metals, and you can use these materials to mount smaller sculptures, too.

△ **Focal point**
This abstract piece appears to hover over the surface of the pool, which also reflects its image, and makes an eye-catching focal point in this small garden.

△△ **Gazing skyward**
John O'Connor's bronze child takes your gaze up to decorative fretwork on a pavilion roof above, while the colour blends harmoniously with the timber frame.

△ **Space to perform**
The tall, cartoon-like figure of a girl striding briskly across the garden creates focus, but needs a large area to convey her energy and momentum.

Scale and proportion

A small piece of sculpture may be lost in a large, open site, but bring it into an intimate courtyard and you'll find that it's in perfect proportion to its surroundings. Try "anchoring" small ornaments by placing them next to a solid piece like a boulder, a hunk of driftwood or an oversized vase. Alternatively, mount decorative objects and plaques, fit them into alcoves in walls and hedges, or raise them closer to head height on plinths. To gauge the size of sculpture required for a site – when planning a focal point at the end of a formal path or at the side of a pool, for example – use piles of cardboard boxes or plastic refuse bins to help you visualize how the sculpture will fit into the proposed setting.

Commissioning a piece

You may discover someone whose work you admire by visiting national or regional gardening shows, dropping in at an artist's studio open day, or checking sculpture and land art websites. Help your chosen artist to visualize what you have in mind with rough sketches and photographs and, if possible, organize a site visit for them. Agree at the outset on the design, its dimensions, and the materials to be used, as well as confirming a price and delivery date for the work.

Materials and cost

There are often less expensive alternatives to traditional sculpture materials. Reconstituted stone, terracotta, or ceramic ornaments, for example, are far cheaper than carved stone, and bronze resin costs less than cast bronze, while lead statuary reproductions are relatively inexpensive. You may also find artists working with driftwood or reclaimed wood, rather than expensive hardwoods.

Hidden torso
Half-hidden by foliage, this weathered terracotta torso appears to grow out of the landscape, and would be a fraction of the cost of a bronze piece.

THEFT AND PROTECTION

Use common sense when placing your sculptures: try to keep them out of sight of passers-by, and consider using alarms or security fixings. For a front garden, choose pieces that are too large and heavy to be carried off easily, and keep them close to the house. Ensure that garden sculpture is covered by your home and contents insurance, and let your insurer know about new purchases.

Garden gallery

As well as providing a decorative focal point, sculpture can transform your garden more generally into a space for art, imbuing it with deeper meaning. In this garden themed around healing, the sculpture of a woman in a striking pose looking upwards, set against a dark yew hedge, could be interpreted as a symbol of hope.

DESIGNER Ruth Wilmott SCULPTOR Rick Kirby

Designing with lights

The beauty of installing creative lighting is that you can design an entirely different look for your garden at night. Soft, subtle lighting, bringing just a few choice elements into focus, is relatively straightforward and makes the most of differing textures and contours. More theatrical styling is possible with the wide range of specialist lighting equipment available. There are important aspects of safety and security to be considered, and you should always discuss your plans with an electrician.

Lighting in the garden

Flooding the garden with light from above creates too harsh an effect, and can cause nuisance to neighbours and add to the problem of light pollution. Avoid strong lights that may shine directly into the eyes of an onlooker. By maintaining areas of shadow you can accentuate the theatrical effect of any garden illumination, and make the night-time experience all the more enchanting. Draw up a plan, taking into account the type of lighting required in each area, such as recessed lighting for a deck, directional spotlighting for a barbecue, or underwater lighting for a fountain. Work out cabling circuits and plug points, and talk through your ideas with a qualified electrician or lighting engineer, preferably before completing any new landscaping work. You can experiment with different lighting effects by simply using a powerful torch, or torches, held at different angles.

Nightlife
Outdoor rooms used for relaxation and entertaining can be lit in a similar way to indoors with low-level lamps, and mini spots to highlight decorative elements.

Ways with water
Moving water features such as cascades are easier to light than static pools, as the surface disturbance masks the light source, while planting can hide cables.

Coloured glow
In contemporary settings, restrained use of coloured lights can create stylish effects. Programmed, colour-changing LEDs are an option for dynamic shows.

Practical considerations

Unless you plan to use solar-powered lights, you need a convenient power supply. Special waterproof outdoor sockets must be installed by a qualified electrician, and any mains cabling needs armoured ducting to prevent accidents. When using low-voltage lights that run from a transformer, house the transformer in a waterproof casing or locate it inside a building. A transformer reduces the voltage from the mains to a lower level at which many garden lighting products work. The size of transformer you will need depends on the power and number of lights you plan to use. Ask your electrician to install an indoor switch so that you can turn the lights on and off easily. LED (light-emitting-diode) lights are both energy efficient and create no heat, making them particularly safe to use in the garden; you will find a huge selection available. If an area is sufficiently sunny, solar-powered lighting is another good option.

Safe passage
If you plan to use the garden at night, illuminate pathways, steps, and changes in level using low-level lighting, and angled recessed lights to avoid glare.

Path lighting
Post lights come in a wide variety of designs, including many solar-powered models, and sets that run from a transformer. Position in the border to light pathways.

Flickering flames
Candles, lanterns, and oil lamps create a magical atmosphere. Never leave them unattended, and take care to keep naked flames away from flammable materials.

Lighting effects

Tiny LED fairy lights running from a transformer are simple to install, and create a romantic ambience when woven through climbers on a pergola. Mini spots are great for uplighting an architectural plant or a piece of statuary, or for highlighting textured surfaces. Recessed, low-level lighting in steps, walls, and decks casts gentle light without glare, and coloured lighting can be used to create contemporary effects, floodlight trees or rendered walls, or to light pools. For a contemporary look, try small white or coloured LED spots set into a decked area or a few underwater lights to illuminate a clear, reflective pool.

△ Mirroring
A single source of illumination bathes this poolside terrace in soft light and produces a perfect reflection in the black, unlit surface.

▷ Uplighting
Matt black mini uplighters are inconspicuous during the day, but can be angled to reveal the shape and texture of plants, decorative elements, walls, and screens at night.

▽ Floodlighting
Bright, even lighting is mainly used for security and can be triggered by infrared sensors. LED spotlights can also be used for dramatic up- or downlighting.

△ Spotlighting
Using a directional spotlight mounted high on a wall and angled in and down towards the subject, you can highlight an area without creating irritating glare.

△ Backlighting
Low-level backlighting throws the foreground elements into relief and creates dramatic shadow patterns on the wall behind. You can also backlight decorative screens.

△ Grazing
This term refers to the effect achieved by setting a light close to or along a wall or floor. It can be angled to illuminate an area, and reveal texture and form.

Choosing lighting and heating

With such a wealth of creative garden lighting now available, it can be difficult to decide what's right for you: this section looks at the relative merits of each option. Heating systems are becoming increasingly popular and allow you to make more use of your garden in the evenings and during cooler weather. However, some heaters and fires are not energy-efficient, so choose carefully and use them in moderation.

Types of lighting

Garden lighting has been revolutionized by the introduction of efficient LEDs, and more reliable and sophisticated solar-powered units. LEDs offer all kinds of "designer" effects, including lights that change colour and systems that can be controlled via a smart phone. While DIY stores carry an increasingly wide range, the largest choice can be found online and via specialist companies.

With the exception of solar-powered lighting, and candles and oil lamps, all other illumination devices need to be connected to the mains. Lights either work directly from the mains or through a transformer that provides a low-voltage current – ideal for a garden, as water and direct current are a lethal combination. Always employ a qualified electrician to install mains lighting, to make connections to mains power, and to fit new switches and plugs.

◁ **Light show**
This garden by Janine Pattison Studios is bathed in light in the evenings, with subtle LEDs grazing the walls and illuminating the modern water feature.

TYPE OF LIGHT

This table shows the pros and cons of the main forms of lighting, but for most types it is also best to discuss your requirements with an electrician or lighting engineer.

WHERE TO SITE

EXTENT OF ILLUMINATION

EXPENSE

INSTALLATION

MAINTENANCE

Heating in the garden

Introducing some kind of environmentally-friendly heat source into the garden extends the use of the plot into the cool of the evening or in spring and autumn. Wherever possible, burn logs and prunings cut from your own garden. Never use treated or tanalized timber, and make sure you read the instructions on appliances to check the type of fuel you can burn. Safety gloves are a must as fire grates get very hot, and make sure you allow chimeneas to cool before covering them. Keep a fire extinguisher handy, and use fireguards.

TYPE OF HEATING	PROS	CONS
FIRE PIT	DIY build possible. Some designs portable. Focal point, with potential for 360° seating. Heats and cooks. Burns garden prunings.	Needs space and safety screen. Ash may stain light surrounds. Poses a danger to children and pets – do not leave unattended.
FIREPLACE	Many different models including cast-iron stoves. Stone and brick styles can form a major garden feature. Burns logs.	Larger models, including those made from stone, take up space and are permanent fixtures. Cast iron rusts.
CHIMENEA	Fits into a small space. Clay designs often very decorative. Easy to cover and protect from weathering.	Both clay and metal types can crack. Clay may start to crumble after absorbing a lot of moisture. Tricky to clean out ashes.
GAS/ELECTRIC	Convenient and no cleaning up afterwards. Instant heat and/or cooking with flexibility; easily controllable.	Burns fossil fuels. Very inefficient considering amount of energy used and heat produced. Heavy cylinders for gas heaters.

LED	LIVE FLAME	ELECTRIC	SOLAR-POWERED

Almost anywhere in the garden. Can be used as pool lighting, recessed lighting, fairy lights, spots, or for security.	Candles, oil lamps, and lanterns may be placed on the ground, in wall niches, on tables, hung from hooks, or floated.	Fluorescent and halogen lights are used for security, spotlights, and lamps, although less extensively – LEDs are favoured now.	Edge of pathways/patios; in ponds (floating/rock lights); on walls; by plants. Some types suitable as spotlights.
Very bright for the size of unit. Casings can enhance and focus light output, while diffusers help to soften it.	Low-level, atmospheric lighting. Candelabras and lanterns are suitable for outdoor dining.	Varies according to fixture – halogens can illuminate entire garden. Coloured fluorescents are for special effects.	Units fitted with modern solar-powered LEDs can be quite bright. Strength of illumination depends on battery type.
Initial costs of units vary considerably, but the running costs are very low and the bulbs can last for years.	Candles, gel and oil lamps are inexpensive compared to electric fittings, but do not offer comparable lighting.	Relatively inexpensive to buy but running costs add up, and the bulbs will need to be replaced more frequently than LEDs.	Costs vary considerably depending on quality. Lights require no mains power installation and running costs are zero.
The same as conventional bulbs – running off mains power or transformer. Useful for hard-to-reach areas.	Take care to site live flames safely on a non-flammable, level surface in shelter. Never leave a candle or lamp unattended.	Lighting can run off mains power or transformer. Consult a qualified electrician for installation (*see opposite*).	Safe and easy DIY lighting. Needs airy spot to operate well. May not light the garden for as long in winter.
LED bulbs last many times longer than other types, and once installed require very little or no maintenance.	Trim wick to keep candle flame low and efficient. Extinguish with a snuffer. Do not move candles when wax is liquid.	Replace bulbs when they burn out. Keep wall lamps and infrared sensors clean.	Photovoltaic cells need regular cleaning. Good quality rechargeable batteries can last up to 20 years.

△ Fire pit
An updated version of the campfire, fire pits are a draw at social gatherings and may also be used for cooking.

△ Chimenea
The chimenea, originally a Mexican device for heating and cooking, comes in several different designs. Ensure that the fire is just below the opening to prevent smoking.

◁ Fireplace
This grand fireplace dominates the garden, creating a dramatic outdoor dining area. Simpler, smaller models for average-sized gardens are widely available.

Designing with plants

Plants perform at their best when provided with the correct combination of growing conditions, and learning about their needs and the kind of soil they prefer will help you devise the right planting scheme for your plot.

Including examples from a range of plant groups should ensure interest all year round. Trees and shrubs give height, depth and shade, as well as the essential framework. Evergreens retain their leaves, so are useful for all-year interest, and the shimmer of frost-covered deciduous plants is one of the pleasures of a winter morning garden. Scented climbers, grasses, perennials and annuals all have their part to play, while spring bulbs and biennials bring a seasonal burst of colour, just when fresh novelties are most needed in the garden.

Plants are very versatile. A structural plant can be a single specimen, such as a stunning cardoon taking centre stage in a border, or a group of plants, perhaps a box hedge clipped to enclose a parterre. Focal plants attract and guide the eye. They don't have to be long-lasting: a lovely individual specimen with vivid flowers or leaf tints works as well as an evergreen spiky *Phormium* or sculptural tree.

Select plants like dahlias for shape as well as colour.

Midrange plants include shrubs, grasses, and herbaceous perennials, and they can help define the style of your garden. Mix strong leaf shapes and flowers and foliage with different colours and textures for a dynamic display. Ground cover is another potential element; choices range from a neat, evergreen carpet to a blowsy show of flowers or scented drift of herbs.

From the heart-lifting first bulbs of spring, through to summer blooms, and on to autumn foliage and scented winter-flowering shrubs like *Mahonia*, seasonal planting is a constantly evolving delight. You can stick to your chosen style, or throw in the odd surprise for fun. Designing with plants is the exciting – and never-ending – pleasure of gardening.

Use layers of plants to create stunning effects.

Understanding plants

Garden plants come from a great number of different habitats around the world and vary in their needs. Providing them with the same conditions in which they grow in the wild is the best way to ensure that they will thrive in your garden. A plant's appearance – the leaves, in particular – can give you a basic understanding of its requirements, but it is best to read the plant label carefully, too. Remember that plants which share a natural habitat will also look good together in the garden.

Shade- or sun-loving?

Imagine the conditions in which a shade-loving plant grows. Light levels are low, so it probably has dark green leaves full of light-catching chlorophyll. Protected from damaging drying winds and scorching sun, it can also afford to have large leaves. Now imagine a plant that has to cope with sizzling midday sun and buffeting winds. Silver or grey leaves with reflective surfaces and protective hairs are less likely to dry out. Leathery or succulent leaves also indicate good tolerance of heat. Many plants fall between these two extremes, but, in general terms, leaves are a useful guide.

Shade-tolerant plants
Moist and shady, sheltered conditions allow large-leaved plants, such as Rheum, Darmera, and Rodgersia, to thrive. Most shade lovers tolerate some full sun during the day, but leaves may scorch with too much exposure.

Sun-tolerant plants
Full sun and dry soil make a testing environment for a plant. Heat- and drought-tolerant plants may have silver, heat-reflective leaves (Artemisia), or narrow grey ones (lavender), which minimize the exposed surface area.

Plants for different soils

It is easier to match your plants to your soil than to try to change the character of your land. Heavy clay can be cold and wet, but it is fertile and productive once plants are established. Sandy soils can be worked all year round at almost any time but will dry out fast in summer. Soil acidity is important if you want to grow ericaceous (acid-loving) plants such as *Pieris*, *Camellia*, or *Rhododendron*. Be aware that labels don't always state whether plants need acid soil conditions. (*For more information on soil types, see p.102.*)

Clay soil
Plants such as Berberis that like fertile moist conditions grow well on heavy clay soil.

Sandy dry soil
If soil is too wet, bulbs, such as alliums, may rot. Free-draining sandy soils suit them best.

Alkaline soil
Soil with a pH value over 7 is considered alkaline – if it is also fairly fertile, roses will love it.

Acid soil
Azaleas are ericaceous plants that require acid soil with a pH value below 6.5.

PLANT GROUPS

ANNUAL
A plant with a life cycle of one year. Usually very floriferous because of the number of seeds it needs to yield in order to reproduce.

BIENNIAL
Plants with a two-year life cycle, producing foliage the first year and flowers the next. Canterbury bells and wallflowers are biennials.

PERENNIAL
Non-woody plants that can live for years. Most die down to the ground in winter and come up again in spring; some are evergreen.

EVERGREEN
A plant that retains its leaves all year round.

DECIDUOUS
A plant that loses its foliage during winter, then produces new leaves in spring.

GRASSES AND SEDGES
A mix of evergreen or deciduous plants with grassy leaves. They can be clump-forming or spreading, and range in height from a few centimetres to two or three metres.

SHRUBS
Evergreen or deciduous plants with a permanent, multi-stemmed woody framework from 30cm–4m (1–12ft) tall.

TREES
Large evergreen and deciduous plants, which usually have a single trunk and are capable of reaching great heights. Trees need careful siting due to their longevity and size.

CLIMBERS
Deciduous and evergreen climbing plants useful for their foliage and flowers. Most need wires or trellis to cling to walls or fences, and can grow to a height of several metres.

AQUATICS
Plants that grow in wet ground or in water fall into three groups: those with leaves held above the water, those that lie on the surface, and those that stay submerged (see p.98).

Growth habits

Understanding a plant's habit helps you to place it in the garden. It also ensures you get the planting density right, so you achieve a balanced border that isn't overwhelmed by plants of unexpected vigour. Height and spread are usually marked on the plant label, but expect some variation due to different growing conditions.

Mat-forming
These plants spread by sending out shoots which then put down roots. Mentha requienii (Corsican mint) will steadily creep over gravel and paving.

Upright
As they often have little sideways spread, upright plants like Verbascum can be planted quite densely. They also provide useful vertical accents in the garden.

Fast-growing
Plants such as Lavatera need space when planted to allow for rapid spread. Plant labels give the size after 10 years, but check with other sources for growth rates.

Clump-forming
Over a few years, plants such as the non-invasive grass Pennisetum alopecuroides form a good-sized clump without threatening to swamp their neighbours.

Climbers
Climbers, including most clematis, take up little horizontal space as they want to grow up rather than out. Train them through shrubs and to clothe vertical structures.

Slow-growing
Many slow-growers will eventually become big, but it can take years. Buxus sempervirens 'Suffruticosa' has a slow growth rate that makes it ideal for low hedging.

Plants in containers

There is no reason why a container garden can't be as well planted as a border. It is an intimate and very flexible form of gardening that allows an almost continual mixing and matching of your plants. However, growing plants in pots can affect their growth rates and restrict their size, since compost, water, and nutrients are limited.

◁ **Big bonus**
A wide range of plants will grow successfully in large containers since they can accommodate more roots, water, and nutrients than small, narrow pots.

△ **Tight squeeze**
The restricted size and volume of compost in small pots limits your plant choices. You must water and feed plants regularly when grown in these conditions.

Mirroring nature

If you bring together plants from different parts of the world but from a similar habitat, it is possible to create a planting design that is both botanically and aesthetically pleasing. Seeing the plants *in situ* in their natural environment will inspire you – and give you a feel for the conditions they require.

Coastal survivors
A plant's ability to cope with gale-force winds and salty spray will govern your choice for a seaside garden. Luckily, there are some beautiful plants that are perfectly adapted.

Woodland effects
You don't need to be a botanical purist to create a woodland garden. You can combine plants from different countries, so long as they all enjoy cool dry shade in summer.

Alpine inspiration
A rock garden is designed to emulate the free-draining dry conditions of an alpine meadow. This image of the real thing shows the effects you can aim for.

Selecting plants

At this stage of the design process you should be getting a clearer idea of the look you want to create in your garden, and thinking about the plants you'll need. Designers often talk about using a "palette" of plants, as if they were paints, and, in many ways, creating a beautiful garden is like painting – except that you are visualizing three dimensions, and your materials, being living, growing things, aren't static. Use the ideas outlined here to help you draw up an inspired planting scheme.

Choosing a planting palette

Focusing your ideas at an early stage in the design process narrows your choices and helps to guide you towards choosing the right plants. It also minimizes expensive mistakes. Sourcing plants is much easier when you have a specific theme, perhaps a favourite colour, or style in mind. A cottage garden, for example, will give you the scope to mix and match a wide range of plants in an informal setting. Something more modern, on the other hand, will demand that you use a limited number of plants in a more organized way. Designing a low-maintenance garden filled with evergreens will, again, focus your choice (see pp.130–233 for garden styles).

△ **Tropical collection**
A flamboyant display of annuals with hardy and tender perennials is high-maintenance, but the results are exciting and worth the effort.

◁ **Easy-care scheme**
The established hardy shrubs and perennials in this formal planting require minimal maintenance. Their structure extends the seasonal appeal right through late autumn and into winter.

Functional planting

Certain garden features design themselves by default. For example, an exposed garden will need a windbreak, while an overlooked plot must have screening for privacy. Other design considerations might include fragrance by the front door, or a tree by the patio to provide shade on a hot sunny day. The design of such schemes is guided by their specific use, and this may limit your choice of suitable plants. The list below details the different design functions plants can fulfil, some of which may be pertinent to your plot.

1 *Provide shelter*
2 *Create a boundary*
3 *Produce food to eat*
4 *Offer shade*
5 *Perfume the garden*
6 *Screen neighbours*
7 *Hide an ugly view*
8 *Provide a wildlife habitat*

Sheltered seating area
Hedges do pretty much the same job as a fence or wall, but they have the edge when it comes to absorbing sound and wind. They also create a much softer effect.

Layers of interest

When space is limited, try to select plants that have a long season of interest. As well as those that flower over a long period, there are also many shrubs and perennials with colourful autumn foliage, structural winter stems, and spring buds. Precious few plants will fulfil all your demands, but look for those that tick the most boxes.

△ **Structure and colour**
The most useful plants here (peonies) work on several levels, providing structure and colour. In spring, their red shoots are followed by lush green foliage, then flowers.

△ **Foliage and form**
A closer look at a peony reveals how its flowers and foliage combine to make it stand out as an individual. Peonies often provide vibrant autumn leaf colour too.

△ **Flower in focus**
Close up you can appreciate the folded and crushed petals of this peony's double blooms. With other plants, such as passion flowers, the detail is in the intricate stamens.

Plant types and their design uses

There is, without doubt, a plant for virtually every situation, be it a tree, shrub, perennial, bedding plant, or bulb. When you're working out a planting plan, consider how best to use each plant, and ask yourself if it will create the look you are after, as well as how it will work next to other plants in the border.

Midrange plants
These make up the majority of the plants in a garden and include perennials and small shrubs. The substance of most plantings, they fill the gaps between bigger, more structural elements.

Structural plants
Plants can be structural on two levels. They can define the limits and framework of a garden, or the term can describe the plant itself, for example, if it has large paddle-shaped leaves.

Focal plants
Like ornaments, these are visual treats for the garden. It could be their distinctive colour, leaf shape, or stature that makes them stand out from other plants in the border.

Ground cover
People tend to think of ground-cover plants as being workmanlike. But there's no reason why they can't do a great job of being ornamental while smothering weeds as well.

Seasonal interest
The changing seasons make gardening a real pleasure. Choosing plants that provide an ever-changing display prolongs a garden's interest, changing its character as time passes.

Using structural plants

Structural plants are the backbone of a garden, forming the framework and helping to anchor other plants within a defined space. A beech hedge encircling a garden works in this way, as does a low box hedge around a border. By their sheer physical presence, individual structural plants – such as a *Gunnera* or *Cordyline* – can give focus to a planting scheme. Identifying key plants and deciding where to position them is the first step towards organizing a planting scheme for any garden.

Creating a framework

Hedging is ideal for defining the boundaries of a large- or medium-sized garden. It also provides shelter and increases privacy. Strike a balance between evergreen and deciduous species: evergreens are effective year-round screens, but because of the low winter sun they can cast a dense gloomy shade, while deciduous hedges allow in some light for most of the year, and can offer seasonal colour, too.

Use structural plants within the garden to frame (or block out) views and to lead your eye around the design. Shrubs in a border, perhaps forming a low hedge, provide a setting for midrange plants, and repeating planting helps to create visual reference points. When planting trees, consider their eventual size and the shade they will cast.

△ **Hedges for definition**
Hedging plants, both small and large scale (in this instance, beech), can be used to define the internal structure of a garden.

▷ **Structure in a border**
Here, green and purple maples (Acer) frame a stone statue, while the sculptural Gunnera at the back forms a focal point.

Temporary structure

While the main framework of a garden should be permanent, much of the planting within it is seasonal, emerging in spring and dying down in winter. Some perennials provide vital structure for all but a few weeks in spring, when, as is the case with many handsome grasses, their stems are cut to make way for new growth. Large, shapely foliage plants, such as *Miscanthus*, act as an anchor for smaller species, or contrast with leafy flowering shrubs like *Deutzia*. Airy plantings also benefit from the occasional strong shape as a visual counterbalance to their wispy forms.

△ **Structural accents**
Clumps of bold foliage (here cannas) in a busy planting scheme act as a foil for slim-stemmed flowers and provide structural accents in a border.

▷ **Reconstructing nature**
Using plants in broad interlocking swathes prevents an over-fussy effect, and the resulting planting, although strongly structured, looks natural.

Year-round interest

While evergreens may seem the obvious choice for year-round interest, visually they can be leaden and static. Deciduous trees and shrubs, on the other hand, may perform for several seasons, with new foliage in spring, followed, perhaps, by flowers, and then berries in late summer and vibrant leaf colour in autumn. In addition, trees often have a beautiful winter silhouette. Many species of *Sorbus* offer these benefits, and are ideal four-season trees for a small garden.

A winter garden may not offer the obvious charms of summer, but there can still be sufficient interest to draw your eye into the garden – perhaps even enticing you to pull on a coat and venture outside.

▷ **Colour and form**
If you mix deciduous and evergreen species, the garden in winter can be both structurally interesting and surprisingly colourful.

▽ **Spring offering**
Trees form an important element of the spring landscape, some offering blossom, others vibrant green new growth.

▽▽ **Formal topiary**
Formal planting is the ultimate in structural design. This row of clipped evergreen trees is balanced and restful, and the effect can be enjoyed during all four seasons.

Using midrange plants

Midrange plants belong to a broad group that includes bulbs, some small shrubs (often called subshrubs), grasses, and most herbaceous perennials. Their great range of shapes, colours, and textures gives you huge scope for creativity, and you'll find plenty to define your chosen garden style. They are also invaluable as gap fillers between structural specimens, and since many flower and reach their full height in their first season or two, you won't have to wait long to enjoy the full effect.

Shape and texture

Some of the best midrange plants rely on their shape and texture for interest more than their flowers. Those with strong leaf shapes, such as *Acanthus*, *Hosta*, *Ligularia*, and *Rodgersia*, can be grouped together for bold shapely plantings; or they can be used to separate plants with frothy flowers or foliage. Using contrasting shapes and textures throughout a planting design creates visual excitement, with no shortage of interest. Imagine the fine leaves of fennel (*Foeniculum vulgare*) against the large sculptural foliage of the globe artichoke (*Cynara cardunculus* Scolymus Group), or the delicate but busy fizz of gypsophila against bold round *Bergenia* foliage. Grouping plants with similar soft textures creates a different, much gentler, effect: try fennel with *Anemanthele lessoniana*, or *Molinia caerulea* subsp. *arundinacea* 'Windspiel' with *Aruncus dioicus* 'Kneiffii' or *Thalictrum delavayi*.

◁ **Multi-layered texture**
This sloping site features layers of beautiful foliage textures and colours, including pompon alliums and feathery fennel.

▽ **Spiky foliage**
The structural leaves of crocosmias give season-long interest; the late summer flowers can almost be seen as a bonus.

Shrubby structure

Many small shrubs are useful additions to a herbaceous planting because they add a degree of permanence and a change of character. Plant short shrubby evergreens at the front of a border to act as a foil to the procession of perennials that come and go as the seasons progress. Good front-line plants include *Teucrium chamaedrys*, *Lotus hirsutus*, *Hebe pinguifolia*, and *Iberis sempervirens*.

◁ **Staying power**
Once its small trumpet-like flowers fade in late summer, the silvery evergreen subshrub Convolvulus cneorum *remains as a foil for other perennials.*

◁◁ **Good mixers**
Low subshrubs, such as Helianthemum, *provide useful low level structure and mix well with perennials, but they also make a reliable display on their own.*

▽ **Foreground interest**
Block plantings of low evergreen hebes provide a weighty foreground that contrasts well with the lighter, airy grasses planted behind.

Flower and leaf colour

Perhaps the most exciting aspect of gardening is the chance to play with colour. If you include herbaceous perennials, the range of leaves and flowers can provide you with almost any tone or shade for your planting palette. When designing a scheme, consider the effect each plant has on its neighbour and decide if you want to use complementary or contrasting colours (see pp.46–47).

In general terms, a mix of colours generates an exuberant, slightly wild feel to a planting. Single-colour-themed borders look more sophisticated and have a cohesion that is satisfying to the eye. The restricted choice of plants also makes designing that much easier. Don't forget that just a hint of a matching shade in a flower or its foliage can be enough to link two plants.

Within a bigger border, colour combinations using two or three plants are effective. These can be timed for seasonal display, say, yellow wallflowers with the near-black tulip, 'Queen of Night'; or for something less transient, pale yellow *Anthemis tinctoria* 'E.C. Buxton', fronted by purple-leaved *Heuchera* 'Plum Pudding', surrounded by the leaves of *Hakonechloa macra* 'Aureola'.

▷ **Early summer border**
A jumble of flower colours and textured foliage injects this border with a huge amount of energy. Adding some summer bedding will add to the overall excitement.

▽ **Focus on foliage**
While still providing a perfect backdrop for other plants in the border, the large ribbed leaves of this luscious blue-green hosta make it a star in its own right.

Shady refuge

Planting choices in this compact courtyard garden are informed by the dappled shade of silver birch trees. Shade-loving perennials, such as aquilegias, *Alchemilla mollis*, and geraniums, vie for attention among leafy ferns and hostas, while the eye is drawn to patches of blue irises and orange geums in the sunnier spots.

DESIGNER Jo Thompson

Using ground cover

Ground-cover plants are used primarily to swamp weeds by creating a densely knitted blanket of leaves, stems and flowers that exclude light and use up all available moisture. The best examples are also decorative features in their own right, offering a tapestry of colour, texture and form, and providing a foil for other plants. Ground cover does not have to be restricted to very low-growing plants, and can include a variety of shapes and sizes, as long as they form a smothering canopy.

Dry sunny sites

Free-draining soils are "hungry"; you can feed them with organic matter but it usually breaks down quickly and its effect is short-lived, so it is best to choose plants suited to the conditions rather than to try to change the soil. Flowering ground-cover plants that thrive on sunny sites include *Helianthemum*, dwarf *Genista*, and low growing shrubby potentillas, such as *Potentilla fruticosa* 'Dart's Golddigger'. For leafy ground cover, try plants with grey leaves, such as *Hebe pinguifolia*, *Santolina chamaecyparissus*, and sage (*Salvia officinalis*). Several plants suited to hot dry conditions are also aromatic and include lavender and thyme. These conditions are the natural habitat of many bulbs, too. Small irises, such as *Iris reticulata*, and smaller species tulips, such as *Tulipa kaufmanniana* and *T. linifolia* Batalinii Group, can be dotted among the ground cover to add extra colour.

▷ **Tough plants for tough sites**
This gravel border features mostly Mediterranean-style ground-cover plants, including thyme, and catmint (Nepeta).

▽ **Sun protection**
Perfect for a hot spot, the silvery leaves of Stachys byzantina *reflect the heat of the sun and prevent the plant from drying out.*

Cool shady sites

Ground shaded by a leafy tree canopy is often extremely dry throughout the summer and provides the biggest challenge for both the plants and the designer. Reducing a tree's crown allows more light and moisture through to the plants below, and adding organic matter to the soil also helps to retain moisture. For dense spreading cover, try *Aegopodium podagraria* 'Variegatum' (variegated ground elder), *Asperula odorata*, *Cornus canadensis*, *Geranium macrorrhizum*, *Pachysandra terminalis*, or *Hedera* (ivy) species.

When shaded by buildings, the soil is usually slightly damper, making it easier to establish ground-cover plants. Shade-loving *Bergenia*, *Epimedium*, *Helleborus orientalis*, hostas, and many ferns, especially the dry-tolerant *Dryopteris* species, all produce a lovely effect.

▷ **Under a light canopy**
Semi-shaded conditions suit a wide range of leafy ground-cover plants, including Asarum, Carex, *and* Rodgersia. *This mix of green shades has a naturalistic quality.*

◁ **Dense shade**
Many colourful hardy geraniums are tough enough to cope with the difficult conditions under a tree canopy.

▽ **Twice the value**
Plants with long-lasting foliage make good ground cover; if, like these astilbes, they also offer flowers, their value is doubled.

Easy-care plants

In large gardens, where you can give them the space they need, vigorous spreading plants, such as *Hedera helix*, *Lonicera pileata*, *Trachystemon orientalis*, and *Vinca major*, make ideal low maintenance ground cover. In smaller gardens, however, giving over large areas to a single species is not always appropriate or practical; it can also be a waste of a good planting opportunity. Where space is limited, it is far better to use a mix of leafy plants, such as *Astilbe*, *Astrantia*, *Bergenia*, and *Geranium endressii*, planted close together. You will achieve the same effect, but it will be more ornamental and can be achieved with very little effort.

▷ **Carpet of colour**
Low-growing Lysimachia *and* Ajuga reptans *suppress weeds while also providing a colourful foil for larger plants.*

▽ **Mat-forming ground cover**
Vinca minor *puts down roots from spreading shoots to form a dense mat. Its small leaves contrast well with those of the* Bergenia.

Using focal plants

Focal plants work on several levels: they can entice you into a garden, distract you from ugly views beyond the boundary, or provide an eye-catching feature within a border. Most focal plants are evergreen or have strong shapes or colours, and offer a long season of interest, but don't dismiss those that perform for only a few weeks each year. Allow them their brief, glorious time in the limelight, and plan the rest of the garden around the show. Remember that focal and feature plants are the same thing.

Visual trickery

In much the same way as you would use a statue or an attractive container, you can site focal plants to lead the viewer's eye to a particular area of the garden. Positioned strategically, they can also distract attention from unsightly objects or views. Their presence not only makes someone shift their gaze, but can entice them to take a stroll around the garden too. When focal plants are repeated throughout a long border they act like visual stepping stones, helping to carry the eye along its length. They also hold the planting together, giving it an essential cohesion. Finally, using a clever trick of perspective, when planted in the foreground, focal plants make the garden behind seem like a separate area waiting to be explored.

◁ **Handle carefully**
Take care that a plant does not overwhelm the garden by grabbing all the attention and becoming an unplanned focal plant.

▽ **Scene stealer**
Pampas grasses have considerable stature, even when they are not in flower. Their late summer display makes them the natural focus of attention.

△ **Worth the wait**
A single plant's display (here a Yucca) can be the raison d'être *and seasonal climax of a whole section of a garden.*

Striking shapes

Many plants have naturally architectural or sculptural shapes: *Acer palmatum* var. *dissectum*, *Cornus alternifolia*, *Phormium*, and *Yucca* all make great focal plants. Many more, however, can be enticed over time with pruning and training to take on striking forms. This can be through traditional topiary, using slow-growing evergreens such as box, yew, *Ilex crenata*, or *Ligustrum delavayanum*. (Avoid fast-growing plants such as *Lonicera nitida*, which needs clipping several times over the summer to stop it losing its shape.) In addition, the adventurous gardener may like to experiment with other creative pruning techniques. By trimming off the lower branches of shrubs and trees you can make standards that produce lollipop shapes, or you can manipulate the branches to form tiers or cascading stems. *Carpinus betulus*, *Cotoneaster frigidus*, *Thuja plicata*, and *Viburnum plicatum* f. *tomentosum* 'Mariesii' are just four that respond well to this type of pruning. When trained, the skeletal winter outlines of deciduous plants can be as interesting as their leafy summer profiles.

Using colour

Very few plants can offer season-long colour, but you can still achieve some great effects with even just a short burst of activity from foliage or flowers. The following are all good candidates for focal plants: the autumn foliage of Japanese acers, azaleas, *Fothergilla*, and larch; the flowers of *Hamamelis*, *Laburnum*, and *Viburnum plicatum* f. *tomentosum* 'Mariesii'; and the winter stems of many of the birches, dogwoods, and willows.

Plants that provide dramatic colour, however, need careful handling. Remember that bright reds or yellows planted at the furthermost corners of the garden have a foreshortening effect. On the other hand, using paler colours at the end of the garden visually lengthens your plot (see p.48).

▷ **Colour care**
Acers are real scene stealers when their foliage fires up in autumn. Position them carefully among more subdued colours so that they can really shine out.

▷▷ **Second innings**
Hydrangea flowers are great value: colourful when fresh in summer, ethereally beautiful when faded in autumn, and stunning in winter with a dusting of frost.

▽ **Come closer**
The vibrant pink, pea-like flowers of Cercis siliquastrum appear before the leaves in early spring. The tree's form provides a focus at other times of the year.

△ **In the limelight**
Large scale centrepieces, these birch trees are made all the more arresting with dramatic winter sunlight.

◁ **Have fun with topiary**
Extravagance and humour are two ingredients that turn a feature into a great focal point. Here, yew is being trained through a giant topiary frame.

Seasonal planting

Designing a garden that offers a continuing series of delights throughout the year is both challenging and highly rewarding. Anticipating the emergence of new shoots, flowers and foliage in spring brings a huge amount of pleasure, which is then matched by the abundance of the summer, followed by warming autumn colours and the stark beauty of winter. With careful planning, you can use plants to decorate your garden 365 days a year with their colour, scent, shape, and form.

Spring awakening

Spring brings welcome colour and energy after the gloom of winter. Nature designed early flowerers for high impact, with brilliant displays from *Amelanchier*, cherries, magnolias, rhododendrons, and *Viburnum*. Bulbs are also keen to impress: flowers of blue (anemone, hyacinth, *Muscari*), yellow (daffodils, tulips), purple (crocus), and red (tulips) all add to the season's vibrant spirit. If you prefer a more subtle effect, choose some of the softer coloured spring-flowering shrubs and smaller plants, such as *Epimedium*, *Fritillaria*, *Helleborus,* and *Primula.* And nearly all spring bulbs have a white selection to temper a colourful display. However, it is often best to give full head to the season and simply enjoy the exuberance – just remember to plant your bulbs in the autumn or you'll miss the show.

▷ **Woodland setting**
Plants and bulbs that thrive beneath trees make use of available light and moisture by flowering before the leaves appear.

▷▷ **Natural drifts of bulbs**
Yellow daffodils and pink magnolia capture the freshness of spring. For naturalistic drifts, throw handfuls of bulbs across the ground and plant them where they land.

Summer profusion

In summer, the emergence of bees and other pollinating insects coincides with the majority of plants coming into flower. This natural abundance offers a huge choice of colours, heights, and shapes, which makes designing for a specific effect relatively easy. Check flowering times and choose a wide range of plants to prolong the display right through the summer months. Select perennials with beautiful foliage, so that when they have finished flowering they still contribute to the overall luxuriant effect, and set out each type of plant in bold groups of at least three for the greatest impact. Finally, to add to the richness, dot summer-flowering bulbs, such as *Allium*, *Gladiolus*, lilies, and *Triteleia,* throughout the border. Keep the display fresh by removing spent flowers and brown or damaged leaves.

◁ **Fiery mix**
The variety of plants available in summer makes a colour theme a much easier option – here a "hot border" of sizzling hues creates a unified display.

Autumn colour

In sheltered gardens, many half-hardy and tender plants, such as dahlias and *Canna*, will continue to flower until the first frosts. Hardy perennials, such as asters, *Aconitum*, and *Actaea* (syn. *Cimicifuga*), flower very late, too, and together with forms of *Fuchsia magellanica*, make good companions for a range of shrubs with fiery autumn leaves. Several summer-flowering perennials, including some peonies and hostas, provide a brief season of autumn leaf colour, but the main stars are the trees and shrubs, such as *Acer*, *Cornus*, *Prunus*, *Rhus*, and some *Berberis*, *Cotoneaster,* and *Viburnum*.

◁ **Seasonal transition**
The overlap between fading perennials and the onset of luminescent autumn foliage colours is a delightful twilight period in the gardening year.

▽ **Borrowed views**
This border has been designed as a stage set for the magnificent beech wood behind, but as the fiery autumn colours of Cotinus, Prunus, *and grasses ignite, all eyes are on the foreground.*

One garden, four seasons

By underplanting a wide range of shrubs and perennials with naturalized spring bulbs you can achieve year-round interest without the need for bedding plants. The unsung heroes of winter are deciduous trees – without the distraction of foliage you can better appreciate their attractive bark and shapely forms.

Spring: fresh and vibrant

Summer: lush and leafy

Autumn: fiery colours

Winter: stripped to the bare bones

Winter interest

There is no shortage of plants to provide colour and interest during the colder months. Winter-flowering honeysuckles, *Fothergilla*, *Hamamelis*, *Mahonia*, *Sarcococca*, and *Viburnum* offer flowers and scent, and the berries or catkins of *Corylus*, *Cotoneaster*, *Crataegus*, *Garrya*, and *Sorbus* add colour and texture. Evergreens and their variegated forms deliver winter foliage, while the bare bones of dormant perennials, such as *Rudbeckia* and *Sedum*, and the stems of grasses, such as *Miscanthus sinensis*, all add to the beauty of the winter garden. Trees also make stunning contributions to a wintry scene: birches with their stark white trunks; the twisted silhouette of *Corylus avellana* 'Contorta'; and the flowers of *Prunus x subhirtella* 'Autumnalis'.

Eyes down
An underplanting of snowdrops brings a glimmer of light to the dark base of shrubs, like this Cornus *(dogwood).*

Planting water features

Water fascinates and captivates like no other garden feature. Its movement, reflections, and sound bring an appealing mix of new sensations to a garden. Water also offers the chance to grow a different range of plants that can attract insects and other wildlife to the garden, whether you are planting up a natural pond, or complementing a modern installation.

Positioning your feature

For a natural look, small features like spouting figures and heads or an overflowing urn can be placed among the planting in borders. Ponds do best where there is good light, away from trees and falling leaves, which will rot and pollute the water. Also site them away from service pipes, such as electricity cables. All features should be viewed as an integral part of the design and placed where any filters and pumps can be hidden by plants, rocks, or decking. Child safety is also a prime consideration.

Choosing plants

Plan your waterside plantings exactly as you would your garden border, taking height, colour, and seasonal interest into account. Plants carry a label that show their preferred water depth – the distance from the crown of the plant (or top of their pot) to the surface of the water – and your choice is governed by the size and depth of your pool. Choose a mixture from the four main groups of water plants: oxygenators to keep the water clear; aquatic plants that grow in the water; and marginals and bog plants to soften the edges.

Siting a pond
Check first that the site does not carry main sewers, drains, or utility pipes. Choose a sunny position with some shade during the day, away from overhanging trees.

Trees are far enough away to prevent pollution from leaves

The pond is the focus of the overall design

Service pipe is a good distance from pond

The view from the house allows you to enjoy the feature

Bog plants
These plants thrive in a moist or wet soil. There is a wide range available, which includes some of the most colourful waterside plants, such as several irises, primulas, Lythrum and evergreen Lysimachia.

Marginal plants
Growing in a few centimetres of water at the margins, these plants soften the line between water and land. As well as colourful or interesting flowers (Saururus, Orontium), many have dramatic foliage (Sagittaria, Pontederia).

To reduce algal bloom, plant marginals in a low-nutrient compost

Aquatic plants
These deep-water plants root on the bottom of the pond, 50cm (20in) or more beneath the water. There are relatively few plants in this group, but it does include water lilies, which grow in water 50cm (20in) to 1.2m (48in) deep.

Sink aquatic plants in their baskets to the correct depth, as marked on their labels

Oxygenators
An essential element in a pond, oxygenators provide oxygen and absorb the nutrients otherwise used by algae. Some, like Ranunculus aquatilis, flower above the water surface.

▽ **Planning ahead**
Making planting ledges and boggy ground part of the initial design of a pond, allows you to grow plants with different depth requirements.

marginal plant depth

aquatic plant depth

Modern water features

In a contemporary setting, water is often used for its reflective properties and movement, rather than as a place to grow plants. However, several water plants, including species of *Juncus, Carex, Cyperus,* and *Equisetum* complement a modern, architectural style. A clean and unfussy look is important, so limit the variety of plants and use those with strong shapes for the best effect. Evergreens work particularly well in a modern setting.

△ **Dramatic statement**
The primitive-looking Equisetum hyemale *(horsetail) is invasive on land, but contained in a pond planter, its stiff, upright shape is very useful to the modern designer.*

◁ **Symmetrical planting**
The round leaves of water lilies emphasize the squareness of this formal pool, while the dramatic foliage of Zantedeschia *adds some exuberance and links the pool with the surrounding planting.*

Small pools

If space is limited, a small fountain, bubbling millstone, or half-barrel or trough filled with water and aquatic plants can give great pleasure. Place your feature by a seat or close to the house where it will be visible from a window. If you cannot plant into the feature itself, position it among plants (*Hosta, Astilbe, Primula, Myosotis, Filipendula,* and *Iris*) that often surround a pond or pool.

Mini oasis
When planting a miniature pool, take care to avoid vigorous plants and rely on subjects like Nymphaea tetragona, *a small, compact water lily.*

Wildlife ponds

The combination of water and a wide variety of aquatic plants creates an attractive habitat for frogs, dragonflies, and aquatic insects, as well as offering cover for fish. Native plants will attract local insects, but any exotic, non-invasive water plants will be beneficial to frogs, toads, and newts. If there is room, introduce a small waterfall to create the splash and moisture ideal for growing ferns and mosses at the pond edge. Also, provide both deep and shallow water for diverse planting and a more natural look.

Natural habitat
Even a small pond will attract a surprising amount of wildlife, and is a useful way of increasing children's interest in nature and the garden.

OTHER PLANTS TO CONSIDER

FOR MODERN WATER FEATURES
Cyperus alternifolius
Equisetum scirpoides
Isolepis cernua
Juncus patens 'Carman's Gray'
Schoenoplectus lacustris subsp.
   tabernaemontani 'Albescens'

FOR WILDLIFE WATER FEATURES
Butomus umbellatus
Caltha palustris
Iris pseudacorus
Myosotis scorpioides
Ranunculus flammula

FOR SMALL WATER FEATURES
Juncus effusus f. *spiralis*
Orontium aquaticum
Primula vialii

Assessing your garden

If your plot isn't a blank canvas, take the time to look carefully at what is already in place before you begin work on a redesign. If you have just moved into a property, it is worth waiting to see what plants emerge and how the garden looks at different times of the year. When planning a makeover of an old garden, cost may be a factor, and you may want to retain and incorporate favourite features.

Get to know your garden soil, too, and notice how much sunshine and rainfall the plot receives. This will tell you what plants will thrive in your particular growing conditions, and help you to avoid costly mistakes. Improving drainage by digging in grit, or adding plenty of compost to poor soil, will also broaden your choice of suitable plants.

The drawbacks of a sloping garden can be turned to an advantage by the use of terraces, steps, raised platforms, or suspended decking. Introducing these elements can revitalize a tired garden, giving it a new lease of life. The same is true of an area that stays constantly damp: transform it into a bog garden or pool and enjoy the pleasures of a wide variety of moisture-loving plants and the ensuing wildlife they attract.

Assess the soil and feed with compost if necessary.

Privacy is important, but it is wise to consider your neighbours' needs before making any major changes to a boundary. A tall, vigorous conifer hedge may shield you from view, but does it also cast a long shadow over their patio for most of the day? Legal obligations may come into play, too, so check first before you finalize your design or begin construction around a shared boundary.

Perhaps the most important piece of advice is to take your time before launching into a garden redesign and new landscaping. And if bare or ugly patches are inevitable while work is carried out, remember that strategically placed containers make a quick and effective screen.

Choosing the right plants for your site is an important first step.

Assessing your soil and aspect

Find out as much as you can about your site before you plan a garden. If you ignore the local environment and specific soil and drainage conditions, you could waste money on unsuitable plants, or discover that your planned seating area is in a wind tunnel, or that the lawn turns into a lake in winter.

Identifying and improving soil

Garden soils range from sticky clays to free-draining sands. Clay soil is prone to waterlogging in winter and dries hard in summer, while sandy soil warms up early in spring, but is a challenge to keep moist in summer. Clays can be very productive and rich in nutrients if manure and grit are dug in, but sands are typically poor and, without adding manure or garden compost mulches, won't retain moisture or nutrients. The ideal "loam" soil contains a mix of clay and sand plus organic matter. Loams are dark and fertile because of the organic content, form a crumb-like structure when forked over, and have good moisture retention. Test your soil (*above right*) before designing planting areas; loams when rolled hold together to form a ball, but crumble under pressure.

Testing clay soil
As clay content increases, you can form it into a ball or sausage, then a ring.

Testing sandy soil
This soil crumbles under light pressure, won't form a ball, and feels gritty.

△ **Grit improves drainage**
Large quantities of coarse grit worked into the top layer of soil (to fork depth) improves the drainage of heavy clay, but drains may also be necessary on waterlogged soils.

◁ **Well-rotted manure benefits all soils**
Manure causes fine clay particles to clump together, improving soil structure and drainage. It also helps sandy soil retain water and nutrients, but use it only as a mulch.

Testing acidity

The soil pH is a measure of acidity and alkalinity – 7 is neutral, below 7 is acid, above 7 is alkaline. Acid soils suit ericaceous plants while many Mediterranean herbs, shrubs, and alpines will grow happily in alkaline, lime-rich conditions. You can pick up clues about your soil by looking around the neighbourhood to see what plants are thriving. Soil type can also vary around a garden due to local anomalies, so carry out several pH tests using an electronic meter or simple chemical testing kit (*right*).

Determining your soil type
Taking samples from around the garden, use a test kit to check acidity/alkalinity.

Checking the aspect

The direction your garden faces has a marked effect on how much sun it receives and how exposed it is to wind. To work out your garden's aspect, stand with your back to the house and use a compass to check the direction you are facing.

Typically, south- and west-facing plots are warm and sunny while north- and east-facing gardens are cooler and shadier (*right*). Filtering the gales on an exposed site reduces wind-chill, and limits damage to structures and plants. As altitude and distance from the sea increase, temperature and exposure can be adversely affected, whereas urban areas produce and hold heat, keeping gardens artificially warm.

Windy sites
Exposure can restrict your choice of plants as well as your enjoyment of the garden. Provide shelter with deciduous hedging, which will help reduce wind speeds without creating turbulence, or use other permeable windbreaks (see also p.61).

Frost pockets
On sloping sites, cold air rolls down to the lowest point and pools there if its path is blocked. Less hardy plants here can suffer frost damage.

MORNING

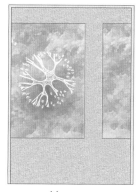

South-facing garden
Gentle sunshine across the garden from the east first thing creates pleasant conditions for summer breakfasts on a patio on the west side of the house.

House

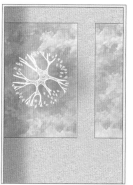

North-facing garden
Morning sun from the east soon disappears behind the house. Plant camellias, and other plants sensitive to morning sun after frost, on the shady east side.

House

East-facing garden
Enjoy breakfast on a patio by the house, but avoid planting wall shrubs here that are sensitive to morning sun after frost. Cold east winds can scorch tender foliage.

House

West-facing garden
The area near the house is shaded for most of the morning and a cool retreat in hot weather, but for early sun, design a seating area at the end of the garden.

House

MIDDAY

South-facing garden
In the height of summer, walls reflect the sun's heat and the whole garden is exposed to the sun, so you and your plants will bake without additional shade.

House

North-facing garden
The area next to the house is completely shaded, but the top end of a longer garden could be in full sun – perfect for a seating area and some sun-loving plants.

House

East-facing garden
Sun filters across the garden from the south but disappears behind the house in the afternoon. Cool after midday, this is a good aspect for a shady conservatory.

House

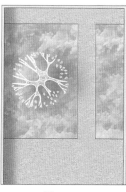

West-facing garden
Most of the garden is in sun at midday, especially in summer. Tender wall shrubs thrive on the house and north and west boundaries. A patio to the south offers shade.

House

EVENING

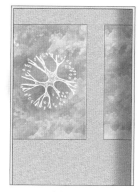

South-facing garden
Heat radiated from walls keeps the patio warm into the night. Most areas of the garden are ideal for frost-tender plants since the garden is warm all day.

House

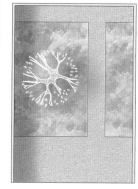

North-facing garden
Gentle light from the west offers an ideal aspect for woodland plantings. A patio on the east side of the garden will capture evening sunlight in summer.

House

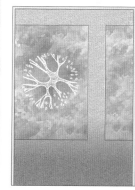

East-facing garden
The area by the house is shady. It can feel chilly sitting out because walls haven't absorbed heat during the day; make a patio at the far end of the garden for evening sun.

House

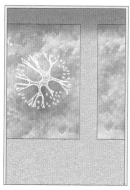

West-facing garden
A dining area by the house allows people to bask in late evening sun, but you may need some shade. Walls absorb sufficient heat to keep the area warm on summer nights.

House

Managing slopes and drainage

Predicting how water moves around, and how it can be directed out of harm's way, is the basis of drainage design. As a general rule, all man-made surfaces should be on an incline and water must flow away from buildings. In most cases, the water runs off hard surfaces, such as terraces or steps, into the soil where it is absorbed. However, sites on hills or with heavy, compacted soil can present drainage problems, and you may need to seek specialist help to avoid waterlogged conditions or flooding.

Drainage issues

All waterproof surfaces (roofs and paved areas) prevent water from draining naturally, and require attention; the water must be channelled to flow into municipal drains, or to run into soakaways or, if in small quantities, directly on to planting beds. The type of soil in a garden will affect drainage, with heavy soils (clays and silts) causing more problems than free-draining types (sands, gravels, and sandy loams).

On a steep site, water will flow quickly, seeking a low point and, eventually, an underground pipe, open ditch, or stream. Particular attention needs to be paid to water moving over bare soil or sparsely vegetated surfaces where it will cause gullies and erosion. However, if the landscape is undulating or contained, water will gather in the dips and in larger wet areas, such as bogs or ponds, and will need an overflow.

If you have a difficult site, determine the upper level of the groundwater (water table) as it may affect where you decide to position your drains or soakaway.

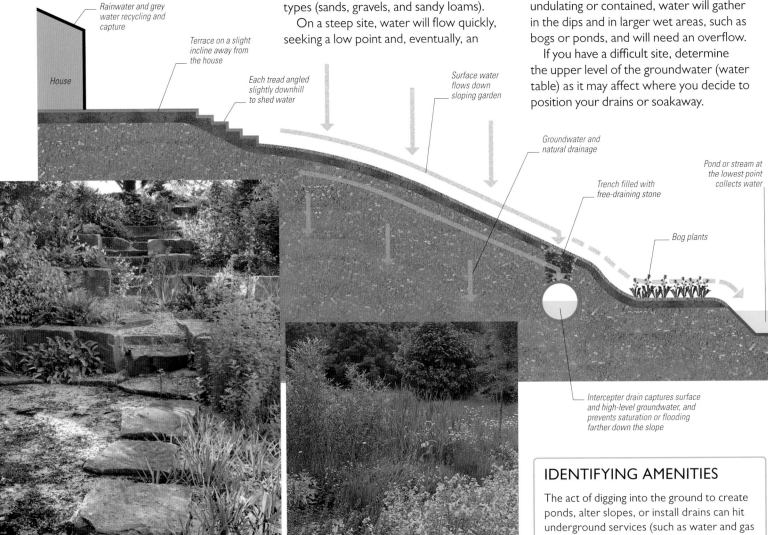

Rainwater and grey water recycling and capture

House

Terrace on a slight incline away from the house

Each tread angled slightly downhill to shed water

Surface water flows down sloping garden

Groundwater and natural drainage

Pond or stream at the lowest point collects water

Trench filled with free-draining stone

Bog plants

Intercepter drain captures surface and high-level groundwater, and prevents saturation or flooding farther down the slope

Sloping garden
All rainwater falling on this garden will eventually find its way into the ground or to the pond, which is located at the lowest point. An overflow may be needed to channel any excess water into an underground drain or soakaway.

Moisture-loving plants
Groundwater may be a problem, but it is also an opportunity. A naturally high water table or a butyl-lined bog garden can make an ideal place for growing a range of beautiful moisture-loving plants.

IDENTIFYING AMENITIES

The act of digging into the ground to create ponds, alter slopes, or install drains can hit underground services (such as water and gas pipes, and electrical cables) or existing drains and sewers. Never excavate the site unless you know what is directly below, and do not presume that amenities are in the exact locations shown on local council plans. Take your time to identify problems, and employ a specialist surveyor if you are in any doubt.

Rainwater collection
This recycled barrel holds enough rainwater to cover a short period of dry weather, and makes an attractive addition to the overall appearance of a garden.

Reduce flooding risks

Where drainage is not managed carefully, it can cause flooding, both in your garden and in the local neighbourhood, if storm drains are unable to cope with the excess. In the UK, there are regulations about paving over front gardens, so check before any redesign. To prevent flooding, install a Sustainable Urban Drainage System (SuDS) by creating areas where water can collect, and then be absorbed slowly into the ground, following heavy rain. Planted areas absorb large quantities of water, helping to mitigate flooding. You can also include small depressions that act as temporary ponds, filled with plants that thrive in wet and drier conditions. The aim is to retain all the water that falls on the garden in the garden. Also install water butts and use the captured rainwater on your plants.

Aquatic plants

Flow diagram
Where waterlogging is not severe, excess surface water can be directed into a drainage ditch or pond. If the water table is high, you will need to install an underground drainage system, preferably using a specialist contractor.

Garden pool
An informal pool can be used to capture excess water and will serve as a perfect habitat for wetland and aquatic plants and animals.

Design considerations

If your garden is on a sloping site, you will need to create flat, usable surfaces. Often this requires construction work so, when drawing up plans, consider budget and time constraints, the overall size and shape of the proposed spaces, and possible access for earth-moving machines. More complex solutions may be required for steeper sites and slopes that are less stable, or where especially large level areas are required.

Decking and platforms

To construct flat platforms or walkways on a slope with minimal disturbance to existing ground levels, it is best to use timber. Decking is especially useful where access for earth-moving is difficult, when slopes are too steep to alter, and on undulating surfaces around wetlands. However, it is short-lived compared to other landform solutions.

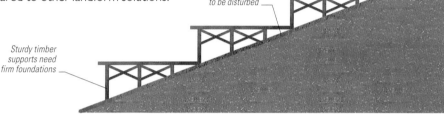

Stepped platforms could be linked by steps

Decking must be well constructed, so seek professional help

Original slope does not need to be disturbed

Sturdy timber supports need firm foundations

Terracing

Small-scale terracing can be used to make horizontal planting beds on a slope. A series of retaining walls, set one above the other, provide structure, then soil is cut away from the slope for backfilling. Work can be done by hand or with a mechanical digger. Any large-scale terracing will require the advice of professional designers and engineers.

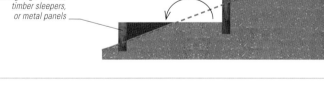

Original slope

Level surface for planting, lawn or patio

Balance the amount cut from the slope, with that required to fill behind the wall

Retaining edge of brick, timber sleepers, or metal panels

Creating gentle slopes

Undulating land can be landscaped into gentle slopes or flatter areas. Excess soil or hardcore may be generated, or more required to achieve the desired levels and, in both cases, this may increase the cost. Any changes will destroy existing vegetation and cannot be carried out beneath the canopies of trees that you want to retain.

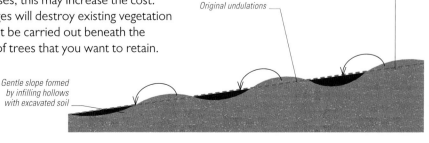

Soil to be removed for infilling

Original undulations

Gentle slope formed by infilling hollows with excavated soil

Down to the ground

Before selecting plants, test the soil in your garden to ensure your favourites will thrive there. Mediterranean-style gardens, such as this one, require free-draining soil and a sunny position, mirroring the conditions found in the plants' native habitats. A gravel mulch helps to prevent stems and leaves from rotting.

DESIGNER Martin Royer

Assessing your garden options

When thinking about a new design for your garden, first ask yourself which elements you like and want to keep, and which you dislike. Next, consider your budget – does it allow you to add a new feature, adapt the existing garden, or will you decide to go for a wholesale makeover, with a new planting design and landscaping? If money is tight, it may still be possible to rejuvenate a tired mature garden simply by taking a fresh approach and injecting some new ideas.

Degrees of change

Before you start designing, think about whether you'd like a completely new look, a new feature such as a patio or a pond, or whether you would prefer to keep the same layout but overhaul the planting. If your garden is small or seen as one space (rather than a series of connected spaces), you may want to rethink the entire area; larger plots will take more time and money to redesign from scratch. List the features you consider important and bear in mind that your needs may change in the future, as your children grow, for example.

A COMPLETELY NEW LOOK

Wholesale change can be hard to visualize, and often means removing existing structures and mature plants. However, it gives you the chance to do something radically different with a garden, and create an innovative space personal to you.

PROS
• An exciting blank canvas upon which to create whatever you want.
• The end result will be more coherent and integrated if you do not have to make compromises around existing elements.

CONS
• Loss of mature trees and shrubs.
• New plants take time to fulfil their potential.
• The reality may not match your vision.
• Short-term loss of wildlife habitats – although, depending on your new design, these should return over time.
• Sometimes a completely blank canvas can be more daunting than adapting an existing layout.

COST CONSIDERATIONS
• Potentially expensive – hard landscaping, and mature plants, if you don't want to wait for plants to grow, are costly.

DEVELOPING AN EXISTING PLOT

This is the most common approach and, even though you will be working with existing elements, it is still possible to refresh the look. List the features you plan to keep. With multi-level or sloping gardens, a site survey may be needed.

PROS
• This approach is usually less time-consuming and costly than a total makeover.
• You can work in stages and tackle different areas of the garden in sequence.
• You can make use of the existing mature planting, so there is no need to wait for everything in your garden to grow.

CONS
• The end result may lack cohesion. It is important to make sure that the features you add are complementary to existing ones.
• The renovations may not have the dramatic impact you are looking for.

COST CONSIDERATIONS
• Working with the current layout is less expensive than a complete makeover, and makes sense if you want to undertake changes in stages as money becomes available.

ADDING A NEW FEATURE

Making a change to just one part of your garden is the simplest option, but take care to integrate a new feature sympathetically. Pay particular attention to choosing materials and colours that blend in well with the existing design.

PROS
• Adding one new feature should be a straightforward change to manage.
• The rest of your garden will still be usable while this feature is being installed.
• Focusing on just one project means you can concentrate on getting the details right.

CONS
• Making sure that your new feature fits visually with the rest of your garden can be difficult.
• You can't let your imagination run free.
• You may damage other areas of the garden while building the new feature. Lawns and existing plants are particularly vulnerable.

COST CONSIDERATIONS
• This is the least expensive option – unless, of course, you are planning something very glamorous. The budget should be relatively straightforward to manage.

Case study: a new family garden

Every garden overhaul begins with a series of questions, and even when you have made a list of desirables and undesirables, you also need to consider the pros and cons of keeping or removing significant elements. For example, if you are thinking of taking out a mature tree because it casts summer shade, check that this disadvantage is not outweighed by its benefits: it may also provide shelter from wind, or privacy and screening from neighbouring buildings. Or, perhaps it adds height to your garden. It is also worth checking if your trees are protected by a tree preservation order (ask your local council).

Making decisions about your garden will be easier if you are very familiar with the plot. If your garden is new to you, be patient and live with it for several seasons to see what appears and what changes, before you make any dramatic alterations.

In the case study discussed here, a family garden is the subject of a renovation. The pictures below show some of the options open to the owners, depending on how much change they want.

The original plot
The way you use a typical family garden, and the amount of time you spend in it, will inevitably change as children grow. Design play areas so that they can be adapted.

INTRODUCE

MORE STRUCTURE
New hard-landscaping elements, such as paths, patios and walls, have immediate impact.

PLAY AREAS
Lay an appropriate surface and add structures that can be changed in the future as needs alter.

OUTDOOR LIVING ROOMS
Extend your living space by creating areas in the garden for eating, entertaining, and relaxing.

ADAPT OR REMOVE

BEDS AND BORDERS
Planting areas can be adapted and new shrubs and perennials added, or they can be totally replanted.

PONDS
Ideal for older children, but fit a grille if you are concerned for the safety of young ones.

UNSIGHTLY PATIOS
It is easy to distract attention from an unattractive terrace with tubs of plants, and garden furniture.

KEEP

OUTBUILDINGS
Sound, useful structures, such as greenhouses, can be integrated into your new design.

MATURE TREES
Try to work around mature, slow-growing trees if possible; they offer valuable structure and height.

PERENNIALS
Keep established plant communities where they are evidently thriving and suit the conditions.

Designing boundaries

Boundaries create a frame for your outdoor space, and are among the most important elements in a garden. They may indicate legal ownership, help to create a microclimate, and provide privacy. Most disputes between neighbours concern boundaries, and there are many legal regulations governing them, so before making any changes, first check who owns yours. If your neighbours have ownership, consult with them first and discuss any proposed changes to avoid conflict later.

Evaluating privacy

Before making changes to a boundary, especially if it is to be higher or removed, take time to evaluate the impact of the changes on your own and your neighbours' privacy and light. Check from all doors and windows, in particular upstairs windows, and assess what you can see now and what you will be able to see once the change has been made. Bear in mind that deciduous trees lose their leaves in the winter, which will mean more light but a less secluded garden. Also, raising the ground level on your side – with a deck, for example – may intrude upon your neighbours' privacy.

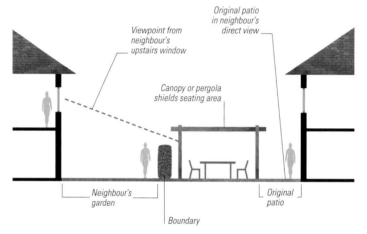

Viewpoint from neighbour's upstairs window

Original patio in neighbour's direct view

Canopy or pergola shields seating area

Neighbour's garden

Boundary

Original patio

Neighbours' views
Carefully chosen structures can create sheltered areas in your garden, reducing the need for high fences or hedging. A patio or seating area can be screened off from your neighbours' view with a canopy or pergola, allowing you to retain your privacy without loss of light to either side.

△ **Pergola cover**
Combined with climbing plants, this is an attractive way to create privacy without blocking light to the rest of your garden.

Increasing privacy

Increasing the height of boundaries may be illegal, so check with your local planning office first. However, it is possible to increase the privacy within your own garden without altering the boundaries themselves. Strategic positioning of new trees can help, but they will take time to grow. Tall, fast-growing evergreen hedges are now subject to planning control, as well as being high-maintenance, and should be avoided. Consider using trellises, which can support climbing plants and also help to create a sheltered microclimate by allowing air to pass through them (see p.61). Best of all, create spaces in your garden that are not overlooked by your neighbours (see diagram above).

◁ **Sheltered patio**
Well-placed planting forms a secluded site for seating areas – an umbrella can give additional privacy when the table is in use.

▷ **Temporary screen**
A makeshift cover like this one creates shelter and privacy wherever it is needed, and can be conveniently packed away.

Keeping in with neighbours

Although we all want some privacy, it is important to establish good relations with neighbours. You could place tall screens around your patio area, and lower fences elsewhere to encourage conversation. When planning your garden, consider anything which could irritate your neighbours, intrude into their space, or block their light.

Communal gardens, on the other hand, are designed to encourage friendship and cooperation. They need careful planning, and you should also consider who will be responsible for the garden's long-term maintenance.

△▷ **Friendly divide**
Low fences encourage communication and friendship between neighbours while also allowing more light into both gardens.

▷ **Shared space**
Communal gardens encourage community spirit and work well where there is shared responsibility for their care.

BOUNDARY REGULATIONS

Planning permission is needed to build a fence or wall over 1m (3ft) high next to a public highway or footpath, and over 2m (6ft) high on other boundaries, so check with your local planning office first. Fence posts should be on your side to ensure that the fence does not intrude on to your neighbour's property, and plant hedges at least 1m (3ft) away from the boundary, on your land. Your title deeds will show you where your garden boundaries lie.

Considering neighbours' light

There are laws governing an individual's right to light. Most light is blocked from gardens by trees, although garden structures and poorly planned building layouts can also create dark zones. Before taking the law into your own hands, seek expert advice. It may be possible to remove part of an offending tree, or to negotiate changes to boundaries to allow your neighbours more light. When planning changes to your own garden, consider the impact they will have on neighbours' light at different times of the day and year, both now and in the future. This particularly applies to trees and hedges, as they will grow in height and width, and could potentially cause problems.

Security issues

Boundaries provide security, but it is best to strike a balance between imprisoning yourself and opening your garden to your surroundings. Police recommend that fencing, walls, or hedges at the front of your house are under a metre (3ft) in height, so your doors and windows are visible from the street. Use lights to illuminate your space, but ensure that you do not floodlight your neighbours' property. Spiky evergreen shrubs, such as *Pyracantha*, holly, or blackthorn can be grown to form attractive barriers that will deter most intruders.

▽◁ **Thorny shield**
Pyracantha is a good choice for a burglar-proof screen, but will take time to grow; combine it with a simple post and wire fence until it matures, then keep it to under 2m (6ft) in height.

▽ **Automatic protection**
Electronic gates maximize security for large properties, or where burglary rates are high. They can be unattractive, so look for well-designed gates that blend in with your garden.

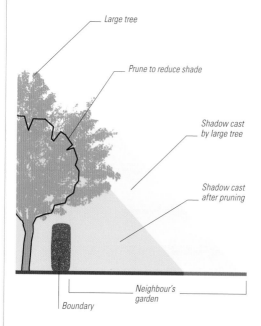

Large tree

Prune to reduce shade

Shadow cast by large tree

Shadow cast after pruning

Neighbour's garden

Boundary

Light idea
Think about how your boundaries, or elements within your garden, will cast shade on to your neighbours' plot. Here, a large tree could be pruned to allow much more light into the adjacent garden.

Creating a plan

Drawing up accurate site and planting plans is a crucial stage of any garden design. By bringing all your ideas together on paper you can see if they are viable within the space available, and get a clear visual image of what you want to achieve. Detailed plans also help prevent any costly mistakes before you buy materials and plants, or employ contractors.

With a few basic tools, and an assistant to help take measurements, you can draw up a site plan yourself. The process is explained over the next few pages, and includes a few tricks of the trade to make it easier. There is also a variety of computer software packages available for this purpose. However, if you have a difficult site or the prospect of drawing a plan is too daunting, you may prefer to employ a surveyor to help you.

When the site plan is complete, and you have decided on the structural elements and plants you intend to keep or introduce, you can start to play around with different design options. Even if you have an idea of the basic shapes you intend to use, it is always interesting to see how redirecting a sightline or introducing a small grove of trees or a collection of containers would change the mood of the garden.

A separate planting plan is also a good idea. Apart from helping you to assess the number of plants needed for your scheme, it will also clarify

A detailed plan, drawn to scale, brings ideas to life.

whether they work well in the overall design and fulfil their intended function. For example, you can use your plan to design a herbaceous bed in a sunny corner, or mark out an area for plants with winter interest that can be seen easily from the house.

Above all, study your plot from all angles and vantage points before you begin. Get to know your soil type and the path of the sun, then relax and enjoy this part of the creative process.

Plan planting carefully so your schemes work as intended.

Creating a site plan

Now that you have mastered the basic theories of garden design, it is time to put your ideas on paper. There are several different types of plan (see *pp.22–23*), but before creating your final design, you need to draw up a site plan, which shows all the basic measurements in your garden, as well as the position, shape and size of elements that you intend to keep. You can then use this plan to develop new layouts and planting designs.

Measuring up
Use the right equipment to ensure measurements are accurate. Get it wrong at this stage and your site plan could be rendered useless.

Getting started

The idea of creating a site plan can be a bit daunting if you haven't put one together before, but most plans are easy to produce, especially if you have a small- to medium-size, fairly regularly shaped garden with straightforward topography. However, if you have a large, irregularly shaped or hilly plot, or even one that is very overgrown, it may be wise to employ a land surveyor (see *opposite*).

When drawing up a site plan for your plot, first take a pencil and sketch pad (A4 or A3 are best) out into the garden and study the boundary and position of any elements you plan to keep, such as outbuildings, hard landscaping, and planting. It is also important to take note of the position of your house, including the doors and windows – not only

because their location will directly affect your ideas and design, but also because your house is one of the best points from which to measure other features, such as trees, sheds, and so on.

Now, roughly sketch the outline of the garden and the position of the relevant elements within it. Refine your sketch until it is clear enough to mark up with measurements. Then start measuring up (see *below and pp.116–117*). Even if you are only planning minimal changes to your plot, it is worth taking a few basic measurements, such as the length and width of the boundaries, to give you a sense of scale for new features, such as flowerbeds or a water feature. Whatever the size and shape of your garden, you will also find it easier with the

help of a family member, friend or neighbour. Take measurements in centimetres, rather than feet and inches, as the metric system makes it simpler to convert sizes to create a scale plan (see *p.118*).

ESSENTIAL EQUIPMENT

To measure up accurately you need the right equipment; most items are available from DIY stores and art suppliers. You can use a digital laser measure instead of tapes.

- Spirit level
- Tape measures of varying lengths – e.g. small, medium, and extra-long – or digital laser measure
- Pegs and string
- Sketch pad

Measuring a rectangular-shaped plot

Rectangular and square gardens are the easiest to measure. Ask your assistant to help you measure all four sides of the garden with a long tape measure, and add the measurements to the corresponding boundaries on your sketch. Then measure the length of the garden's two diagonals and mark them up on your sketch, too. To ascertain the position of features, measure at right angles to the house the distance to the feature/plant you want to keep. Do the same from a boundary, as shown below.

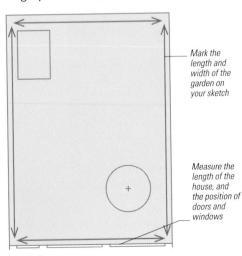

Mark the length and width of the garden on your sketch

Measure the length of the house, and the position of doors and windows

Boundaries
Carefully measure all four sides of your plot. Also measure the house and the distance from the house to the boundary.

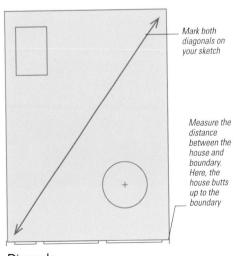

Mark both diagonals on your sketch

Measure the distance between the house and boundary. Here, the house butts up to the boundary

Diagonals
Diagonal measurements help to create an accurate plan of the plot if it is not a perfect square or rectangle.

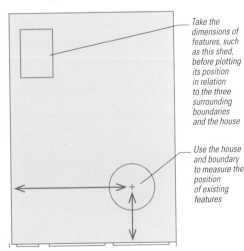

Take the dimensions of features, such as this shed, before plotting its position in relation to the three surrounding boundaries and the house

Use the house and boundary to measure the position of existing features

Features
Plot the position of features that you plan to keep by taking measurements at 90° from the house and boundary.

Site plans for rectangular plots

When you have decided which scale you are going to use, convert your measurements accordingly (see p.118). For large- or medium-sized plots you may want to create more than one plan for different areas, or use different scales to focus on a planting bed or similar feature that requires more detail. When drawing up your plan, use an A3 pad of graph or squared paper; you can use plain paper and a set square, but it is more difficult and the results may not be as accurate. Then, using a sharp pencil and ruler, plot the measurements on the paper and draw out your scale plan. You can then go over the pencil lines in pen.

You will need

- Metric, A3, squared or graph paper, or plain paper
- Set square
- Scale rule and/or clear plastic ruler
- Pencil and pens
- Rubber

1 Start in the bottom left-hand corner of your page. Draw the wall or walls of your house – including the positions and dimensions of the doors and windows.

2 To draw in the boundaries, mark the length and width on the plan, and add the diagonals. Diagonals show if the plot is a perfect square or rectangle, or slightly off.

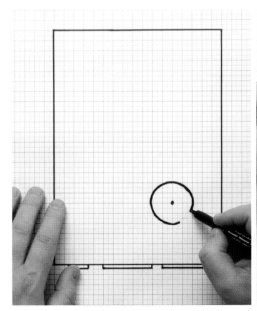

3 Use the measurements you took from the house and the boundaries with a tri-square to add trees and major planting – don't forget to include their canopies.

4 Lastly, plot all other features on your site plan. Carefully draw on sheds, greenhouses, patios, pools, paths, and outbuildings, if you are planning to keep them.

Measuring gradients

This method is only suitable for small inclines. It is useful if you want a couple of steps or terraced flowerbed and need to calculate the required heights. For more complex works or difficult sites, employ a land surveyor.

You will need

- 1 length of wood just over 1m (3ft) long
- Spirit level and tape measure
- 2 or 3 wooden pegs

Lay the wood from ground level to the top of the peg

Use a spirit level to ensure that the wood is exactly level

Measure the height of each upright peg from ground level

20cm — 1m (3ft)
50cm — 1m (3ft)
35cm — 1m (3ft)

1 From a specified point on the slope, measure 1m (3ft) down the hill, and hammer in a peg. Check it is vertical using a spirit level.

2 Lay the wood from the soil surface at your original point to the top of the peg, and use a spirit level to check it is horizontal. Measure the height of the peg.

3 Then, 1m (3ft) further down the slope, hammer in a second peg, as before. Lay the wood from the bottom of the first peg to the top of the second.

4 Measure the height of the second peg. Repeat these steps as necessary until you reach the bottom of the slope. Next, calculate the "fall" or drop.

5 To do this, add up the heights of all the pegs. Here the calculation would be: 35cm + 50cm + 20cm = 105cm over 3m (14in + 20in + 8in = 42in over 9ft).

EMPLOYING A SURVEYOR

You may wish to employ a land surveyor to produce a site plan for you if you have a difficult site. Surveyors in your local area can be found online. Land surveyors come under the jurisdiction of the Royal Institution of Chartered Surveyors (RICS), and it is advisable to check with them that the person you plan to employ is a member.

The cost of employing a land surveyor will depend on the size and complexity of your plot, but expect to pay between £800 and £1500. This fee will pay for a topographical survey, but a cross-section may cost more. Not all land surveyors are used to surveying gardens, so explain your needs carefully to ensure you employ the right professional for the job.

Measuring an irregularly shaped plot

If your plot is large, has an irregular boundary, is hilly or undulating, or very overgrown, it may be best to pay a surveyor to measure it accurately and draw a site plan. However, the methods shown here are not especially difficult, so try one and see how you fare before calling in the experts.

Advanced techniques

Although the measuring techniques shown here are slightly more involved than those used on page 114, they are still relatively straightforward. There are two methods to choose from: "taking offsets" and "triangulation". Start with an outline sketch of your garden on an A4 or A3 sheet of paper (see p.114). Then choose the technique you find easier, but do not use a combination of the two, as this will make the process more complicated, especially when you come to transfer your measurements to a scale plan (see p.118). For both methods, start by taking measurements of the façade of your house, including windows, doors and gaps between the house and boundary, and mark these on your sketch.

Taking offsets

To take offsets, you need two tape measures – one long and one shorter, to measure the length and width of your plot – and a giant tri-square, essentially a huge set square. Use the tri-square to help you to lay the long tape measure along the full length of the garden on the ground at exactly 90° to the house. Use the second, shorter tape to measure at 90° (again, use the tri-square to ensure the accuracy of your right angles) the distances from this main line to points along the boundary and to relevant features you want to keep. Clearly mark these measurements in centimetres on your initial sketch.

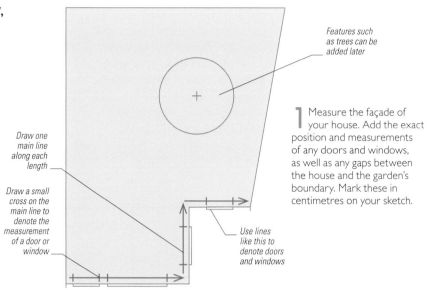

Features such as trees can be added later

Draw one main line along each length

Draw a small cross on the main line to denote the measurement of a door or window

Use lines like this to denote doors and windows

1 Measure the façade of your house. Add the exact position and measurements of any doors and windows, as well as any gaps between the house and the garden's boundary. Mark these in centimetres on your sketch.

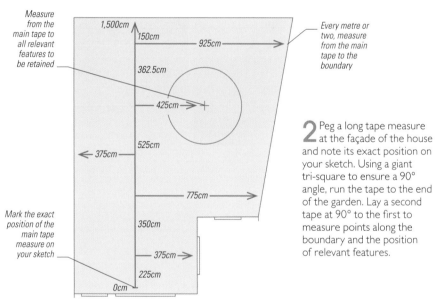

Measure from the main tape to all relevant features to be retained

1,500cm
150cm
925cm
362.5cm
425cm
525cm
375cm
775cm
350cm
375cm
225cm
0cm

Every metre or two, measure from the main tape to the boundary

Mark the exact position of the main tape measure on your sketch

2 Peg a long tape measure at the façade of the house and note its exact position on your sketch. Using a giant tri-square to ensure a 90° angle, run the tape to the end of the garden. Lay a second tape at 90° to the first to measure points along the boundary and the position of relevant features.

Getting some perspective

Whether you want to redesign part or all of your garden, site plans are an indispensable tool. However, unless you have at least some experience in reworking spaces or are naturally adept at imagining change, they may not help you to visualize how your new garden will look in three dimensions.

However, this simple idea will help to convey a sense of scale and proportion. You will need several bamboo canes, each just over 1m (3ft) in length, a tape measure and a giant tri-square. Form a square grid by pushing the canes into the ground at 1m (3ft)

intervals, and so that they are 1m (3ft) high (you can clip off the tops with secateurs if necessary). Take a photograph of your garden with the bamboo grid and print it out. Then enlarge it – to A4 or A3 size – on a colour photocopier. Lay a sheet of tracing paper over the photocopy and then use the canes to help you draw your proposed new features in perspective (see p.22). Use the grid to block in areas of planting, or to design screens, using the vertical canes to judge the heights.

Mapping your garden
This visualization technique works best in open spaces. Take an initial photograph of the area you want to design from the spot where you will be viewing the garden.

Using triangulation

On paper, this advanced measuring technique looks slightly more complicated than taking offsets, but in practice many garden designers consider triangulation easier and favour it over the offset method.

Triangulation involves marking two spots on the house – usually 1–2m (3–6ft) apart, but they could be further apart on a larger property – and then measuring from each of these spots to one point on the boundary, or a relevant feature, to form a triangle. This triangle and its measurements should then be marked on your sketch. Repeat this process at several points along the boundary – or the edges of a feature, such as a shed or a tree and its canopy. The more measurements you take, the more accurate your site plan will be.

You can then use these measurements to plot points on a scale plan and reproduce the exact dimensions of the garden and position of the boundaries, and any additional structures and key plants (see p.119).

Triangulation is a good method for measuring the position of curved boundaries accurately

1 Measure the façade of your house, and the doors and windows, and mark these measurements on your garden sketch (see *Step 1* in "Taking offsets", opposite).

Draw in the house, windows and doors (see Step 1 in "Taking offsets", opposite)

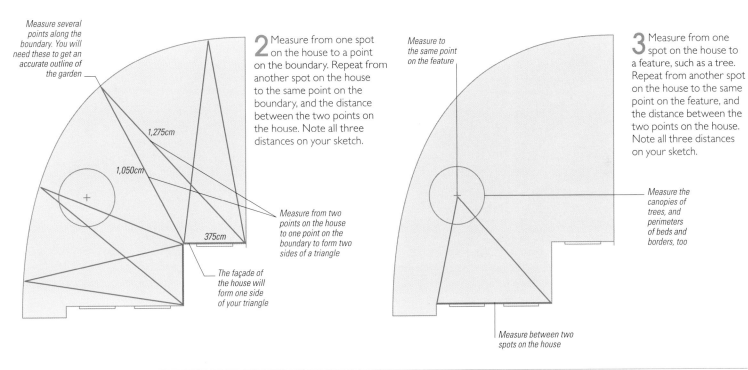

Measure several points along the boundary. You will need these to get an accurate outline of the garden

1,275cm

1,050cm

375cm

2 Measure from one spot on the house to a point on the boundary. Repeat from another spot on the house to the same point on the boundary, and the distance between the two points on the house. Note all three distances on your sketch.

Measure from two points on the house to one point on the boundary to form two sides of a triangle

The façade of the house will form one side of your triangle

Measure to the same point on the feature

3 Measure from one spot on the house to a feature, such as a tree. Repeat from another spot on the house to the same point on the feature, and the distance between the two points on the house. Note all three distances on your sketch.

Measure the canopies of trees, and perimeters of beds and borders, too

Measure between two spots on the house

1 Place the bamboo canes 1m (3ft) apart to form a square grid over the whole area – use a tape measure and giant tri-square to ensure accuracy.

2 Make sure that the bamboo canes are the same height, 1m (3ft) is a good choice, or the sense of perspective will be lost. Take another photograph of the garden.

2m

3m

3 Print out the photograph and enlarge it on a colour photocopier. Lay tracing paper over the image, then use the canes as a guide to draw your proposed features.

Using scale and drawing more complex plans

Essentially, a scale plan is a proportional visual representation of your garden, and you can draw one easily by converting the measurements you took of your garden (see *p.114 and pp.116–17*) to one of the scales outlined below. It is also worth investing in a scale rule (a Toblerone-shaped ruler with scales such as 1:10, 1:20, and 1:50 marked on it) for this job, as it dispenses with the need for calculations. When your site plan is complete, use it as the basis for your design and planting ideas.

Choosing a scale

There are several scales to choose from, including 1:10, 1:20, 1:50, 1:100 and 1:200. Put simply, a 1:1 scale shows an object at its actual size; on a 1:10 scale plan, 1cm on paper represents 10cm measured in your garden; on a 1:20 scale, 1cm on paper represents 20cm on the ground; and on a 1:50 scale, 1cm on paper represents 50cm in your garden. For small domestic gardens, it is best to use scales of 1:20 or 1:50; for a larger plot, you may want to use a 1:100 scale, or even a 1:200 scale for an extensive country garden.

Designers often draw more than one plan, and use different scales to show different details. For example, a 1:50 scale can be used for planting plans, and a 1:20 or 1:10 scale is best for structural features, such as a pond.

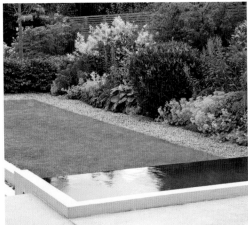

Whole garden 1:100
This is the best scale for an overview of medium-sized to large gardens. If your garden is particularly big, you may have to draw your site plan on an A1 sheet of paper.

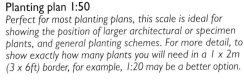

Planting plan 1:50
Perfect for most planting plans, this scale is ideal for showing the position of larger architectural or specimen plants, and general planting schemes. For more detail, to show exactly how many plants you will need in a 1 x 2m (3 x 6ft) border, for example, 1:20 may be a better option.

Architectural details 1:20
This scale allows you to work out quantities of hard landscaping materials, such as pavers. Use it to calculate the exact numbers you will need if building garden features yourself, or supply building contractors with a 1:20 plan to enable them to make these calculations.

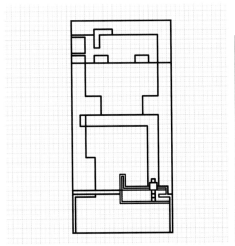

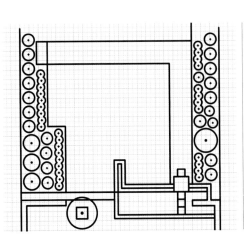

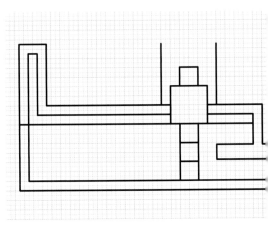

Drawing a plan for an irregularly shaped plot

You will need

- Metric, A3, squared or graph paper, or plain paper and set square
- Large pair of compasses (for triangulation)
- Scale rule and/or clear plastic ruler
- Pencil and pens, and rubber

Regardless of the method – triangulation or offsets – used to measure your irregular plot and its features, start by drawing your house and the doors and windows on your plan.

If you used offsets, draw a line at 90° to the house to represent the tape measure. Using the graph paper's grid and a ruler or a scale rule, plot the boundary measurements at 90° to this line; join the dots to form the boundary. Then add features, also plotting measurements at 90° to the central line.

If you took measurements using triangulation, use the method on the right to draw up your scale site plan.

TOP TIPS

- Use Google Earth to check the shape of your plot. On larger or more open plots you may even see trees, features and sheds.
- Don't over-complicate your sketch. If necessary, use more than one sheet to record dimensions of the main garden, and a separate sheet for details, such as planting plans.
- If an impenetrable area of vegetation gets in the way, estimate its dimensions from the measurements around it.
- When drawing your site plan, use metric graph paper for a more accurate result.

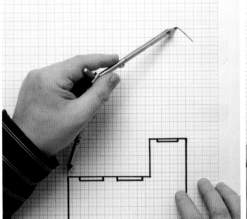

1 Draw the house, doors and windows, and then set the compasses to the first scaled measurement you took from the house. Place the point where you measured from on the house, and draw a small arc.

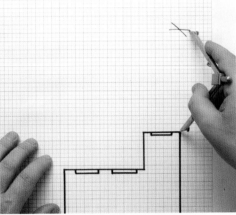

2 Reset the compasses to the second measurement you took from the house to form the triangle. Place the point where you measured from on the house, and draw a second arc to cross the first.

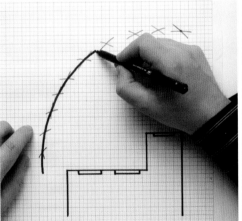

3 Repeat Step 1 and Step 2 for all of your boundary triangulation measurements. With a pencil, join up the centre point of each of the crosses to plot your boundary. You can then go over the line in pen.

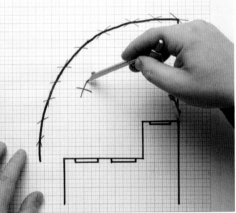

4 Use the same technique outlined in Steps 1 to 3 to plot the position of the garden's features – such as outbuildings, trees, plants, or water features – to create your scale site plan.

The finished site plan

You've taken all the necessary measurements, converted them to your chosen scale and drawn up your scale site plan (or plans, if you chose to use more than one). This accurate representation of your garden's boundary, and any existing features that you intend to work around, is an important design tool. Take photocopies of your plan, scan it onto a computer and print out copies, or make a few tracings. You can then use these copies or tracings to sketch shapes and ideas that will fit the plot.

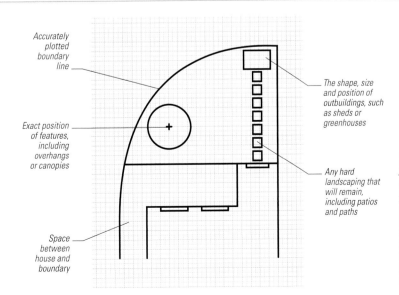

Accurately plotted boundary line

Exact position of features, including overhangs or canopies

Space between house and boundary

The shape, size and position of outbuildings, such as sheds or greenhouses

Any hard landscaping that will remain, including patios and paths

Using your working plan

As well as creating your own design, you can use a scale site plan to show builders the size and type of surfaces and features you want. Also some design companies offer postal services, particularly for planting schemes, and ask for a site plan to help them produce an accurate plan.

Experimenting with plans

More accurate than a bubble diagram or sketch, a scale drawing enables you to experiment with different layouts in enough detail to ensure that the design fits and works well. Although all proposed elements, such as paths and planting, must be drawn to scale, the drawing does not need to be too technical. Here, designer Richard Sneesby explores four ideas for one simple plot.

One garden: four solutions

This simple plan (*see right*) shows a rectangular plot, with the rear elevation of the house located along the bottom line. Adjoining the house is a patio, and the garden includes an existing tree and shed. There is also a rear access gate in the top-right corner.

Each of the four plans shows different design options for this site. All feature a lawn, pond, paving/deck area, as well as access to the back gate, and three include a shed. The tree has been removed in two schemes, as it would compromise the suggested layout.

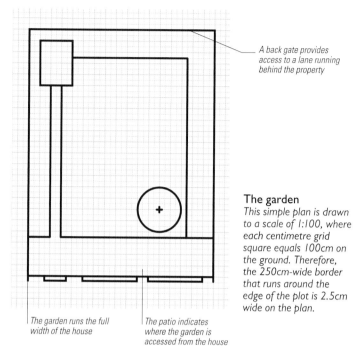

A back gate provides access to a lane running behind the property

The garden
This simple plan is drawn to a scale of 1:100, where each centimetre grid square equals 100cm on the ground. Therefore, the 250cm-wide border that runs around the edge of the plot is 2.5cm wide on the plan.

The garden runs the full width of the house

The patio indicates where the garden is accessed from the house

Option one

By positioning rectangular areas diagonally, the corner-to-corner orientation of this garden gives it a dramatic appearance. The design provides planting areas that are deep enough for larger specimens, and a triangular pond that can be appreciated from the nearby seating area. This is a garden of two halves, with a hedge dividing (and possibly screening) the two lawn areas, allowing each section to be given a distinct character.

Option two

The garden here is divided by a series of hedges that create a visual and physical chicane, keeping views short and varied; they also act as a unifying element across the plot. The hedges would be grown to different heights to allow or inhibit views, giving visual variety. Rows of trees reinforce the division created by the hedges but would allow views beneath their canopies. The design also includes rectilinear flowerbeds, a formal pond, and a shed hidden behind a high hedge.

Shed or summerhouse adds height and structure

Path and patio are laid with the same material to show consistency across the garden

A wide, formal lawn provides plenty of space for family recreation and socializing

Growing large shrubs or small trees in pots allows bolder planting near the house

Angles at work
Diagonal alignments work well in rectangular plots, especially in urban areas. They create generous planting beds and throw the eye to the corners, helping to make full use of the space available.

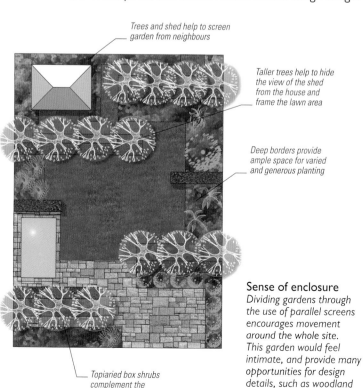

Trees and shed help to screen garden from neighbours

Taller trees help to hide the view of the shed from the house and frame the lawn area

Deep borders provide ample space for varied and generous planting

Topiaried box shrubs complement the formality of the pond

Sense of enclosure
Dividing gardens through the use of parallel screens encourages movement around the whole site. This garden would feel intimate, and provide many opportunities for design details, such as woodland areas and sculpture.

Option three

With its strong diagonal axis, this design works in a similar way to Option one. The oval-shaped lawn provides a central space, further defined by a low, flowering hedge. The trees also help reinforce the geometry and partially enclose the central area. The summerhouse is a focal element here, while a decked area and pool overlap on to the lawn to provide opportunities for attractive detailing. The planting beds are deep and generous.

Option four

This curvilinear plan would be more complicated to set out on the ground than the other designs, but would accommodate existing features and levels more easily. The lines are sweeping organic curves, the pond much less formal, and there are two distinct seating areas. Planting beds vary in width to allow a wide variety of plants and combinations to be grown. However, as there are no hedges, taller plants would be needed to prevent the garden from looking and feeling too open.

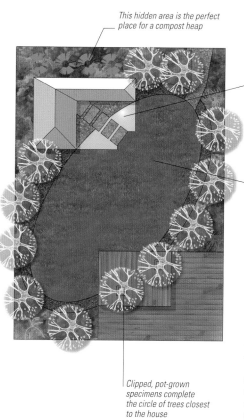

This hidden area is the perfect place for a compost heap

The pond, crossed via a small bridge, provides a restful setting for the summerhouse

The oval-shaped lawn makes full use of the site, and is kept private by the surrounding trees

Clipped, pot-grown specimens complete the circle of trees closest to the house

Oval approach
Central circular zones can help to unify a space and bring the garden together. Using an oval shape, in particular, gives the garden a sense of direction, and leads the eye across the spacious lawn.

A limited range of materials adds interest without clutter

Decked seating area acts as a focal point and provides space for seasonal containers

Larger trees give shelter and privacy, and help to define the view through the garden from the house

Informal gravelled area offers easy access and long views up the garden

Flexible design
Curved, organic shapes can be used to create a more relaxed feel, and the layout can be adapted to accommodate larger plants as they grow. Such shapes are difficult to build using paving materials.

Using design software

To create a plan on your computer, you can choose from a wide range of garden design software packages available. Look for options appropriate to your level of skill and the amount of detail you want to include. Most are quick to learn and some are free to download, although the price you pay generally determines the quality of the plan you can produce. Some packages also show how selected plants will develop over time; check that the package you choose provides a searchable database of plants suitable for your soil and climate conditions.

Professional designers use specialist computer-aided design (CAD) software to design accurate 2D layouts for contract drawings and commercial tendering, often combined with SketchUp illustrations to create 3D visuals of their ideas.

Bird's-eye view
Using SketchUp, designer Richard Sneesby shows a formal layout of terraces adjacent to the house. Shadows are geo-located to show the sunny areas throughout the year.

View from the terrace
3D models, such as this, allow you to see the design from eye level. Simple modelling, without planting details, helps develop an understanding of the space and structure.

Planning your planting

A combination of practicality and artistic flair is required to plan a planting scheme. The practical considerations include soil type, aspect, and the amount of shade and sun the site receives. You may also want to consider using plants to offer shelter, structure, or scent close to a seating area. Your ideas and inspirations inject the all-important artistic input.

First steps

Before planning your planting, draw up a site plan (see pp.114–119). You can then start thinking about the whole design of your garden, and how planting fits into the overall look. Sketch in the shapes and sizes of proposed beds and borders, and take photographs of the garden, too – either an aerial shot from a bedroom window, or from the area from which your planting will be most often viewed. You can then use these to help judge the scale of planting you need.

Visualization technique
You may find it easier to visualize your planting if you dummy it up by using garden objects of similar sizes, such as bamboo canes, buckets, cardboard boxes, and pots.

Choosing the right plants

You can either start with a list of your favourite plants and work them into your scheme, or decide on the look you want and then find plants to fit the heights and shapes required on your site plan. In reality, though, a planting plan usually ends up being a combination of both.

Whichever approach you take, bear the following points in mind. First, make sure the plants you choose will cope with the site and soil conditions; then when arranging plants on your plan, check their height, texture and shape in relation to those you will be placing next to them. Flowering period is important if you are looking to highlight a particular season; otherwise focus on foliage attributes first. In a small garden, a planting palette limited to relatively few different types of plants will have the greatest impact. For inspiration, go to the garden centre and group your chosen plants together. Or search online: Pinterest, Instagram, and Houzz offer lots of planting ideas.

△ Habitat match
In this naturalistic planting, drought-tolerant succulents and alpines, which require free-draining conditions, are planted in a bed of gravel and pebbles.

△ Balanced forms
Choose a range of marginals with different leaf shapes, such as these irises and astilbes, for a balanced poolside display.

▷ Consider the seasons
Make the most of the available light and moist ground in late winter and spring when planting under deciduous trees.

Plants with design functions

It is easy to become fixated on flower and, to a lesser extent, leaf colour, but many plants offer other equally attractive attributes that will add an extra dimension to your planting. Perfume is an obvious one and is a must near patios and around doors and windows, while structure – for example, the domed hummocks of *Hebe* and the sword-like leaves of *Phormium* – can be used to give visual emphasis to a planting. Many climbers can be trained over trellis to disguise an ugly view, and tough hedging plants, such as hornbeam or yew, make perfect windbreaks.

◁ **Fill the gaps**
Bulbs provide seasonal colour and can be squeezed between permanent plantings. Spring bulbs will cheer your border before most perennials appear, and Allium bulbs (left) in early summer are followed by colourful Gladiolus *and* Nerine.

▽ **Year-round interest**
Flower colour is often a transient feature, but foliage has long-term impact and should be seen as the mainstay of any border throughout different seasons.

▷ **Winter colour**
Winter flowers are a treat, so make sure you can see them from a path or the house. Several Hamamelis *have the bonus of scent.*

▷▷ **Scented plants**
These are best planted and enjoyed in warm sheltered areas of the garden where strong winds won't dissipate their perfume.

Coastal retreat

The drought-tolerant planting scheme in this garden is
designed to evoke the landscape along the Mediterranean
coast. A sunny site and free-draining soil provide the
perfect conditions for salvias, verbascums, *Centranthus
ruber* 'Albus', and wildflower *Jasione montana*, while pine
and tamarisk trees, typical of the region, offer cool shade.

DESIGNER Robert Myers

Drawing up a planting plan

Planting plans don't have to be complicated, but they can be a great aid, helping you to organize your ideas and calculate planting quantities. Just measure your garden fairly accurately and produce a simple scale plan (see *pp.114–119*), then use this to outline areas of plants and, in more detail, the shapes of planting groups and individual specimens.

Grouping plants

The lure of an instant effect often tempts new designers to cram too much into a small space, but overcrowded plants tend to be unhealthy, so always bear in mind their final spreads when drawing up your plan. You can achieve a fuller look by grouping plants together. With perennials, larger groups of three or more of a single species will have a stronger, more substantial effect than single plants dotted around, which can look messy. Grouping plants in sausage shapes (which works well for cottage- and prairie-style plantings), or triangles, is satisfying to the eye and makes it easier to dovetail disparate groups. Also, try placing the occasional plant away from its group to suggest it has self seeded for a naturalistic look. With shrubs, you can either plant in groups for an instant effect, or singly and wait for them to fill the space. Plant trees at a good distance from your property to prevent subsidence, and give them plenty of space to mature.

A formal planting scheme near the house will create a contrast with natural plantings elsewhere. Try a simple parterre formed of squares or rectangles enclosing a cross, and outline your design with box hedging. Avoid making the beds too small, because once planted up they could look cramped and overly fussy.

Prairie-style drift planting
Interlocking sausage-shaped drifts of plants give a less contrived look. Make the shapes a good size for maximum impact.

Modernist blocks
Strong geometric shapes are emphasized and complemented by bold blocks of planting, such as cubes of hedging.

Parterres
The symmetry and formality of a parterre makes planning fairly simple. Start with the outline hedging, then add the infill plants.

Random planting
To recreate a natural habitat, place plants in random groups. To avoid a chaotic design, use a limited colour palette.

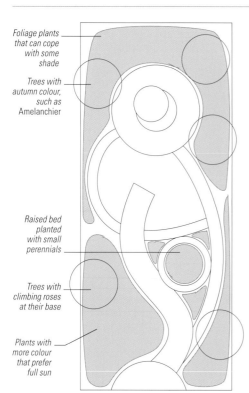

Foliage plants that can cope with some shade

Trees with autumn colour, such as Amelanchier

Raised bed planted with small perennials

Trees with climbing roses at their base

Plants with more colour that prefer full sun

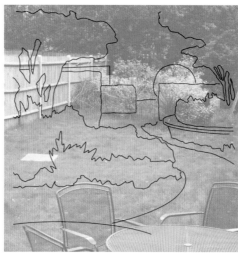

△ **Sketching on a photograph**
If you find drawing difficult, doodling over a photograph will help you visualize the design in the context of your garden and get a sense of perspective.

◁ **Bubble diagram**
This type of simplistic sketch, with rough shapes and annotation, will help you to position larger plants, such as trees, and pinpoint key areas of full sun or shade.

Sketching ideas

One of the simplest ways to visualize a planting plan for a small garden is to sketch the view from an upstair's window. Give full rein to your imagination and don't worry about accuracy at this stage. Next, identify the views from the house at ground level (stand by the back door) and consider whether you want planting to enhance, frame or block them. Finally, walk around the plot visualizing the overall layout, and the shapes and positions of structural plants, such as hedges and shrubs. Mark these on your sketch as simple shapes.

Take photographs as well, so you can refer to them when you come to draw your plan. If you feel confident, you can sketch your ideas directly on to photographs; if not, work on a sheet of tracing paper laid on top. You may find that black and white print-outs are less distracting to work with than colour pictures. Use your rough sketches as the basis for preparing a more organized planting plan.

The final planting plan

If you are preparing a plan for your own use you will not need fancy graphics, but if it is for a client a professional-looking plan (see *symbols on p.22*) is appropriate.

On your scale plan, first draw the outlines of the areas you want to plant, then add specific plants. To help you position trees or shrubs, draw circles to scale, depicting their likely spread. Mark perennials in as freehand shapes. To help you calculate planting densities, mark out a square metre on the ground and work out plant spacings for different species using their final spreads. Keep a note of them for future reference.

Draw your plan on graph paper, or on paper marked with a pencil grid of 1cm squares – you can then erase the latter when you ink in your final design. The scale you choose for your plan depends on the size of the beds or borders you are designing, but for a detailed plan, a scale of 1:50 or 1:20 is appropriate (see *p.118 for more on scale*).

Use acrylic tracing paper to copy your final sketch and produce a clean, finished drawing. Office suppliers sell tracing paper on rolls or as large sheets. Architect's offices often offer a copying service for large plans. You will need at least two copies: one for best and one that can be taken out into the garden at planting time. Consider laminating plans to make them weatherproof.

▷ **The finalized plan**
This is a planting plan for the border shown below. The shapes indicate the position and number of plants within each group. The plan also shows their final spreads, so you can see how they will fit together.

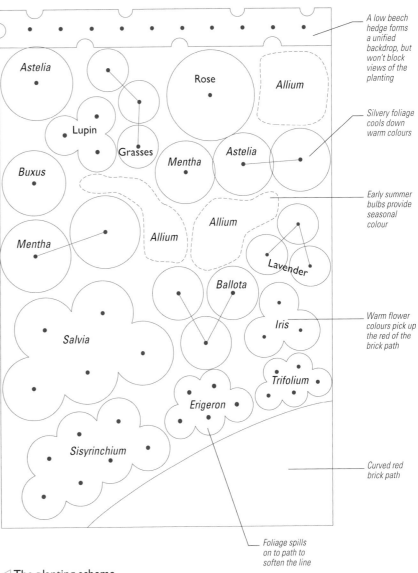

A low beech hedge forms a unified backdrop, but won't block views of the planting

Silvery foliage cools down warm colours

Early summer bulbs provide seasonal colour

Warm flower colours pick up the red of the brick path

Curved red brick path

Foliage spills on to path to soften the line

◁ **The planting scheme**
Successful plantings, such as in one of Cleve West's RHS Chelsea Flower Show gardens, will inspire your own creations, helping you to visualize how plants look in situ. Make a note of combinations that work well and use your smartphone or digital camera to take snaps of plantings that catch your eye.

COSTING UP YOUR PLANTS

If you can afford large shrubs and trees you can produce an instantly mature look; a smaller budget means younger plants and patience while you wait for them to grow. Perennials flower and reach their maximum height in the first couple of years, so don't spend a fortune on big plants.

It is worth asking garden centres and retail nurseries if they give discounts to designers; some also offer a plant sourcing service. If you can show you are a trade customer, wholesale nurseries allow you to buy plants in bulk.

Examples of planting plans

Irrespective of the style of garden you're designing, whenever you're putting together a planting plan check first that the plants you choose suit the site, soil and climate. If working on a design for a client, it is vital that you talk through your planting ideas with them before committing to a final design, not only to help them visualize the finished garden, but also to agree on a scheme that they can easily maintain.

A divided garden

Unless you divide it up in some way, a rectilinear garden holds no surprises. To avoid the "what you see is what you get" effect, designer Fran Coulter created a visual break between a decked terrace along the side and back of the house and the rest of the garden.

Plants used include:
1 *Rosa* 'New Dawn'
2 *Clematis* 'Pink Fantasy'
3 *Trachelospermum jasminoides*
4 *Lonicera nitida* 'Baggesen's Gold'
5 *Buxus sempervirens*
6 *Weigela* NAOMI CAMPBELL ('Bokrashine')
7 *Nepeta nervosa*
8 *Vitis vinifera* 'Purpurea'

Design in focus
When a garden is overlooked by neighbours, especially from an upstairs window, a climber-clad pergola provides privacy for seating or dining areas. However, in this design – the area shown is approximately 3.5 x 2.5m (11 x 8ft) – the pergola is used as a colourful boundary between a decked terrace and the garden beyond. The wood is painted a matt red to match the Scandinavian-style property. In Sweden, the paint is traditionally made with iron and copper ores, and these tones are picked up in the planting: the purple grapevine, wine-red Weigela, *and the pink rose and clematis.*

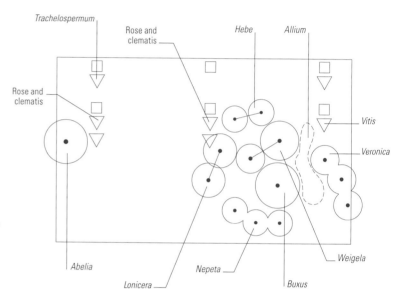

Shady area

This 3.5m (11ft) square border is backed by a high stone wall and cherry laurel. The owners asked designer Paul Williams for shade-tolerant planting that would mirror the formality of the adjacent garden. The plants here are mostly green with the odd splash of colour.

Plants used
1 *Dryopteris affinis* 'Cristata'
2 *Gazania*
3 *Prunus laurocerasus*
4 *Hosta* 'Krossa Regal'
5 *Taxus baccata*

Design in focus
To emphasize the formality of the garden on the other side of the path, this border (of which this is one section) is broken up with yew "buttresses" every three metres. Each section contains a simple planting and an urn or feature plant. Foliage is important: the plants need to be shapely and shade tolerant. Seasonal plants in the stone urn can contrast with or complement the surrounding plants.

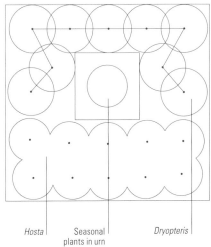

Hosta | Seasonal plants in urn | Dryopteris

City garden

Adam Frost designed this small city garden filled with romantic cottage-style planting. Soft red bricks are the perfect foil for the colour-themed planting, which is a sumptuous mix of crimson, pink, and mauve.

Plants used
1 *Salix elaeagnos* subsp. *angustifolia*
2 *Persicaria bistorta* 'Superba'
3 *Rosa* 'Souvenir du Docteur Jamain'
4 *Heuchera* 'Chocolate Ruffles'
5 *Astrantia major* 'Roma'

Design in focus
At the centre of this border, which measures roughly 1.2 x 2m (4 x 6ft), is a highly fragrant, dark crimson cup-shaped rose, its glossy green leaves forming an open framework for the slim stems of the Persicaria and Astrantia to grow through. These pale pink perennials complement the rich tones of the rose and help reflect light into the scheme, and are fringed at ground level by a wine-coloured Heuchera. The Salix, with its pale green filigree leaves, provides the perfect neutral backdrop to the warm colours.

Persicaria | Salix | Persicaria

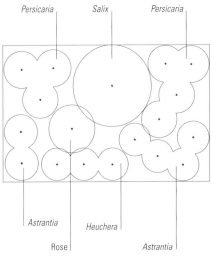

Astrantia | Heuchera | Astrantia
Rose

CHOOSING A STYLE

Garden styles explained

In design terms, style refers to the way in which we express ideas and organize materials, plants, colours, and ornaments to create a composition that can be understood and appreciated. While some garden styles are short-lived fashions, others represent major movements, each with their own aims and motives. In classically inspired formal design, order, repetition, and axial symmetry are used to create strict visual and spatial balance. This style dates from antiquity, and even when interpreted for modern gardens, the basic design principles still apply. In contrast, Modernism, which developed as an influential garden design movement in the early 20th century, uses asymmetry to create more complex views through the space, and many designers today have adapted elements of this approach to achieve stylish, crisp gardens. Others have taken a more relaxed approach, creating their own set of rules and evolving new ways to achieve harmonious designs.

The power of plants
Prairie-style planting is a dramatic way to create naturalistic swathes of colour over a large area.

Space to play
The open space and minimal planting in this garden offer the flexibility to accommodate many uses.

Eclectic influences
Combining different design elements can create a space that equals more than the sum of its parts.

EXTERNAL INFLUENCES

Garden styles commonly draw inspiration from cultural or historic reference points, which give them a particular theme. The aim is to create a stylized interpretation of reality, rather than an accurate representation. Japanese-style gardens, for instance, often lack the original philosophical and religious meaning but are nonetheless atmospheric. Similarly, the traditional cottage garden is a highly romanticized view of the simpler artisan model.

Broader issues and lifestyle changes have also helped to shape garden design. The increase in foreign travel has given gardeners a taste for the al fresco life (as seen in places like the Mediterranean), and for more exotic planting, which is being used increasingly in city gardens where warm microclimates allow a broader range of plants to thrive. Meanwhile, concerns about the environment are driving the use of sustainable materials and gardening for wildlife.

Minimalism updated
Modern materials, strong lines, and understated planting give this design a bold, contemporary edge.

Ideas explored
Garden style takes ideas and inspiration from around the world that can be easily adapted and recreated.

Leafy mix
Plants with colourful and attractively shaped leaves lend a lush, exotic look to contemporary designs.

FUNCTIONAL SPACE

The idea of the working garden has long been a recurrent feature of garden history, where the focus has involved growing food for the table. While the current trend for healthy eating has put home produce at the heart of many gardens once more, the functional requirements of gardens today are far broader, and reflect individual lifestyles more closely. Hence, families commonly require space for leisure, play and socializing, while other gardeners seek refuge from daily pressures in a calm space, ideal for rest and relaxation.

THE WAY AHEAD

As population densities increase, the urban garden is coming under ever greater pressure, diminishing in size but increasing in value. A century ago, a one-acre plot would have been considered quite small, but now people fill balconies, roof terraces, and postage stamp-sized gardens with vibrant ideas, creating a new idiom in direct contrast to much larger and expansive country gardens in which abundant space is the key characteristic.

Cottage dream
Generously filled borders and a haphazard approach to planting are typical of the cottage garden style.

Wildlife habitats
Even small garden ponds and boggy areas provide an excellent habitat for a wide range of wildlife.

Urban living
Ever-decreasing outdoor space is forcing gardeners and designers to develop creative new solutions.

Just as the form and function of gardens are changing, new styles are also being developed. Cutting-edge gardens often celebrate the man-made, creating dramatic and sometimes thought-provoking gardens that can be humorous or whimsical, philosophical and profound, short-lived or permanent. Designers of these conceptual or non-conformist spaces have thrown out the rulebooks to make cutting-edge gardens for a future generation. The cultural connection in many of these designs is strong, with some offering social commentary or presenting a reflection of modern society. Other designers mix up styles to create a fusion of the old and new, perhaps weaving cottage-style planting into a Modernist-inspired ground plan, or employing modern materials, sculptures, and technology in a formal, symmetrical layout.

As styles and references merge, so innovative ideas, fresh possibilities, and new idioms arise. Where once garden style was seen as conservative and predictable, it has now been rejuvenated and celebrates change. In addition, new links with architecture and art are being forged, and garden design is now considered a dynamic and socially relevant discipline.

Formal rules
A parterre planted with box hedging illustrates the symmetry and geometry of the formal garden style.

Blue sky thinking
Modern garden designers are constantly pushing the boundaries to create and develop new styles.

Productive patch
Attractive vegetables and herbs integrate easily into most garden styles, even where space is limited.

Formal gardens

Designed as expressions of man's dominance over nature, the features and natural elements in formal gardens are contained in an imposed geometry and structure. This idea is rooted in classical architecture and design, and many of the best examples of this type of garden can be seen in France and Italy.

A successful formal garden has a balanced design, achieved through symmetry and a clearly recognizable ground plan or pattern. Organized around a central axis or pathway, formal plans often focus on a key view through the garden from the house. In larger gardens there may be space for several axial routes that cross the central path, and sometimes reach out into the wider landscape. Sculpture, water, or decorative paving are also used to punctuate the areas where these routes intersect.

The geometry of the formal garden is clear and easily identifiable, but generous scale and balanced proportions are key considerations. Rectilinear shapes and forms feature most commonly in this type of garden, but any regular symmetrical shape can be used, as long as it sits on at least one axis. Circles, ovals, ellipses, and equilateral triangles are all options.

The material palette tends to be kept to a minimum, with gravel and regular paving stones most frequently seen. However, decorative elements, such as cobble mosaics or brick designs, are also popular. Water is employed either as a reflective surface or used for jets and fountains.

Dynamic water features provide movement.

Lawns and hedges are key planting features, the latter helping to define space or views, while dwarf hedging can be used to edge borders, create parterres, or form knot gardens. Pleached trees help to add height, and where space allows, avenues of trees line paths to accentuate vistas and draw the eye to a focal point in the distance.

Symmetry about a central axis attracts attention to focal points – such as sculpture or water features – in a formal garden.

What is formal style?

Formal garden design relates directly to the classical architecture of Greece and Italy. Ordered gardens originally provided a setting for the villas of the wealthy or powerful across Europe, echoing the symmetry of their grand houses. Known as "power gardening", it was seen as the ultimate in garden-making, embodying a sense of control. Although famous formal gardens, such as Versailles, are vast, the basic principles of the style can be applied to gardens of any size, even tiny urban spaces, where ordered, balanced designs work very well.

Symmetry about a central axis is crucial to emphasize the focus of the garden. Planting and construction are geometric and simple, with lawn, clipped hedges, and avenues forcing planting into order, and balustrades, steps, terraces, and wide gravel pathways all conspiring to unify the garden space.

Formal gardens in detail

Formality demands an axis, or central line, which is the basis of the garden plan. This could be a pathway or lawn, or even a central planting bed. Generally, the axis focuses on a dominant feature, such as a sculpture, statue, or ornament.

If space allows, cross-axes can be created; some larger gardens have multiple axial routes that create views along and across the garden. A dramatic sense of scale and proportion is essential as planting and paving are often kept simple – one reason why many modernists and minimalists appreciate this style.

The space should initially be divided into halves or quarters. Larger gardens can be partitioned further, but divisions should be as sizeable as possible to maximize the impact of long vistas, or the repetition of topiary or trees. Parterres, water pools, and expanses of lawn are typical of classical formality; examples by contemporary designers may also feature decorative borders that soften the garden's structure.

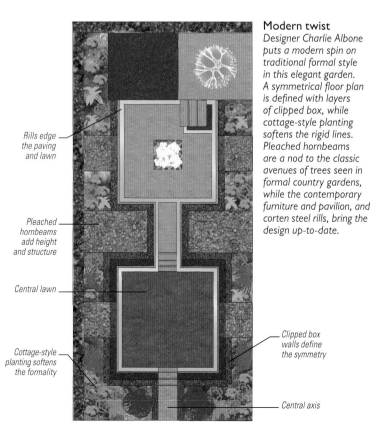

Rills edge
the paving
and lawn

Pleached
hornbeams
add height
and structure

Central lawn

Cottage-style
planting softens
the formality

Modern twist
Designer Charlie Albone puts a modern spin on traditional formal style in this elegant garden. A symmetrical floor plan is defined with layers of clipped box, while cottage-style planting softens the rigid lines. Pleached hornbeams are a nod to the classic avenues of trees seen in formal country gardens, while the contemporary furniture and pavilion, and corten steel rills, bring the design up-to-date.

Clipped box
walls define
the symmetry

Central axis

DESIGN INFLUENCES

Vaux le Vicomte by André Le Nôtre.

Although some of the earliest Islamic gardens were formal in layout, often divided by rills into quarters, classical and Renaissance influences have come to define this style. The doyen of the formal garden is André Le Nôtre, one of a long line of gardeners turned designers who found fame in France under the reign of Louis XIV. The gardens he designed at Versailles and Vaux le Vicomte are his most famous legacies. The false perspectives, level changes and reflective pools of both gardens are typical of Le Nôtre's approach to design, which won him the affection of the King.

Hedges, vast lawns, water features, and parterres of box and cut turf, often decorated with coloured gravel, as seen in Le Nôtre's work, set the tone for all formal gardens that followed, with views and perspectives manipulated for the best theatrical effect.

Key design elements

1 Symmetry
The symmetrical balance of a formal design can be achieved at any scale. Here, an olive tree and a parterre form a focal point in a circle that intersects the pebbled and paved central path.

2 Statuary
Gods and mythological creatures were the original subjects of statuary in formal gardens. In modern designs, contemporary figurative subjects and abstract works function well as focal points.

3 Topiary
Clipped hedging, typically box or yew for evergreen structure, is used to define space. Topiary provides architectural definition, and dwarf box hedges are used to form patterns in parterres.

4 Ornament
Large, ornate urns, often on plinths or balustrades, provide focal points or punctuation. Modern formal gardens use the same technique, although elaborate decoration is reduced.

5 Natural stone
Paving provides an architectural element for pathways and terraces. Sawn and honed natural stone slabs can create regular patterns, or they can be used to edge lawns and gravel paths.

Interpreting the style

Although the rules of formality are simple and clear, it is still a remarkably versatile style. The overall layout can be completely symmetrical and axial, or you can select just a few formal elements. One axis can be more dominant than another, for example, or a series of balanced, rectangular beds can be veiled by soft, romantic planting. You can also experiment with the style and opt for a traditional look or bring formality right up to date.

△△ **Contrasting elements**
An overflowing bowl creates a focus at the centre of this parterre in an enclosed corner of the Alhambra, bringing a dynamic quality to the formal planting.

△ **Contemporary order**
A simple rectangular lawn, elegant pleached hornbeams, and a pale paved surface create restrained formality. The three plinths and subtle lighting lend focus.

△▷ **Urban formality**
Limestone paving creates a crisp, formal edge to this lawn, offering clear definition. Pleached lime trees provide increased privacy in this urban space.

▷ **Ornamental hedging**
A parterre-style panel of box cartouches makes a decorative statement of light, shade, and texture. The pattern will read particularly well from the first floor.

"Set the geometric rules of formality, then decide which ones to break"

◁◁ **Aquatic symmetry**
Pools and a connecting rill form the focus of this formal arrangement, with the sculpture and fountain on the central axis. The planting is then arranged symmetrically.

◁ **Sculpted greenery**
Here, the tightly clipped topiary supports the axial layout. The mossy path itself breaks the rigid formality, with lawn softening the edges of the rustic paving slabs.

▽◁ **Softer planting**
Steel edging evokes a sense of formality in this grid-pattern garden, and is in stark contrast to the soft, light-catching grasses and perennials that fill the borders.

GARDENS TO VISIT

VAUX LE VICOMTE, Seine-et-Marne, France
Designed by Le Nôtre using false perspectives and axial layout. vaux-le-vicomte.com

VERSAILLES, Yvelines, France
André Le Nôtre's best-known garden. chateauversailles.fr

VILLA GAMBERAIA, Settignano, Italy
Garden of allées and formal compartments that radiate around the house. villagamberaia.com

ALHAMBRA & GENERALIFE,
Granada, Spain
Evidence of the Islamic influence on formal design in Europe, with water as a central theme. alhambra.org

DUMBARTON OAKS, Washington DC, US
Originally designed as a series of formal spaces and vistas, but with some naturalistic planting. doaks.org

CASE STUDY
BALANCED VIEWS

A symphony of classic formal style and contemporary features, this elegant garden is orderly and calming, providing beautiful views from the terrace over lawns, topiary, fruit, and flowers, while the gentle sound and twinkling reflections of a water wall soothe the spirits.

Simple shapes

Laid out on a symmetrical floor plan, the garden features a central rectangular lawn flanked by paths and pebble-shaped box (*Buxus*) topiaries. A smaller terrace on the right breaks the formal pattern, but identical stone links the two paved areas.

Italian influences

Inspired by Italian Renaissance gardens, the terrace features a water wall made from grey-green marble and travertine limestone. The soothing sound of gently flowing water sets the mood, bringing a sense of calm to this formal space.

Designers **Tommaso del Buono and Paul Gazerwitz**
Show **RHS Chelsea Flower Show**
Award **Gold Medal**

Citrus scents

The garden includes many Mediterranean influences, such as the lemon trees in large terracotta pots that flank the terrace on both sides, augmenting the design with their scented flowers and bright fruits.

Flower forms

To temper all the straight lines and geometric forms, the designers have included areas of soft planting that feature a range of herbs, perennials, and grasses, including *Stipa gigantea, Gladiolus byzantinus*, and *Anchusa azurea* 'Loddon Royalist'.

Green corridors

An avenue of pleached lime trees (*Tilia* x *europaea* 'Pallida') have been trained to form an elegant green canopy. Working in perfect harmony with the other clipped forms, they also have a practical use in the shade they provide to the terrace.

Formal garden plans

Although formal design follows specific rules, there is, as these three gardens show, plenty of scope for interpretation. Here, the designers Charlotte Rowe, who usually produces more contemporary works, and George Carter have both merged formal lines with classical details, yet two very different gardens have emerged. At Port Lympne, the early 20th-century layout proves that formal designs can be timeless.

Classic lines

In this small space, designed by Charlotte Rowe, the simplicity of design works well: the beds retain a mix of just a few species. The urn and *Ligustrum* topiary add height and a sense of scale to the scheme, while the *Hydrangea* provides an elegant focus to the central axis.

Key ingredients

1 *Ligustrum jonandrum*
2 *Hydrangea macrophylla*
3 *Artemisia* 'Powis Castle'
4 *Geranium sanguineum*

Charlotte says:

"My simple, understated design for this front garden in Kensington had to fit in with the regulations of the local conservation area. I used Yorkstone and bricks to match similar detailing on the house façade and evergreen screening for privacy but kept the overall design simple and understated.

"I'm often influenced by Modernist designers, such as Luis Barragán and Dan Kiley, so it was interesting to retain a sense of precision here in such a classical format. I think of the hard landscaping materials as the bone structure of the garden, which the planting can soften and enhance."

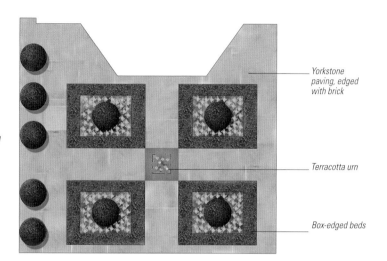

Yorkstone paving, edged with brick

Terracotta urn

Box-edged beds

Statement piece

This eye-catching chessboard at Port Lympne in south-east England is one of a sequence of formal "rooms" created in the 1920s by Philip Tilden for Sir Philip Sassoon, MP. Former head of gardens, Jeremy Edmond managed the site for many years.

Key ingredients
1 *Taxus baccata*
2 *Verbena venosa*
3 *Begonia semperflorens*

Jeremy says:
"This garden is one of a pair – the other, the Striped Garden, is on the other side of the main walkway. This one was designed to be looked at from the terrace above, and the pattern of lawn and bedding reads well from this position. We use annual bedding to add colour; usually pansies and polyanthus in winter, and Begonia and Verbena in summer. The changing view within the garden is its most majestic feature. Maintenance is difficult, but the graphic impact makes it worthwhile."

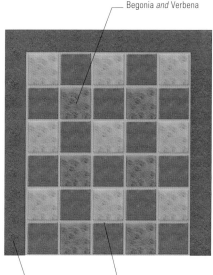

Summer chessboard of Begonia and Verbena

Grid of single bricks, laid to divide the planting, dates from the garden's inception

Garden enclosed within yew hedge

Below ground

The basement garden of this London mews house, designed by George Carter, is meant to be viewed from above. The minimal planting is architectural, to complement the property's classical focal points, such as the door frame at the end of the plot.

Key ingredients
1 *Cupressus arizonica* var. *arizonica*
2 Portland stone paving
3 *Cyclamen coum* subsp. *coum* f. *albissimum*
4 *Hebe* 'Pewter Dome'
5 *Festuca glauca*

George says:
"This is typical of my work – especially in smaller London spaces, where I think simplicity and order help give a sense of spaciousness. The garden was quite shaded, which led to the use of water to add sparkle and movement. The design was influenced by the work of the 18th-century architect James Gibbs – this is reflected in the door frame on the boundary wall. After dark, lighting creates the effect of an additional room."

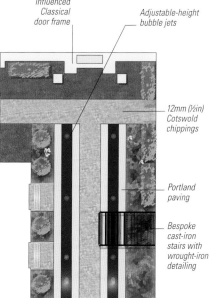

James Gibbs-influenced Classical door frame

Adjustable-height bubble jets

12mm (½in) Cotswold chippings

Portland paving

Bespoke cast-iron stairs with wrought-iron detailing

Cottage gardens

Celebrated for their abundant planting and apparent confusion, cottage gardens are traditionally simple and regular in layout, with a path to the door, and rectangular beds on either side. They were first used as productive spaces in rural locations, created to supplement the diet of the peasant, with the focus on food rather than flowers.

The cottage garden idyll that came to the fore in the late 19th century was, in fact, largely an urban invention – a reaction to the unrelenting cityscape, where people were more concerned with colour and scent than growing produce. Traditional cottage gardens were also championed by the famous garden designer Gertrude Jekyll, who refined them to form the basis of her Arts and Crafts planting schemes, which we now regard as typical of this style.

Jewel-like aubretia cascades over a weathered stone wall.

The scale of cottage gardens is generally intimate, sometimes even restrictive to movement, as dense planting is allowed to spill across pathways. Self-seeding is encouraged, as are plants that can colonize gaps in paving. Hedges are frequently used to divide the garden into a series of enclosed spaces with different planting schemes and atmospheres. The combination of soft and riotous planting with formal clipped hedges and decorative topiary results in one of the most successful contrasts in this design style. Away from the house, in larger gardens, there may be room for meadow planting and native hedges that create a wilder impression.

The most appropriate hard materials for use in cottage gardens are natural stone or brick, with weathered or rescued materials favoured for their aged and subtle appearance. Gravel is also used for pathways, partly because it allows easy self-seeding, and simple post-and-rail or picket fences also suit this naturalistic design style.

Decorative produce in a working garden.

While many cottage gardens adhere to simple patterns, others are more free-flowing, with sinuous pathways carving up the space, although any geometry is often blurred by the abundant planting and only revealed in winter when it dies down.

What is cottage style?

The romance of the cottage garden wins the hearts of many designers across the world. This is mainly due to the dominant force of the planting, profusion of colour, and the sheer variety of species used in this quintessentially English style. At its best, a cottage garden uses thematic or coordinated flower and foliage colour within small compartments or "rooms", as seen to great effect in the gardens at Sissinghurst or Hidcote Manor.

Cottage gardens in detail

The layout of a cottage garden should be simple and geometric, yet many diverge from this pattern into more idiosyncratic twists and turns, especially as the design moves further away from the house where wilder planting dominates. Pathways are often narrow, so that the plants partially obscure a clear way through. This romantic planting softens the appearance of a garden, and brings you into close contact with scent, foliage textures, and spectacular blazes of colour.

The paved areas are constructed from small-scale units, such as brick, gravel, setts or cobbles, which allow mosses, lichens or creeping plants to colonize the joints and surfaces. Simple seats, old well heads, tanks, pumps, and local "found" materials make interesting focal points and create a serendipitous quality, while arbours or arches decorate the thresholds between the various garden spaces.

Lawns are used, but it is the planting beds that are considered most important. Elsewhere in the garden, fruit and vegetable beds retain the simple geometry of the earliest cottage gardens, with brick or compacted earth paths providing access to these working borders.

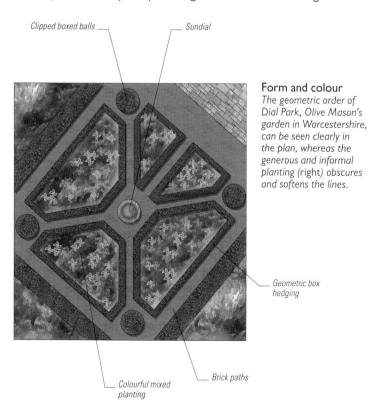

Clipped boxed balls — Sundial

Geometric box hedging

Colourful mixed planting — Brick paths

Form and colour
The geometric order of Dial Park, Olive Mason's garden in Worcestershire, can be seen clearly in the plan, whereas the generous and informal planting (right) obscures and softens the lines.

Summer colour in a garden for all seasons

With its wide range of foliage textures, tumbling climbers, colourful perennials and perfumed flowers, Olive Mason's garden is planted for year-round interest. In spring, green and white foliage predominates, interspersed with subtle drifts of daffodils, tulips, hyacinths, and forget-me-nots. The colours intensify in early summer (*above*) to warm pinks and mauves, with roses, geraniums, delphiniums, clematis and centaureas. As summer progresses into autumn, the palette deepens to the cerise, deep blues and purples of asters, phlox, dahlias and aconites, and in winter everything is cut back to reveal the simple pattern of the box hedges, enhanced by a bark mulch spread over the bare beds.

Key design elements

1 Profuse planting
Cottage gardens require intensive maintenance due to the complex planting. The art lies in the skilful association of planting partners, and the selective editing of species that become too dominant.

2 Rustic furniture
The patina of timber garden furniture changes organically over time; plants can be encouraged to weave through it to create an impression of apparently natural, but actually cultivated, recolonization.

3 Rose arbours
These make pretty shelters for seating, and can also be used to link different areas. Here the intense colour and delicate scent of a pink rose help to awaken the senses on a walk through the garden.

4 Weathered paths
Brick, stone sett, and gravel pathways provide textured surfaces as a foil to the complex planting on either side, allowing plants to seed and soften the boundary between path and border.

5 Vegetables and herbs
Productive borders are often seen in cottage gardens, with cut flowers and herbs used in association. This attractive mix softens the functional appearance of these areas, and may also help to control pests.

DESIGN INFLUENCES

Munstead Wood designed by Gertrude Jekyll.

The modern interpretation of the cottage garden is based to a great extent upon the work of Gertrude Jekyll and her architect partner, Edwin Lutyens. They created many outstanding designs in the 1890s under the auspices of the Arts and Crafts Movement. Jekyll used local cottage gardens around Surrey as the inspiration for her planting schemes, teamed with elements from her Mediterranean travels and colour theories developed during her fine art training.

Together, Jekyll and Lutyens designed and planted enormous borders in a luxuriant and romantic style, which brought timeless cottage-garden qualities to the estates of some of the wealthiest Edwardian families. Their approach set the agenda for the English garden over the next century.

Interpreting the style

A profusion of plants disguises the underlying geometry of this garden style. Plan simple-shaped beds and make sure they can accommodate a good depth of planting. The repetition of plants, colour themes, and hedging can bring some order to the borders, which are primarily created for variety and complexity.

△△ **Sunshine and flowers**
The late-summer colours of dahlias and cosmos ramble through shrubs, splashing their warm tints close to the incidental seat and almost smothering the path.

△ **Corner for reflection**
A old rustic seat, surrounded by soft drifts of pink perennials and a delicate white rambling rose, provides a quiet place for rest and contemplation.

△ **Decorative food crops**
Purple-flowered lavender echoes the vivid cabbage leaves in this potager. The lively mix of produce and ornamental planting is typical of the cottage garden style.

▷ **Underlying framework**
The rectangular beds and pathways can just about be seen beneath the warm-toned perennials and the searing carmine spikes of Lythrum virgatum 'The Rocket'.

"A sense of discovery, curiosity, and mystery is central to the cottage garden experience"

GARDENS TO VISIT

EAST LAMBROOK MANOR, Somerset
A cottage garden for modern times, planted by Margery Fish. eastlambrook.com

HIDCOTE MANOR, Gloucestershire
Celebrated Arts & Crafts masterpiece.
nationaltrust.org.uk/hidcote

MUNSTEAD WOOD, Surrey
Gertrude Jeykll's house and garden.
munsteadwood.org.uk

SISSINGHURST CASTLE GARDEN, Kent
Vita Sackville-West's 20th-century garden.
nationaltrust.org.uk/sissinghurst-castle-garden

△△ **Restricted palette**
The cottage garden is reinterpreted by the design company Oehme, van Sweden in this border in Virginia, US, where shrubs and perennials are intricately woven together.

△◁ **Framing vistas and views**
This rose-covered pergola provides height and enclosure, as well as rich colour and perfume. Use various structures to define the entrances linking different spaces.

△ **Simple restraint**
Low box hedging contains the unstructured border planting of poppies, salvia, and foxgloves; a technique appropriate to front gardens, where greater order may be required.

CASE STUDY

FLOWERING GLORY

Cottage gardens are all about the plants, shown in this contemporary design, which blends a profusion of blooms in a medley of colours and forms, while the natural tones of the timber decking and stone sculpture ensure the plants are never upstaged.

Woodland edge

The twining wisteria and river birch, *Betula nigra*, with its peeling shaggy bark, punctuate the garden space with their structural presence, rising up above a sea of colourful perennial plants, edible herbs, and dainty annuals.

Water for wildlife

Water often plays a part in cottage gardens, whether half barrel, pond, or natural pool (as here). As well as offering a relaxing space to swim, the water in a natural pool is cleansed by a range of plants that attract many forms of wildlife.

Designer **Jo Thompson**
Show **RHS Chelsea Flower Show**
Award **Silver-gilt Medal**

Organised chaos

The overall look of a cottage garden is never too contrived. Here, the dense, slightly unruly planting scheme and open spaces designed for relaxation are perfectly balanced to create a sense of natural abundance.

Escape to nature

A timber-framed, two-storey retreat, with nods to country vernacular style, is reached via a path of raised decking through lush planting and over water, allowing visitors the fantasy of escaping to a rural idyll that is the essence of cottage design.

Leafy seclusion

Boundary edges are blurred by a wall of trees and shrubs planted around the perimeter, which also helps to shelter the garden and create a private space. Native trees, such as the field maple, provide food and habitats for birds, insects, and other wildlife.

Cottage garden plans

Abundant planting and a mass of flower forms, textures, and colours define a cottage garden, with the hard landscaping – usually narrow paths of stone, brick, or gravel – taking a back seat. In the design by Gabriella Pape and Isabelle Van Groeningen, the lively soft planting comes in many colours, while Jinny Blom has opted to celebrate bright pinks and rich reds in a limited, warm palette.

Sea of plants and flowers

This garden was designed by Gabriella Pape and Isabelle Van Groeningen for the RHS Chelsea Flower Show as an homage to Karl Foerster, a great nurseryman who experimented with perennial plants. It creates the sensation of swimming through the foliage and flowers.

Key ingredients

1 *Digitalis purpurea* 'Alba'
2 *Hakonechloa macra* 'Aureola'
3 *Hosta* 'Sum and Substance'
4 *Veronica* 'Shirley Blue'
5 *Paeonia lactiflora* 'Duchesse de Nemours'
6 *Aquilegia chrysantha*
7 *Hosta* 'Royal Standard'
8 *Achillea* 'Moonshine'

Isabelle says:

"This layout was based on Karl Foerster's own garden in Potsdam, Germany, so it's not typical of our work. The planting, however, is. Influenced by the English style, it incorporates colourful matrix planting, and drifts of plants and flowers are reminiscent of Edwardian woodland gardens. These themes recur a lot in our work.

"Our influences are varied and we often bounce ideas off each other to develop design solutions. English garden designers, such as Vita Sackville-West, Geoffrey Jellicoe, and Charles Wade, are a major influence. We also create gardens and their planting around existing elements."

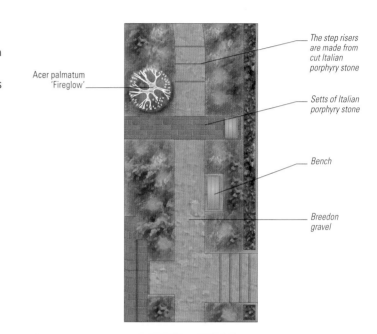

Acer palmatum 'Fireglow'

The step risers are made from cut Italian porphyry stone

Setts of Italian porphyry stone

Bench

Breedon gravel

Restrained palette

Modernist treatments, such as simple, clean paving, provide a cool contrast to the hot-hued palette of plants that tumble and explode around this garden by Jinny Blom. In true cottage style, the seemingly haphazard, densely packed planting pockets soften and relax the more ordered layout. The use of gravel allows plants to self-seed, creating additional random patterns of spontaneous growth. Grasses, seedheads, and bulbs create veils of foliage and texture.

Key ingredients

1 *Betula nigra*
2 *Akebia quinata*
3 *Geranium* PATRICIA
4 *Allium sphaerocephalon*
5 *Verbena bonariensis*
6 *Panicum virgatum* 'Heavy Metal'

Jinny says:

"This view is just one part of a multi-levelled garden – the different parts of which are connected by walkways and steps, so that, overall, the design flows nicely. The clients were a young family, and, as a result, the design needed to be robust, allowing the children to play freely.

"We agreed a strategy of hard-wearing, virtually indestructible materials that would be softened with romantic planting. This seems to have paid off, as the garden has matured well. We have recently added yew hedging in order to create a visual anchor in winter.

"I was inspired by the work of Italian architect Carlo Scarpa. In terms of flow and visual stimuli, his work was very important in creating the design."

White limestone bench top

Crushed limestone gravel

White limestone paving

Mediterranean gardens

Two garden types are associated with the Mediterranean region: informal and formal. Informal gardens tend to feature gravel, with planting arranged in structural groups or masses. This look is inspired by the shrubby vegetation (*maquis*) of the south of France or the more arid regions of southern Italy and Spain. Olives, citrus fruits, vines, lavender, and rosemary thrive in these conditions, as do succulents and grasses, while colours tend to be muted, incorporating soft sage-grey greens and purple-blues. Gravel is used between areas of planting and to create pathways. Drifts of plants appear to emerge spontaneously in the gravel, perhaps punctuated by arrangements of rocks and boulders. Sometimes a dry stream bed is recreated with clusters of informally arranged drought-resistant plants.

For more intimate and often urban spaces, terracotta instantly evokes the style, supplemented by mosaic tiles or features to add splashes of colour. Walls are often white-washed, creating clear backdrops for shadows, but where paint is used, hues are often bold. Rustic containers introduce colourful planting at key points, and may be used as focal features or arranged in informal groups of different sizes.

The formal gardens of the Mediterranean tend to utilize water and stone, often with clipped hedges and specimen trees such as tall, slender cypresses. In some of the gardens of Spain and southern Italy there is a clear Moorish influence, as seen in the courtyards and water features of Spain's Generalife and the Alhambra. Decorative parterre planting is also typical of the formal style, with plants selected for foliage rather than flower colour, and densely planted trees such as *Quercus ilex* (holm oak) providing cool shade.

Fleshy succulents are ideally suited to a warm, dry site.

Typical Mediterranean courtyards offer seating areas in shade.

What is Mediterranean style?

The popularity of the Mediterranean as a holiday destination has created a thirst for gardens that reflect this region. The mild winters and warm, dry summers favour specific groups of plants, often hardy and low-growing, with olive trees, vines, lavender, various herbs and many succulents combining to produce a distinctive style. These plants are designed to look natural, against a background of textured surfaces such as gravel and scree. Trees provide dappled shade, and water (a precious resource) is used sparingly, if at all. Any outdoor space can reflect a Mediterranean atmosphere, from large, sheltered plots to colourful, decorative courtyards and roof terraces. Across the world, California, South Africa, and parts of Australia and Chile have similar climates to southern Europe and make excellent locations for Mediterranean gardens.

Mediterranean style in detail

In Mediterranean gravel gardens, pathways are not defined by formal paving. Instead, gravel is used across the entire space, serving as both hard landscaping and a mulch for planted areas. This unifies the garden, allowing plants to be grouped informally and leaving smaller areas of paving to provide more stable surfaces for seating.

Pergolas or arbours are used for shade, and when planted with vines and other climbers enhance the Mediterranean atmosphere, providing the perfect location for sharing al fresco meals. Alternatively, plant trees for patterned shade, either in groves or as individual specimens in key locations.

Water is used to create sound or as a focal point, but, as a precious resource in these landscapes, it would not normally be seen in the form of large pools. In courtyard gardens, decorative rills or bubbling fountains echo the Moorish gardens of Spain and southern Italy. Colourful tiles and mosaics provide vibrant patterns while planted terracotta pots add splashes of vivid red or pink.

Californian-style Mediterranean garden
In this Californian gravel garden, designer Bernard Trainor has created a low, curved wall – which doubles as a sinuous seat – close to the house and beneath the shade of some trees. The wall frames the space while providing a backdrop to the water bowl.

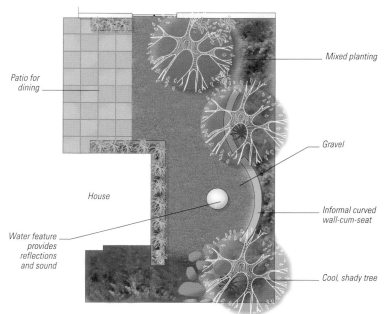

Patio for dining

Mixed planting

House

Gravel

Water feature provides reflections and sound

Informal curved wall-cum-seat

Cool, shady tree

DESIGN INFLUENCES

The dry landscapes of the Mediterranean with their soft colours have influenced many garden-makers. Gertrude Jekyll included Mediterranean species in her planting schemes, mixed with more familiar border plants. In the late 20th century, Beth Chatto created dry gravel gardens inspired by plants of the *maquis* (Mediterranean scrubland), and in France designer Michel Semini took similar inspiration from *maquis*-style planting. Today, James Basson leads the way in Provençal garden design.

Garden by Michel Semini, southern France.

Key design elements

1 Shady seating areas
In these sun-drenched gardens, shade is key, and can be provided by trees planted as individuals or in groups. Timber pergolas and arbours with climbers also provide a shady setting for outdoor dining.

2 Gravel floor
Limestone forms the typical gravel of the Mediterranean, creating a light, textured surface through which plants can grow. Larger boulders can be used as focal points. Landscape fabric below suppresses weeds.

3 Rills and pools
Water is often confined to rills in more formal gardens, and used to refresh the air or to mark spatial divisions. In gravel gardens, overflowing containers or water bowls are used for reflections and gentle sound.

4 Succulents and silver foliage
Many species have adapted to drought with fine, silver, or fleshy foliage. Rosemary and lavender are typical, with *Euphorbia*, *Agave*, *Yucca*, *Bergenia*, and *Genista* providing suitable associations.

5 Terracotta pots and tiles
The Mediterranean is famous for the terracotta pots used in gardens, as focal points or as planted containers. Old olive oil pots make sculptural features. Aim for larger-sized pots where possible.

6 Mosaic features
Floor surfaces in courtyards (or on roof terraces) are created from tiny, coloured cobbles laid out in intricate patterns. Glazed and brightly coloured tiles are also often used to decorate walls and grottoes.

Interpreting the style

This style is often typified by the materials and planting. Gravel gardens recreate dry, sun-baked landscapes, using rustic limestone or terracotta for pattern and decoration, while planting is informal and drought-tolerant. Formal gardens are often defined by cypress or palm avenues, with arbours for shade. Courtyards are often decorated with glazed tiles, and may also be filled with leafy plants to create an oasis with water as a focal point.

◁△▷ **Avenue of cypress trees**
Tall, slim, elegant cypress trees create a formal avenue to frame this walkway, highlighting the gazebo as a shady focal point in this ordered garden.

△ **Provençal landscape**
The wide joints in the pale limestone paths create patterns and allow thymes to colonize. Lavender-blues are virtually the only flower colours.

◁ **Splashes of colour**
Brilliant colour dominates this sun-filled space, the painted wall clashing with the bougainvillea overhead, which offers some shade for outdoor dining.

▷ **Deceptively simple**
The quiet simplicity of this gravel garden is emphasized by the decorative water feature, which reflects dappled light from the vast tree canopy overhead.

◁ Bubble fountain
A tall terracotta pot is lined and used as a bubble fountain, perfect for a terrace feature. Water circulates from a reservoir concealed below.

▽ Moorish look
In this Moroccan courtyard, lush planting forms a backdrop to the elegant tiles and raised water bowl.

"*Create contrasts of sun and shade, bold texture and sizzling colour*"

△ Courtyard calm
Stone and gravel create flexible and functional surfaces in this small urban space, with large pots, architectural foliage plants and seat cushions providing the main drama.

◁ Foliage garden
Simple colour-washed rendered walls provide a coordinating architectural backdrop to textured planting and sculpture, reflected in turn in the pool alongside.

GARDENS TO VISIT

ALHAMBRA, Granada, Spain
Islamic and Renaissance influences combine with water, planted terraces and courtyards.
alhambra.org

BARCELONA BOTANIC GARDEN, Spain
Featuring a huge collection of Mediterranean species from Catalonia and around the world.
museuciencies.cat/visitans/jardi-botanic

JARDIN MAJORELLE, Marrakesh, Morocco
Famed for its planting and deep blue walls.
jardinmajorelle.com

STRYBING ARBORETUM, San Francisco, US
A wonderful collection of native Californian and Mediterranean planting.
sfbotanicalgarden.org

CASE STUDY

SUN-KISSED RETREAT

Mediterranean gardens are famous for their tough yet beautiful drought-tolerant plants, sun-drenched open spaces, and dancing fountains and water spouts. Here, these elements are combined in a modern update of a traditional courtyard garden.

Precious water

Water is present in almost every Mediterranean garden, and here the spouts pour into a cool, refreshing rill, adding movement and sound to the design. The rendered wall links tonally with the informal stone paving that divides the space.

Tapestry of colour

The informal planting scheme cleverly combines a tapestry of different colours and textures, using heat- and drought-tolerant perennial plants, including silvery artemisia, achillea, red *Dianthus cruentus*, and white *Centranthus*.

Designer **Cleve West**
Show **RHS Chelsea Flower Show**
Award **Gold Medal and Best in Show**

Cracked terrain

The rocky terrain of the Mediterranean coast is echoed in the irregular stone paving. Mortar joints between the stones allow rain to slowly percolate into the ground, ensuring that any available moisture is not lost.

Form and shade

The pagoda tree (*Sophora japonica*) in the centre and yew hedges beyond provide much-needed shade and natural structure to anchor the design. They also help to convey a sense of enclosure, creating a private area for relaxation.

Ancient origins

Sculpted columns, made from textured concrete and terracotta, are included to evoke the ruins of an ancient temple. They act like a stage set, contextualizing the design and giving the garden a feeling of permanence.

Mediterranean garden plans

There are two Mediterranean garden types: naturalistic and wild, and formal. Each of these designs merges elements of both. Karla Newell's courtyard is a burst of colour set around the rectilinear lines of a Moorish pool, and Michel Semini's relaxed garden in southern France features formal hedging. In Acres Wild's design, the garden is laid out according to a strict grid, and its planting is aromatic and lively.

Moorish design

Colourful tiles and walls add depth and interest to Karla Newell's own garden. Planting is dense and textured, using palms and large-leafed architectural species. The pool, kept clear to reveal the lively mosaic, provides a focal point around which pots and specimens are arranged.

Key ingredients
1 *Fuchsia magellanica*
2 *Euonymus japonicus* 'Latifolius Albomarginatus'
3 *Acer palmatum* var. *dissectum*
4 *Hosta sieboldiana* var. *elegans*
5 *Arum italicum* 'Pictum'
6 *Pelargonium* 'Vancouver Centennial'
7 Italian glass mosaic
8 *Lathyrus odoratus*

Karla says:
"My Brighton garden was inspired by Spanish and Moroccan courtyards – such as the Majorelle in Marrakech, in which intense, painted colour is combined with carefully detailed spaces. I like crafted elements, so I laid and designed the pattern for the mosaic tiles (based on traditional Moroccan designs) myself.

"The garden's not far from the beach, and enjoys a sheltered microclimate, enabling me to introduce a Mediterranean range of plants. The planting palette is varied and relatively high maintenance, which suits me as I have a keen interest in gardening. The space provides an outdoor room."

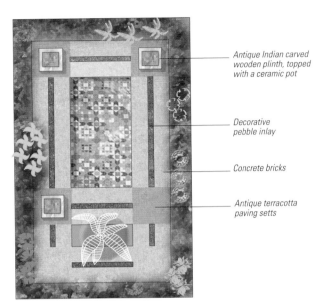

Antique Indian carved wooden plinth, topped with a ceramic pot

Decorative pebble inlay

Concrete bricks

Antique terracotta paving setts

Rustic charm

Key Mediterranean plants are included in this area (the rear entrance) of a large Provençal garden by Michel Semini, with an olive tree taking centre stage and providing essential shade.

Key ingredients
1 *Viburnum tinus*
2 *Nerium oleander*
3 *Olea europaea*
4 *Lavandula stoechas*
5 Gravel

Michel says:
"This plot in Provence was once a derelict sheepfold. It was first cleared and developed as a garden, but as been improved and expanded since. The Alpilles mountains form its backdrop.

"I wanted a sense of mystery, and to link the planting with the landscape using green and silvery foliage.

"The rustic character of the sheepfold was a key consideration when choosing the materials for the garden. I like to mix the influences of the site, my client's needs and my own ideas, and in this garden they all came together well."

Chalk-stone paving slabs

Chalk-stone edging

7–12mm (¼–½in) gravel

Good taste

Debbie Roberts and Ian Smith of Acres Wild tend to work with the prevailing conditions in a garden, and this section of a steeply sloping, well-drained sunny plot with panoramic views lent itself to Mediterranean herbs. The paving creates an informal terrace.

Key ingredients
1 *Origanum vulgare* 'Aureum'
2 *Allium schoenoprasum*
3 *Santolina chamaecyparissus*
4 Terracotta paving
5 *Thymus citriodorus* 'Bertram Anderson'

Debbie says:
"The clients wanted their garden divided into intimate, sheltered 'rooms' and they helped to style these, although it was important to create the right microclimates first. This space, close to the kitchen and with dry soil, made Mediterranean herb-planting appropriate. But it was also a space that people walked through to access the rest of the garden, so had to look good."

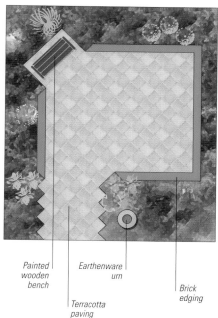

Painted wooden bench

Earthenware urn

Brick edging

Terracotta paving

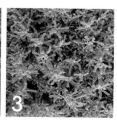

Modernist gardens

Asymmetry is key to Modernist designs, which are also characterized by free-flowing space and the play of light and shade. These gardens are often "pared down" spaces, using quality materials, spatial relationships, and clever styling to succeed.

In many Modernist gardens, one or two views may be emphasized, but the partial enclosure of space within walls or hedges means that they are open to personal interpretation, as the visitor is not forced by the design to experience them in just one way. Sharp lines reinforce the contrast between horizontals and verticals, and water is used architecturally, often as a reflective surface.

The material palette is minimal – smooth rendered concrete is often used for paving and walls, while limestone or slate, with little or no detailing, are other good options for floors. Designers also prefer large slabs that minimize joints and create clean, uninterrupted surfaces.

Planting is restricted too, with many Modernist gardens featuring only trees, hedges, and lawn, punctuated by key architectural specimens.

The geometry of Modernism tends to be rectilinear and emphasizes the horizontal line, although there are examples of garden designs in this style that are based on circles or ovals. Plans are frequently created on grids that relate the house to the garden, helping to blur the distinction between the interior and exterior spaces.

"Green" architecture in a modern courtyard garden.

The Modernist Movement was originally associated with the Bauhaus School of the 1920s and '30s, which embraced new technologies and proclaimed that form should follow function. However, it was not until after World War II that it found favour among some landscape designers, who reacted against the old schools of garden design, and created outdoor spaces that were functional and adapted to human, rather than plants', needs. Modernism continues to influence outdoor space, with some designers combining a broader planting palette, including perennials or wild flowers, with crisp, high-quality landscaping.

A tranquil infinity pool reflects a unified environment.

What is Modernist style?

The creation, definition and celebration of space is crucial to the success of Modernist gardens. Their primary emphasis is leisure and the enjoyment of life outdoors, with planting frequently used as an architectural element. Clipped hedges, specimen trees, and large blocks of planting provide simple, sculptural surfaces or screens, which complement the horizontal expanses of timber, stone, concrete, or water. From the original functional focus of Bauhaus, the Modernist approach flourished in the US, especially in California where the climate encouraged the use of the garden as an outdoor room. The architectural philosophy of Modernism, which views planting as only one element of the whole composition and not the principal reason for the garden's creation, has led to the development of many beautiful, elegant spaces.

Modernist style in detail

Crisp and clean, Modernist designs suit gardens of any size, and can provide an antidote to crowded cities and hectic lifestyles. Relying on scale and proportion to create drama in the absence of decorative embellishments, these gardens focus on open, uncluttered spaces that offer the perfect setting for outdoor living.

Most Modernist gardens are based on a geometric layout, with the horizontal lines of rectangles providing a sense of movement. These dynamic lines contrast with the verticals of trees, hedges, or walls, and slice through space to unite different sections of the garden.

Materials are selected for their surface qualities – decking, polished concrete, limestone, and gravel produce expansive surfaces, often punctuated by reflective water or specimen trees, and this honest use of materials requires stunning high-quality finishes and architectural precision. Fine lawns, clipped hedges, and simple planting are typical of most Modernist 20th-century gardens, but contemporary designers sometimes include a more complex palette.

Inside out
Here the main terrace of Casa Mirindiba in Brazil (right), designed by Marcio Kogan, extends into the garden to create a sheltered space, part interior and part exterior in character. The long, narrow swimming pool reflects the stone wall, and lighting picks out surfaces and tree canopies to create interest after dark.

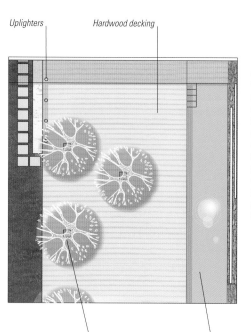

Uplighters   Hardwood decking

Minimalist tree planting

Reflective swimming pool, or "lap" pool

DESIGN INFLUENCES

Thomas Church's iconic pool.

Celebrated as the founder of Modernist garden design, Thomas Church thought that gardens were primarily for people and should reflect their owners' lifestyle and needs. Many of Church's theories are explored in his 1955 book, *Gardens are for People*, and in his iconic Modernist garden, El Novillero in California, designed in 1948.

The garden is based on a regular grid that relates to the adjacent poolhouse and is defined with concrete paving and decks. Into this regular pattern Church wove sweeping curves to create the swimming pool (left) and lawns, echoing the winding river in the salt marshes below, while existing oaks were retained to frame the view.

The simplicity and elegance of the materials, and the overall geometry, result in a composition which confirmed Church as one of the greatest landscape architects of the last century.

Key design elements

1 Asymmetry
Although a central axis may be used in Modernist design it is rarely a dominant feature. Rectangles of lawn, water, paving, or planting interlock more intuitively to create sharply defined but irregular patterns.

2 Modern materials
The clean lines of steel, concrete, glass, and timber emphasize the precision of the manufacturing process. Paving joints are minimized, and subtle lighting is used to enhance the surfaces.

3 Planting in blocks
The variety of species is often limited and planted in large blocks or masses. Grasses and perennials, interplanted to catch the light and create movement, have revitalized the style.

4 Contemporary furniture
Modernist garden furniture is architectural in style. Design classics, such as the sculptural Barcelona chair, set the tone for elegant recliners, simple tables and matching benches (*left*).

5 Reflective water
Reflective pools create unruffled surfaces and bring light into the garden. Modern technology now allows water pools to brim or overflow, maximizing the expanse and impact of the reflective surface.

Interpreting the style

The manipulation of space is central to Modernism, creating gardens free from clutter or fuss. This style demands a clearly defined geometric layout, so that the proportions of the main features can be appreciated. Keep material and plant palettes to a minimum, and pay particular attention to the finer details. Fixings can be hidden to create smooth flowing surfaces.

△ **Pool garden**
Smooth rendered walls surround this garden with a neutral backdrop, allowing the reflective water and planting to take centre stage. Decks overhang the pool to create an impression of floating surfaces. Planting is restricted, but simple blocks of texture create the necessary impact.

▷ **Bamboo screen**
Decking creates a warm, tactile surface, which is ideal for city or roof gardens. Here the planting is contained within simple cube or box planters that screen this private space.

"The play of light and shadow breathes life into the Modernist garden"

Complementary colours
Texture, colour and shape combine to create this small garden. The ochre tones of the brickwork contrast with the warm terracotta-rendered surfaces, while clipped evergreens, grasses, and irises offer natural forms.

Textural composition
Contrasting surfaces of honed limestone, precise dry stone walls, and reflective steel-edged water create the drama here, softened by the dense planting of irises and Stipa beyond.

△ Classic structure
The rectangular pool, deck, and path are classic Modernist features, complemented here by blocks of dwarf hedging and an untamed leafy backdrop.

◁ Geometrical design
The architecture of this garden space is the dominant theme, with the rectangular pool based on the dimensions of the picture window. Repeated cordylines arranged along the balcony above create a sculptural splash.

GARDENS TO VISIT

BURY COURT, Farnham, Surrey
Includes a grid pattern grass garden by Christopher Bradley-Hole.
burycourtbarn.com

ST CATHERINE'S COLLEGE, Oxford
Designed by Arne Jacobsen.
stcatz.ox.ac.uk

VILLA NOAILLES, Hyères, France
Cubist garden designed by Gabriel Guevrekian. villanoailles-hyeres.com

ART INSTITUTE GARDENS, Chicago, US
Designed by Dan Kiley. artic.edu/garden-overview

EL NOVILLERO, Sonoma, California, US
Thomas Church's iconic Modernist garden.
gardenvisit.com/gardens/el_novillero_garden

CASE STUDY

BUILDING BLOCKS

The simple, clean lines of this garden betray an exacting design that has modern, contemporary detailing at its heart. Crisp blocks of planting, paving, and water are set out on an assymetrical floor plan, reflecting Modernist design principles.

Visual play

Contrasts of texture and form are used to great effect in the design. Smooth paving, a reflective water feature lined with flat pebbles, trim beech hedging, and multi-stemmed *Osmanthus* trees, conspire to create bold visual effects.

Ordered space

The space in the garden is set out in rectilinear blocks of paving, planting, and water. Some of the areas are open, others are enclosed, hiding and then revealing aspects of the design as the visitor walks through the different spaces.

Asymmetrical plan

The stone wall panels were inspired by a Piet Mondrian painting. Combining calm, clean lines and strict geometry with an asymmetrical plan, they perfectly represent principles typical of Modernist design and brought strikingly to life in this garden.

Floral contrasts

The different flower shapes provide refined contrasts and accents. The yellow daisy-like flowers of *Doronicum* stand tall above tiny *Euphorbia polychroma* bracts, and contrast in colour and form with red tulips and blue cornflowers.

Clear colours

The colour palette shows a typically Modernist restraint. Shades of green predominate, allowing the primary colours of red, blue, and yellow to shine through.

Modernist garden plans

The Modernist garden has a simple, geometric layout and a balanced design, with the emphasis on sculptural planting and quality materials. The three designs here are perfect examples of gardens that embrace these principles. The planting schemes are simple and bold, allowing space and material texture to be the focus, and they all exemplify 20th-century Modernist architect Ludwig Mies van der Rohe's maxim: "Less is more".

Maximizing space

Planting is restricted in this elegant garden by Vladimir Djurovic, where surface and texture are the highlights. The clever lighting design draws attention to the low bench seats made from the same material as the paving, and to the apparently floating fire cowl, which becomes a giant focal point for the terrace.

Key ingredients
1 Red cedarwood table
2 *Acer palmatum*
3 Lighting
4 Natural stone-honed finish

Vladimir says:
"This garden was developed as a holiday retreat. The space available for the garden was quite restricted, and a major part of the design process was dedicated to creating a sense or illusion of space.

"The brief was quite demanding: the client loves to live outdoors when in residence, and the garden needed to reflect this – with spaces for cooking and dining, relaxing, entertaining large groups of people, and so on.

"The restricted topography and the fact that the house is arranged on split levels also made the connection and sequencing of space more difficult.

"The result is typical of my work – I aim to produce memorable spaces, no matter what their scale. I am inspired by nature, and like to feel that my work brings people closer to the natural world."

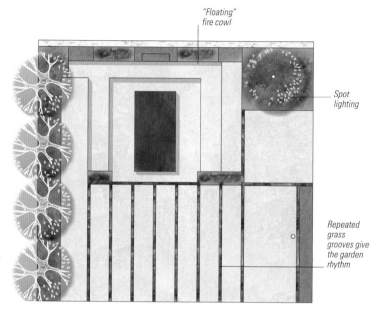

"Floating" fire cowl

Spot lighting

Repeated grass grooves give the garden rhythm

Grid lock

The owners of this property asked Andrew Wilson for a spacious design with a semi-industrial quality, to complement a new, polished, dark-green fibreglass house extension with long curtain walls made of glass.

Key ingredients
1 *Betula pendula*
2 *Stipa gigantea*
3 *Deschampsia cespitosa* 'Bronzeschleier'
4 *Yucca aloifolia*
5 *Ligustrum delavayanum*

Andrew says:
"The long, low roof of the new building extension was echoed in the horizontals of the paving, low walls, and steps. The trees, mainly pine and birch, provide towering verticals that produce the classic contrast central to most Modernist compositions.

"The garden is paved in coloured, poured concrete that appears to float out across a reflecting infinity-edge pool. Darker rendered walls provide subtle screening and a backdrop for uplighting to create an ambient glow after dark."

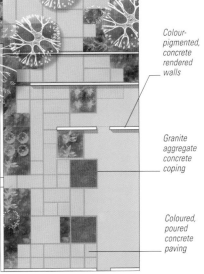

Colour-pigmented, concrete rendered walls

Granite aggregate concrete coping

Coloured, poured concrete paving

Room outside

Created by Wynniatt-Husey Clarke, this London garden was commissioned to complete renovations to the client's house.

Key ingredients
1 Hardwood panel fencing
2 *Carpinus betulus*
3 'Floating' cantilevered hardwood bench
4 Self-binding crushed slate particles
5 *Zantedeschia aethiopica* 'Crowborough'

Patrick Clarke says:
"The garden emerged from a close cooperation between the architect, client, and garden designer. More than anything, it reflects a clear ambition to see the building and garden as a single entity.

"The rendered 'blade' wall, colour-matched to the interior finish, gives the impression that the back wall of the house has been moved to the end of the garden. The threshold between inside and out is seamless, with the same paving used for both, and a frameless door creating minimal intrusion. Asymmetry is used as a way of creating a dynamic quality within the garden as one moves through the space."

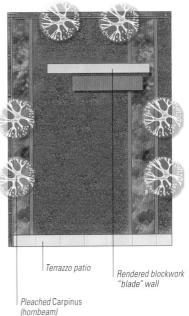

Terrazzo patio

Rendered blockwork "blade" wall

Pleached Carpinus (hornbeam)

Japanese gardens

The Japanese garden is often perceived in the West as a single garden style, when in fact there are many different approaches and philosophies, some of which are based on traditional rituals or have spiritual meaning. These diverse design theories make a definitive translation of this style difficult.

At their heart, however, Japanese gardens share some key characteristics. Symmetry, for example, is eschewed in favour of balanced asymmetry. These harmonious layouts are achieved by the careful placement of objects and plants of various sizes, forms, and textures, frequently contrasting rough with smooth, vertical with horizontal, or hard with soft. The area of the garden is often restricted, but individual elements are not forced or crowded together, and the space between objects is considered essential to the overall design.

Cherry blossom has been celebrated for centuries.

Japanese gardens are appreciated as visual compositions for contemplation, rather than as spaces to be cultivated or enjoyed for leisure. Traditionally, natural stone was used, although many modern gardens feature concrete or stone with different finishes. Bamboo and timber are also popular materials.

The famous dry Zen gardens use fine gravel raked into fluid patterns, and planting in these symbolic gardens is minimal, often limited to mosses and lichens around the base of a group of rocks.

Water is seen as a purifying element, especially important in Japanese tea rituals. Small pools, often in stone containers, or streams, provide reflective details.

A mountain landscape recreated in miniature.

Planting in Japanese gardens is restrained, with bamboo, grasses, and irises providing verticals, and plants such as camellias, cherry trees, peonies, and rhododendrons used for flower and form. The underlying geometry is not easily discernible, but irregular plans may be complemented by paths made from rectangular blocks. Informal stepping stones or meandering pathways are also typical, as the changing views or winding terrain provide an aid to concentration and meditation.

What is Japanese style?

After centuries of isolation, the harmonious asymmetry of Japanese gardens came as a shock to Western travellers in the 19th century, who were used to more formal and geometrical layouts. The balance of hard elements, such as rocks, stepping stones and gravel, with tightly clipped shrubs and trees, created a contrast that still appeals. Meticulous positioning of the main elements to disguise restricted spaces, or to provide links to the landscape beyond, is crucial to the success of many of these sculptural and highly controlled gardens.

Japanese style in detail

Many plants used in Japanese gardens are subjected to tight pruning regimes to maintain or restrict their size, but also to ensure that they remain in proportion to their surroundings; maples, azaleas, camellias, and bamboo are all controlled in this way.

In turn, rocks are selected for their weathered qualities, and their innate characteristics are carefully considered before final placement is agreed. A pleasing contrast between verticals and horizontals is also important to achieve. Gravel is used to symbolize water and provides a neutral but textured foil to the planting and rock formations. In Zen gardens the gravel is raked into precise patterns, and this daily ritual is considered conducive to contemplation and self-knowledge.

In stroll gardens, the route through the space is scrupulously planned, and the winding paths or stepping stones ensure that the visitor stops to experience the views that are revealed along the way.

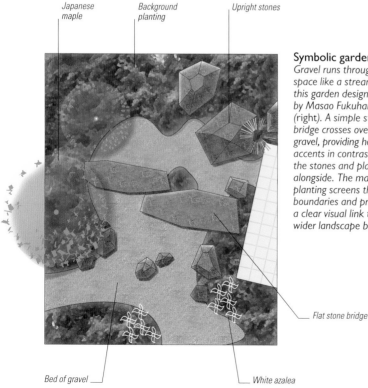

Japanese maple

Background planting

Upright stones

Symbolic garden
Gravel runs through this space like a stream in this garden designed by Masao Fukuhama (right). A simple stone bridge crosses over the gravel, providing horizontal accents in contrast to the stones and planting alongside. The massed planting screens the boundaries and provides a clear visual link to the wider landscape beyond.

Flat stone bridge

Bed of gravel

White azalea

Key design elements

1 Japanese plants
Evergreens are often densely planted and pruned to provide a consistent structure, whereas deciduous species are used for flower colour or seasonal change.

2 Water features and pools
Water is either used expansively as a reflective surface, or in smaller features, such as the stone water basins (*tsukubai*) associated with the tea ceremony.

Religious influences

Japan's rich tapestry of religious belief is fundamental to the design of its exquisite gardens. Both the ancient religion of Shinto, and the Buddhist teachings that were introduced later, celebrate the natural world, and all natural elements are seen as sacred and thus worthy of respect and worship. This philosophical approach is expressed in many Japanese gardens by the sensitive placement of significant rocks, trees, or other natural phenomena, with specimen maples, magnolias or cherries often displayed against a backdrop of dark foliage. The cultivation of beauty as a spiritual activity is also reflected in Zen tea gardens, in which a *roji* (dewy path) lit by stone lanterns leads the visitor through an intimate landscape to the ceremonial tea house.

3 Symbolic ornaments
Stone lanterns, water basins, and buddhas are often placed close to paths leading to the tea ceremony. Pagodas or stupas create focal points in larger gardens.

4 Gravel and rocks
Gravel is used to represent water, with stones symbolizing islands, boats, or even animals. Great care is taken over the placement and orientation of the stones.

5 Bamboo fencing
Fences and gates are often made from bamboo fastened with elaborate ties or bindings. These are used as boundaries and screens, or to direct or control views.

6 Stepping stones
Stepping stones create a heightened self-awareness through the garden. Often used as a route to the tea ceremony, they resemble a dewy path through the forest.

Interpreting the style

Pressure on land means that most Japanese gardens are very small, and designed to be looked at rather than used. Sculptural courtyard gardens, laid out to be viewed from important windows or terraces, focus on a few carefully selected stones or trees. Larger gardens are also highly manipulated, with precisely positioned plants, trained to deceive the eye – here there is more room for a range of trees, intricate pathways, water features, and views into the *shakkei* or "borrowed landscape" beyond.

△△ **Tranquil moss garden**
The uneven and meandering stepping stones stand out against a soft emerald carpet of moss. Exquisite views are created to be admired along the way.

△ **Gravel and stones**
In this contemporary courtyard, carefully chosen rocks and stones form a sculptural route across gravel and moss, punctuated by the verticals of specimen trees.

△ **Miniature landscape**
A typical arrangement of interior, veranda, and garden presents a staged sequence of space. Here a contorted specimen pine provides a magnificent focal point.

▷ **Reflections of autumn**
Japanese maples shade the banks of a pool, dropping their colourful leaves like jewels onto the ground. Stepping stones offer access across the still water.

◁◁ **Falling water**

The placement of vertical and horizontal rocks is key to the success of waterfalls and dry gravel systems alike. This three-step cascade produces a calming water sound.

◁ **Transcendent stones**

Balance is an important attribute of the Japanese garden, emphasized here by this precarious sculpture of flat stones, and echoed by the low hedges and ground cover beyond.

▷ **Illusions of space**

An illusion of distance is created here, by emphasizing the foreground with a stone lantern and balustrade. The autumn canopies can be appreciated from the path.

△△ **Sinuous steps**

Curving stone steps provide an enticing route through the garden, creating a similar effect to winding stepping-stone paths. Subtle layered planting follows the rhythm.

△ **Geometric space**

This modern design uses horizontal and vertical steel panels to form a transparent deck and unified boundary, through which the stems and foliage of plants emerge.

JAPANESE GARDENS TO VISIT

KATSURA IMPERIAL VILLA, Kyoto, Japan
Stroll garden with extensive water and woodland. sankan.kunaicho.go.jp

RYOANJI, Kyoto, Japan
Zen Buddhist raked gravel garden. ryoanji.jp

TOFUKUJI, Kyoto, Japan
Zen temple garden with *Acer* collection.
tofukuji.jp

TATTON PARK, Cheshire
One of the best Japanese gardens in the UK.
tattonpark.org.uk

GOLDEN GATE PARK, San Francisco, US
Japanese stroll-style tea garden.

"Japanese gardens are symbolically and spiritually connected to the landscape"

CASE STUDY

EASTERN INFLUENCE

Key elements of a traditional Japanese tea garden, including the cascading stream, mossy pathway, teahouse, and restrained planting palette, are used in this modern interpretation, where every element is carefully crafted to create a landscape in miniature.

Planting traditions

Many plants associated with Japanese style feature in the design. The rich red colouring of *Acer palmatum* is echoed by the pinkish young *Pieris* leaves, while spiky stems of *Equisetum* and iris foliage shoot up from moss-dotted rocks.

Space to reflect

Despite its limited dimensions, this garden creates a real feeling of space. The teahouse focal point, careful layering of the planting, and the natural slope enhanced by a gently tumbling stream, all work to create an illusion of a bigger garden.

Calming stream

Water is an essential part of
a tea garden, and the stream
symbolizes the renewal of life.
The mossy stone walls that form
the cascade create a visual motif;
water and stone represent yin
and yang, complementary
opposites that create harmony.

Stop for tea

Here made from rough
sawn timber, the teahouse
is a traditional element of
the Japanese tea ceremony.
Inside, visitors are invited
to drink tea and reflect
upon the tranquil scene
and harmonious planting.

Green innovation

The living green roofs of the
entrance arch and teahouse are
a modern addition – they would
traditionally be thatched or tiled.
Succulent planting helps provide
wildlife habitats, while softening the
contrast between the buildings
and the surrounding plants.

Japanese garden plans

These gardens cannot simply be recreated with a haphazard collection of Japanese ornaments and species; successful Japanese designs integrate a careful balance of plants and objects that often have symbolic and spiritual meaning. In these two examples, Maggie Judycki and Haruko Seki have perfectly captured the notion of the soothing, contemplative garden, and the subtleties of natural colours and forms.

Living art

The fish-filled pond is a meditative focal point in Maggie Judycki's own garden. Rocks, ornaments, and planting are carefully arranged around it and a split bamboo fence filters light in horizontal patterns across its surface. The leaves of a *Sassafras* and a *Betula* merge and rustle above.

Key ingredients
1. *Acer* 'Rubrum'
2. *Sassafras albidum*
3. Bamboo fence
4. *Hosta* 'Francee'
5. *Cotoneaster salicifolius* 'Gnom'
6. Japanese bathing stool
7. *Sarcococca hookeriana* var. *humilis*
8. *Betula utilis* var. *jacquemontii*

Maggie says:
"This is my own garden, and it's been a work in progress for many years. I started out as a stone sculptor, which has helped me to use and understand hard materials. I tend to start with them and soften the surfaces with planting.

"Sitting places are important to me too. A favourite is the Japanese bathing stool, ideal for contemplation when I'm feeding the koi carp. Living art and the movement it creates is also fascinating — we can see the pool from the house, and it's a constantly changing view. The garden is typical of my work in that I customize the space for each client."

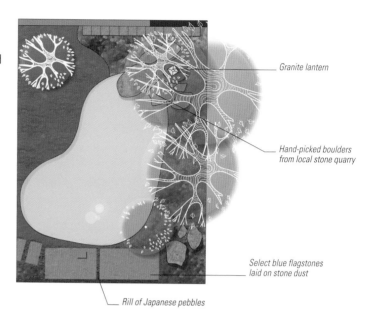

Granite lantern

Hand-picked boulders from local stone quarry

Select blue flagstones laid on stone dust

Rill of Japanese pebbles

  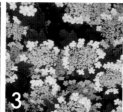   

1   2   3   4   5   6

Capturing movement

This Japanse garden by Haruko Seki of Studio Lasso was designed for the RHS Chelsea Flower Show and it has since been recreated in a private garden in south London. The swirling curves of gravel paving and green mounds create a sense of movement and enclosure, while simple, transparent planting produces a delicate filigree of foliage. Still water reflects the lit glass panels, which give an ethereal glow and are decorated with the silhouettes of bamboo leaves and canes. The contrasting pale raked gravel and grass help define the composition.

Key ingredients
1 Cercis canadensis 'Forest Pansy'
2 Phyllostachys aurea
3 Viburnum opulus
4 Abelia 'Edward Goucher'
5 Spiraea cantoniensis
6 Stipa tenuissima

Haruko says:
"The client who bought this garden leads a stressful life and was attracted to the calmness of the composition. The design encourages a feeling of peace and opens one's senses up to the environment – for example, the whispering of the breeze through the planting is central its success.

"In all of my work, I use the space to enhance the changing character of nature; I believe this is an essential quality in a Japanese garden. I am also influenced by the late landscape architect Sir Geoffrey Jellicoe, who explored the relationship between the landscape and the subconscious."

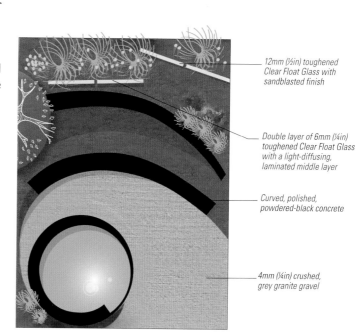

12mm (½in) toughened Clear Float Glass with sandblasted finish

Double layer of 6mm (¼in) toughened Clear Float Glass with a light-diffusing, laminated middle layer

Curved, polished, powdered-black concrete

4mm (¼in) crushed, grey granite gravel

Foliage gardens

This approach to garden-making is seen across the world, but works especially well in warm climates where planting is naturally lush, and a jungle look with tall vegetation is not hard to achieve. Texture and shape drive the design, rather than a season of bloom. Layouts vary in their composition, but all combine areas devoted primarily to foliage, with the emphasis on contrasting varieties and plant forms. Clearings are carved out of dense vegetation, creating a sense of seclusion and separation, with paths winding between. Decorative bark or pine needles are often used to create a jungle-floor softness underfoot.

These gardens are typically organic in shape, without hard edges or a sense of formality, but where man-made structures do encroach, the contrast is often startling, with the bold use of rustic materials such as rough-hewn timber and unworked stone. Interestingly, sleek Modernism also works well with foliage planting. Water is frequently present in the form of energizing waterfalls and streams, or even swimming pools.

Foliage gardens date back to 19th-century European colonial gardens, where the indigenous, richly diverse local flora found in Australia, New Zealand, South Africa, Malaysia, India, and the Caribbean was used to produce a celebration of flower colour and foliage texture. The gardens of the 20th-century Brazilian designer Roberto Burle Marx are modern interpretations of this genre, with huge areas devoted to rich tapestries of foliage.

Exotic bromeliads add colour.

In temperate zones, this approach has been adopted in some urban gardens with the emphasis on architectural plants, such as tree ferns, bamboo, loquat, *Fatsia, Phormium,* and cordylines, which are combined to create a sense of drama. Some designers also experiment with grasses, water, or woodland planting to gain similar effects, but formal lawns are rarely seen in these gardens.

Foliage colour, texture, and shape provide dramatic effects.

What is foliage style?

The jungle-like appearance of many foliage gardens creates an atmosphere of irresistible exuberance. Plants chosen for their interesting leaves dominate and the use of individual specimens and large-leaf perennials *en masse* yields a gorgeously textured landscape with dramatic spots of bright colour. A network of pathways and clearings forges a route through the garden, offering the visitor a close-up view of the planting. Cooler climate foliage gardens concentrate on mass plantings of grasses and woodland glades.

Foliage style in detail

The enjoyment of foliage gardens derives from the sheer volume and variety of planting. In larger gardens there may be space for grassy areas, swimming pools and terraces but, generally, most of the available garden space is devoted to leaves. Flowers are often subordinate and provide stabs of vivid colour among the foliage.

Taller species such as *Eucalyptus*, palms, cordylines, and bamboo provide height and vertical interest, while the space below is filled with lower-growing shrubs, grasses, and perennials. The main emphasis is on structural and foliage planting, but sewn into this rich canvas is a brilliant embroidery of flower colour, with *Strelitzia* (bird of paradise) and *Canna* typical in warmer climates, and dahlias or lobelia more appropriate in temperate regions.

Larger cities form heat islands, in which warmer than average temperatures allow more exotic species to find a home. In the UK, this has led to the phenomenon of urban jungle gardens.

DESIGN INFLUENCES

The most notable name associated with this style is Roberto Burle Marx, the artist/ecologist/designer who worked in spectacular fashion with the rich flora of his native Brazil. His gardens demonstrate a painterly sensibility to landscape design, celebrating foliage pattern and saturated flower colour. In what was formerly known as the Odette Monteiro Garden, huge plates of textured ground cover feature along a dramatic lawned valley. His planting schemes are particularly impressive when seen from above.

The Luis Cezar Fernandes (formerly Odette Monteiro) Garden, Brazil.

The exotic garden
In this remarkable garden (right), created by the late Will Giles at his home near the centre of Norwich, rich planting exploded from the borders over gravel paths. Sparks of colour came from the purple-leaved Canna *and tall yellow sunflowers. Cacti and succulents were brought outside over the summer months, while containers of other plants of differing heights, including grasses and herbs, fringed the vibrant, foliage-rich display.*

Gravel path is a foil for planting

House

A palm (Trachycarpus fortunei) *gives height to the planting scheme*

Exotic banana (Musa) *contributes to the subtropical feel*

Pots form a central island

Canna *'Durban' adds hot colour accents*

Key design elements

1 Bold foliage
The key element is foliage that makes a statement. The plants that dominate demand attention; strappy *Phormium* perhaps, or tall-growing bamboo, or *Musa* (banana) with its fabric-like leaves.

2 Colourful highlights
Bright flower colour lifts the general greenness of these gardens, providing surprises along the way. Here *Dahlia* 'Bishop of Llandaff' adds rich red flowers and dark foliage.

3 Pools and reflections
Clear pools, perhaps edged with lilies or papyrus, create reflective surfaces. Waterfalls add sound and energy, and boulders set by jungle pools provide naturalistic seats.

4 Containers
In cooler climates, planting exotics and tender species in pots offers the designer greater flexibility – they can easily be moved under cover in winter. Dramatic pots can also be used as focal points in a scheme.

5 Materials
Hard materials are often sourced locally. Gravel or stone, often rough-hewn, are used for paved surfaces, but timber and bamboo are also common. Walls covered with whitewash or painted render add intense colour.

6 Height and structure
Tall plants are essential to create jungle-like layering. This banana-like *Ensete, Trachycarpus* (Chusan palm), and *Eucalyptus* give height to the canopy, and offer protection and shade to plants below.

Interpreting the style

Foliage gardens deliberately set out to overwhelm the onlooker with the sheer volume and scale of planting in the jungle-like borders. When grouping your plants, consider details – such as the shape, texture, and colour of leaves – to produce exciting contrasts. Add bright colour with variegated foliage and striking, subtropical flowers to complete the vibrant mixture.

GARDENS TO VISIT

THE EXOTIC GARDEN OF EZE,
Monaco, France
Exotic plants from all over the world.
jardinexotique-eze.fr

TREBAH, Cornwall, UK
Subtropical garden on a Cornish hillside.
trebah-garden.co.uk

WIGANDIA, Victoria, Australia
Garden on slopes of Mount Noorat.
wigandia.com

SITIO ROBERTO BURLE MARX,
Rio de Janeiro, Brazil
The late artist's own large garden.
museusdorio.com.br

JIM THOMPSON HOUSE, Thailand
A lush jungle garden in Bangkok.
jimthompsonhouse.com

△△ **Palm and gravel mix**
An informal clearing is edged with the elegant, fanned leaves of Chusan palms (Trachycarpus fortunei) with vertical jets of brilliant red cannas dotted between. A low mound of dark green planting complements the composition.

△ **Spiky combinations**
The instantly recognizable, sword-shaped foliage and tall flower spikes of Phormium tenax dominate this space – echoed by the sharp points of agaves and the fine-cut leaves of palms.

◁ **Grassy effects**
A basket-weave path meanders through a border of fine textures, which include the repeated arching rosettes of Hakonechloa macra 'Aureola' – a grass that takes on warm orange tones in autumn.

◁ **Hot pot**
This incidental association plays on the similarities between the tones of the glazed pot and the veined Canna *leaves. Carmine-red flowers turn up the heat.*

◁◁ **Verdant enclosure**
Even within the confines of a small and overlooked city garden, it is possible to create privacy and a space to relax. Here, a hot tub is enclosed by hedges of densely planted bamboo and tall hurdles.

△ **Sunset spires**
Phormium *'Sundowner'*, Astelia chathamica *'Silver Spear', and the heads of* Verbena bonariensis *conspire to produce a glorious display of glowing colour in the evening sun.*

◁ **Cool pool**
An array of fleshy foliage closes in to create a secluded swimming pool alongside a sun-filled terrace in this thickly planted jungle garden.

"Foliage gardens are a feast of sculptural shapes and forms"

CASE STUDY

FEAST OF FOLIAGE

Architectural forms and leafy contrasts are key to the success of this garden, which, despite the lack of flowers, is a triumph of sculptural shapes, textures, and colours, created by inspiring foliage combinations from small trees, shrubs, perennials, and grasses.

Hardy exotic

The lancewood *Pseudopanax crassifolius* makes an intriguing statement, with its weird, almost dead-looking foliage and gaunt form. A surprisingly hardy tree from New Zealand, it is guaranteed to create a talking point in any garden.

Leaf combos

Foliage can offer pleasingly bold colour contrasts, as here with the orange-brown leaves of tassel cord rushes (*Baloskion tetraphyllum*) next to silvery *Artemisia*. Elsewhere, bright green tree ferns overhang white-splashed hostas.

Designers **Andrew Fisher Tomlin and Dan Bowyer**
Show **RHS Hampton Court Palace Flower Show**
Award **Gold Medal and Best Summer Garden**

Simple materials

The landscaping materials, such as the grey paving and dark, almost black, boundary walls, provide excellent foils for the foliage, their smooth texture and contemporary colours contrasting with, but never upstaging, the leaves.

Colour spots

The foliage-dominated planting is lifted by spots of flower colour. Most of the blooms are small, such as those of *Canna indica* and *Duranta erecta* 'Geisha Girl' – the restricted palette of orange and blue complements the leaf colours.

Jungle enclosure

The palm, *Butia yatay*, and tree ferns provide a sense of privacy and enclosure, without being too overbearing in this small space. The jungle-like plants are also relatively hardy, and would be ideal for a city garden in a temperate climate.

Foliage garden plans

In two of these gardens, British designers have used a range of tender and hardy plants to achieve a foliage effect in a cool climate. The third, in Florida, is a leafy, tropical extravaganza. In all three, the exuberance of dense foliage and architectural planting needs some sense of control, and this is provided by paving, water and structural elements, such as the screens and boundaries. These also offer contrasts in texture and form.

Layered planting

In designer Declan Buckley's own garden, a rich tapestry of layered planting sits alongside the bold geometry of paving and a pool; the use of reflective water increases textural impact. There is a great sense of contrast here, between the open, light terrace and the narrow pathways.

Key ingredients
1 *Phyllostachys nigra*
2 *Euonymus japonicus*
3 *Fatsia japonica*
4 *Pseudosasa japonica*
5 *Geranium palmatum*
6 *Astelia chathamica*
7 *Buxus sempervirens*
8 *Cycas revoluta*

Declan says:
"After years spent growing plants in pots on a roof terrace, it was a relief to have a garden to plant them in. The site is a

long rectangle, overlooked by five-storey houses, so bold and layered architectural planting helps to screen the site and provides privacy. Conversely, the end wall of my own house is solid glazing, which gives me a dramatic view across the pool and into the luxuriant planting.

"London's warmer temperatures allow more tender and unusual species to thrive, and plants were chosen for their texture and form – flower and colour came second. A strong, simple framework softened by foliage is key to all my projects."

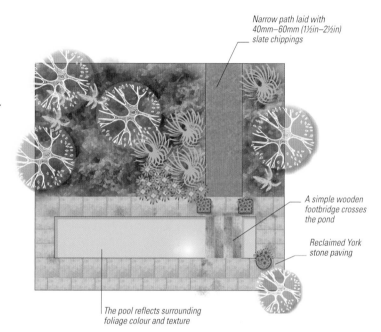

Narrow path laid with 40mm–60mm (1½in–2½in) slate chippings

A simple wooden footbridge crosses the pond

Reclaimed York stone paving

The pool reflects surrounding foliage colour and texture

English exotic

In this small garden by Annie Guilfoyle, a mass of oversized and textured exotic planting hovers over a wooden deck.

Key ingredients
1 *Phormium cookianum* subsp. *hookeri* 'Tricolor'
2 *Eriobotrya japonica*
3 *Euphorbia mellifera*
4 *Musa basjoo*
5 *Polystichum setiferum*

Annie says:
"This garden is close to the River Thames in Kew. It's a tiny space that had to capture the essence of the East, where my clients had spent a great deal of time, yet link seamlessly with the house. To create deeper planting areas, I set the layout at an angle – which also seemed to make the boundaries disappear. This is typical of my work, as I try to maximize usable space in small gardens, balancing room for relaxing and entertaining with rich, full planting.

"The garden is pretty low-maintenance, and it was good to work with a client who didn't demand year-round colour."

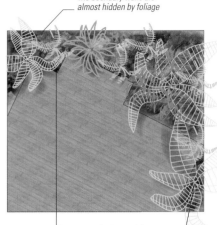

The boundary walls are almost hidden by foliage

The decking's zigzag edging increases the sense of space

Ground-level ferns add another layer of texture

Tropical refuge

Raymond Jungles has used large, fleshy and verdant leaves to create shade in this Florida Keys garden. Glimpses of art lead the eye through the plants.

Key ingredients
1 *Pritchardia pacifica*
2 *Attalea cohune*
3 *Solanum wendlandii*
4 *Areca vestiaria*
5 *Heliconia rostrata*

Raymond says:
"I created this garden for my family and it was a labour of love. I am influenced a great deal by other designers, in particular, Roberto Burle Marx, Luis Barragán, and Richard Serra. In some ways, this garden was a laboratory in which I grew specimens, some of which were collected in Brazil with Burle Marx himself. The result looks typical of my work, but nowadays I tend to use native species whenever possible. We tried to maximize light in the house and garden, and used sliding doors to differentiate between interior and exterior space. Many of the materials are rescued and re-utilised."

Terracotta-coloured wall adds visual drama

Regimented, square paving slabs give a sense of order to the lush jungle foliage

The bench doubles as a work of art

Productive gardens

Historically, two main types of productive garden evolved: the large walled gardens of wealthy Victorian estate owners, which offered exotic fruit, fresh vegetables, and cut flowers for weekend entertainments, and, at the other extreme, allotments, cottage gardens, and areas of private gardens devoted to growing produce as a hobby, or to supplement the diet.

The Victorians elevated productive gardening to a fine art, but they were not the first to mix fruit, vegetables and flowers in the same area. Medieval abbey gardens were typically divided into small herb and vegetable beds with some decorative planting, and Renaissance gardens in France featured ornamental produce in elegant parterres, known as "potagers". This term is still used today to describe an attractive productive garden.

The Dig for Victory campaign during World War II generated a huge enthusiasm for home-grown produce in the UK, but this waned as wealth increased after the conflict. Today, our increasing desire for organic food, and concerns about the carbon footprint of imported goods, is fuelling a revival of the kitchen garden, albeit on a smaller scale.

Most productive gardens tend to be orderly, with geometric beds separated by paths for ease of access and maintenance. However, designs today also include tiny spaces, where fruit and vegetables

A scarecrow protects valuable crops.

are grown informally in pots on a patio or balcony, or even in a window box. Materials for surfaces focus on the utilitarian – concrete slabs, brick paths, or compacted earth are all practical options and suit the look.

Planting varies seasonally, with fruit trees and bushes providing the permanent structure. Low box hedges may also be included, often to contain herbs that tend to flop and spread, while rainwater, required for irrigation, can be captured in butts or other recycling vessels.

Formal potager at Château de Villandry in France.

What is productive style?

In large productive gardens, the layout and surfaces tend to be functional, creating a sense of ordered abundance, while in smaller spaces, the design is often more relaxed, with planters used to squeeze in as many crops as possible. Traditional designs were influenced by early monastic or physic gardens, which were divided into geometric beds filled with herbs and vegetables, punctuated by taller focal plants, such as bay trees or standard roses, in the centre. These simple design plans are used in contempory edible gardens, too, with bed sizes often shrunk to fit smaller urban plots. Functional paths – made of brick, stone, or gravel – allow space to tend the fruits and vegetables easily, while colourful rows of crops, fruitful containers, and decorative interplanting create garden designs that provide a feast for the eyes as well as the table.

Productive gardens in detail

As the 20th century came to a close, productive planting was pushed to the end of the main garden to give flowers, shrubs, and trees pride of place. Today, this approach is changing, as more people realize that growing food close to home is not only fun, but also allows you to enjoy fruit and vegetables that are either not available in the shops or, like raspberries or blueberries, expensive to buy.

Productive gardens need to be planned carefully to make them easy to manage. When planting in the ground, different crops should be planted in different beds each year to prevent the build-up of soil-borne pests and diseases. In small gardens and on patios or terraces, compact crops, such as tomatoes, chilli peppers, aubergines, and leafy salad crops can be grown successfully in pots or larger planters. Cold frames, greenhouses, and sunny windowsills indoors allow you to extend the growing season, while bee-friendly plants, such as lavender and open-flowered dahlias, inject colour and bring in pollinators to guarantee a good crop.

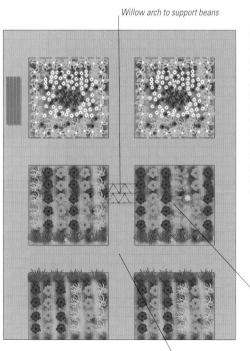

Willow arch to support beans

Square beds with a colourful mix of flowers and leafy crops

Wide paths for access and structure

Colourful potager
Here, the ordered character (left) of the vegetable garden, with its rows of crops and strong rectilinear pattern, makes a beautiful impression (right). Tall supports for runner beans and clipped hedging are used to enclose the space, and red dahlias and lavender add extra splashes of colour.

DESIGN INFLUENCES

While many modern productive gardens are a mix of styles, some still echo the regimental formality of the walled kitchen gardens of the great English country houses. Victorian aristocrats showed off their wealth by serving exotic hothouse produce to guests, but the main function of the garden was to provide fresh food for the whole household.

Crops were set out in orderly lines in geometric beds edged with box and separated by paths made of gravel or beaten earth, or ash produced by the glasshouse boilers. Tender fruit trees were trained along south-facing walls that radiated heat to give them additional protection, while soft fruit bushes were grown under netted frames to prevent birds from eating the harvest.

Large, heated greenhouses were often built into the structure of the wall, allowing early cropping and the cultivation of tender produce, such as peaches and apricots.

Traditional walled kitchen garden.

Key design elements

1 Raised beds
Raised beds were first introduced to improve drainage, but they also provide a sense of order. An increased height of up to 1m (3ft) allows those with a disability to tend their gardens more easily.

2 Wide paths
Pathways should be at least 1m (3ft) wide in order to make the garden easy to navigate. Hard surfaces, such as brick, concrete or stone slabs, or gravel, are ideal since they withstand heavy everyday use.

3 Rustic obelisks
Ornamental features are always put to good use. Trellis and wooden or metal obelisks create height and rhythm in the garden, but also provide support for climbers, such as runner beans or sweet peas.

4 Planting in rows
Crops planted in rows can be easily recorded, cared for, and harvested, and the spaces between rows provide access for weeding. This geometric layout gives these beds their strong character.

5 Practical containers
Pots can be used to grow a wide range of edibles in small gardens and on patios and terraces. Large containers hold more compost and water and require less maintenance than smaller types.

Interpreting the style

When planning a fruit and vegetable garden, you can opt for a formal design with regular pathways, or go for a more relaxed approach, using a series of planters and pots. Low hedges or raised beds give coherence to border edges in larger gardens, and beans, corn, and fruit trees provide height. Introduce colour with flowers that attract beneficial insects, or choose those you can eat.

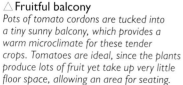

△ **Fruitful balcony**
Pots of tomato cordons are tucked into a tiny sunny balcony, which provides a warm microclimate for these tender crops. Tomatoes are ideal, since the plants produce lots of fruit yet take up very little floor space, allowing an area for seating.

▷ **Olive terraces**
Rows of mature olive trees provide a sculptural element in this elegant scheme for a warm, sunny urban space. A layer of culinary herbs is planted below to soften the architecture.

◁ **Edible windowbox bouquet**
Strawberries have been planted along with edible flowers, including nasturtiums and pot marigolds, in this contemporary windowbox. The marigolds have a citrus flavour and nasturtiums taste peppery.

▷ **Salad in a planter**
Suitable for use in restricted spaces, this stained timber planter contains a mix of salad crops and herbs. Tomatoes or strawberries would also be appropriate.

▽ **Urban kitchen garden**
This small city courtyard has been transformed into a tiny allotment, with baskets of crops and a cleverly designed dining table that doubles as a planter for salad leaves, herbs, and flowers.

▽▷ **Eye-catching gourds**
Productive planting can be included in the design of a main garden. Here, gourds are used as a decorative climber, giving privacy to the seating area. Pink dahlias provide late summer colour below.

◁ **Lettuce and herb mix**
Raised timber planters offer easily accessible beds for herbs and salad leaves. The rough woven rope edging on those shown here helps to combat attacks by slugs and snails.

GARDENS TO VISIT

BROGDALE, Kent
Home of the National Fruit Collection.
brogdalecollections.org

LOST GARDENS OF HELIGAN, Cornwall
Walled garden with many traditional cultivars. heligan.com

WEST DEAN, West Sussex
Beautifully restored Edwardian kitchen garden. westdean.org.uk/gardens

RHS GARDEN WISLEY, Surrey
Includes herb, fruit, and vegetable gardens. rhs.org.uk/wisley

CHATEAU DE VILLANDRY, France
Formal Renaissance kitchen garden. chateauvillandry.fr

"Homegrown produce is one of the joys of a gardening life"

CASE STUDY

EDIBLE EDEN

Productive gardens can be any shape or size, and even in this small plot, the designer has squeezed a wide range of edibles into raised beds and narrow borders, mingled with flowers that attract bees and other pollinators to create a beautiful, bountiful space.

Elegant yields

Rustic materials and a mix of vegetables, herbs, and flowers references cottage style. Every bed is crammed with edibles, from beetroot and lettuces to beans scrambling up wigwams, but the overall look is decorative and orderly.

Practical paving

The red brick pathway marries well with the traditional styling. Both practical and decorative, it lends an old-fashioned look, while allowing plenty of space for wheelbarrows and a hard surface from which to cultivate and harvest the produce.

Herb focal point

A clipped bay tree edged
with a skirt of culinary herbs
– including rosemary, parsley,
and thyme – provides a
beautiful, aromatic focal point
in the centre of the garden,
and a readily accessible source
of fresh herbs for the kitchen.

Crops in close-up

The wide edging on the raised
beds doubles as both work
surface and informal seating
from which to admire the
garden. It also allows crops to
be inspected at close quarters
so that damage from pests and
diseases can be spotted quickly.

Potted extras

In small gardens, compact
crops can be grown in pots
and containers to increase the
growing space. These patio
tomatoes have been bred for
such a purpose and produce
high yields of sweet fruits
on small bushy plants.

Productive garden plans

In a productive garden, function generally wins over style, but the two are not mutually exclusive. These three gardens are packed with delicious edible plants, yet each, in its own way, looks great. Maurice Butcher's design bursts with edible produce; Bunny Guinness's vegetable garden gives a nod to formality with its timber raised bed; and an allotment society has mixed herbs, flowers, and vegetables in a small space.

Wildly productive

Even the paving in this natural-looking productive garden, designed by Maurice Butcher for the RHS Hampton Court Palace Flower Show, is softened by a profusion of planting – in this case chamomile which, when trodden on, releases a scent.

Key ingredients

1 *Chamaemelum nobile* (lawn chamomile)
2 *Santolina rosmarinifolia* (cotton lavender)
3 *Petroselinum crispum* (parsley)
4 *Mentha suaveolens* (apple mint)
5 *Galium odoratum* (sweet woodruff)
6 *Chamaedaphne* 'Cassandra' (leatherleaf lettuce)
7 *Thymus* 'Doone Valley' (thyme)
8 *Salvia officinalis* 'Tricolor' (sage)

Maurice says:

"This small kitchen garden was created for enthusiastic gardeners. The emphasis is on medicinal and culinary herbs for regular harvesting, but the space is for relaxing, too. The clients also wanted something organic and with a low carbon footprint.

"As the design developed it became clear that we were working towards a blend of fruit, vegetables, and herbs, and that they should be the dominant elements.

"I take inspiration from many things – including literature, art and travel. The input and character of my clients are essential ingredients in my work, too."

Grey-green concrete stepping stones surrounded by chamomile

This *Arbutus unedo* (strawberry tree) is the focal point around which the garden is organised

Raising veg

The geometric layout of this garden by Bunny Guinness includes the sort of well-equipped detailing needed in a hard-working space. The raised beds of vegetables are easy to reach and maintain.

Key ingredients
1 *Phaseolus coccineus* (runner beans)
2 *Allium cepa* (garden onions)
3 *Daucus carota* subsp. *sativus* (carrots)
4 *Beta vulgaris* subsp. *vulgaris* (red chard)
5 *Vitis vinifera* (vine)

Bunny says:
"This garden was originally dominated by an overgrown Leylandii hedge. Once this was removed, the space really opened up and a backdrop of native plants was revealed, which help to soften my design.

"The space works hard, which is typical of my approach. The owner is a barbecue enthusiast, so I created a space for entertaining, with a barbecue and built-in sink, and a small greenhouse.

"My influences often come from the architects I work with, and new or interesting ideas I see on my travels."

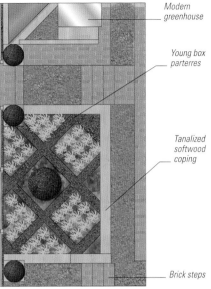

Modern greenhouse

Young box parterres

Tanalized softwood coping

Brick steps

Garden allotment

This garden was designed by the Manchester Allotment Society for the RHS Flower Show Tatton Park and aims to show how easy it is to integrate a few crops into the average domestic garden.

Key ingredients
1 Beehive-style composter
2 Wild flowers
3 *Ocimum basilicum* (basil) and other herbs
4 *Solanum melongena* (aubergine)
5 *Cucurbita pepo* (pumpkin)

Packed with a variety of herbs, including basil, fennel, sage, and parsley, the crops are squeezed into raised wooden beds and small patches of soil in between. French marigolds (Tagetes) are woven through the herb plants, providing colour and helping to deter flying pests.

Tender crops, such as aubergines and tomatoes, are also included. They can be grown outside in a sheltered sunny garden, and ripen towards the end of summer. A few pumpkin plants scramble up supports at the back of the plot.

The white beehive composter creates a decorative yet practical focal point, and wild flowers help to lure pollinating insects to the fruiting vegetables.

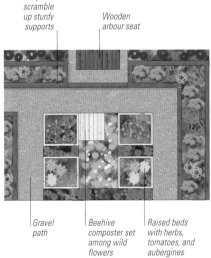

Pumpkins scramble up sturdy supports

Wooden arbour seat

Gravel path

Beehive composter set among wild flowers

Raised beds with herbs, tomatoes, and aubergines

   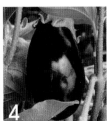 

Family gardens

As leisure time increased in the middle of the 20th century, the concept of a garden shifted from a formal area that was walked through or viewed from the house, to a space that provided a focus for family life. Specific areas devoted to relaxation, children's play, and dining have become increasingly popular, and today these spaces form the template for many family designs.

Family gardens are often a blend of styles. Their layouts can be rectangular or curved, with flexible designs for children's areas that will accommodate their changing needs as they grow. Play equipment helps to introduce strong colour into the design, while planting areas that attract a range of wildlife can also provide entertainment for young ones.

The safety of babies and young children is a top priority in these gardens, with jets and cascades, where the main water reservoir is underground, used instead of open water features. However, naturalistic ponds are perfect for older children, who will enjoy the aquatic creatures and wildlife these features attract.

Natural surroundings can be adapted to create play areas.

Natural or composite stone are popular materials for dining and seating areas, with bark chippings, or other soft yet resilient materials, providing practical surfaces for play spaces. In larger gardens, the transition between the children's and adults' areas can easily be managed with separate, designated areas, but in smaller plots the design may need to be more adaptable, perhaps using play equipment that can be cleared away as night falls. Lighting can also help to create a different ambience for adults to enjoy after dark.

Planting in a family garden needs to be robust and easy to maintain; it should also be free from toxic plants and sharp thorns. Hardwearing turf is the best choice for lawns used by children, or opt for easy-care artificial grass.

A swimming pool provides hours of fun for older children.

What is a family garden?

A family garden can be almost any style that has been adapted to provide a flexible space for games, room for entertainment and play, and an area for dining. The smallest of gardens can accommodate a sandpit or swing, while larger plots have space for separate adult- and child-friendly zones.

Family gardens in detail

The concept of the outdoor room celebrates family life. Terraces need to be large enough to accommodate a dining table and chairs, with space for a barbecue or even an outdoor kitchen.

For play, there are two schools of thought: structured play relies upon equipment, but children have different needs as they grow, so flexibility is important. For example, a small sandpit located close to the house allows parents to watch their young children more easily; then, as they grow and move down the garden to seek more adventure, swings, slides, and climbing frames can be introduced.

Unstructured play provides a rich and interesting environment in which children can be encouraged to take some risks – building dens, pond-dipping, climbing trees, and watching wildlife. This requires a more subtle approach to design and one in which parents cannot be too precious about their gardening exploits, giving preference to the needs of their inquisitive children.

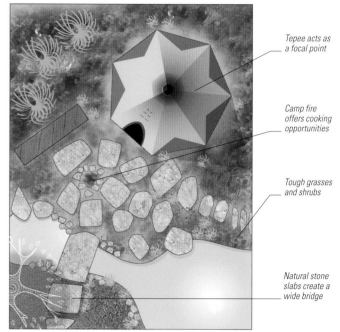

Tepee acts as a focal point

Camp fire offers cooking opportunities

Tough grasses and shrubs

Natural stone slabs create a wide bridge

Natural playground
Designed by Chuck Stopherd of Hidden Garden, this garden (right) for older children offers valuable opportunities for outdoor play. The tepee, firepit, and pool, hidden behind trees, provide a natural setting for children to take risks and explore their environment.

A 1950s family garden designed for play.

DESIGN INFLUENCES

The opening up of the garden as a family facility is a relatively recent occurrence, although outdoor dining *en famille* has always been a tradition in Mediterranean countries. Thomas Church's book, *Gardens Are For People*, first published in 1955, changed perceptions of the garden and signalled a move away from intensive gardening and towards the development of the outdoor room. Later, John Brookes developed these ideas in his designs and 1969 book *Room Outside: A New Approach To Garden Design*. Today, gardens are places of enjoyment, education, and fun for families to share.

Key design elements

1 Play equipment
The children's area can feature large items of play equipment, such as a swing or climbing frame. If space is limited, some items may still be included by adapting a pergola or similar structure.

2 Colourful materials
Splashes of bright, primary colours are an essential ingredient in a family garden. These can be introduced via planting, equipment, or hard landscaping.

3 Dens and tents
Part of the children's area could include a den: a place of their own where they can extend their imagination through play. It may be sited within view of the house or tucked away in a corner.

4 Tough plants
Plants have to be versatile and tough to withstand rough treatment from children and pets. Closely planted, often with some evergreens and seasonal colour, they must also be easy to look after.

5 Wildlife features
Ponds with sloping sides to allow creatures access, boxes for birds, habitats to give shelter to hedgehogs, and plants to attract bees and butterflies, are all ideal for family gardens.

6 Easy-care seating
Seating needs to be suitable for children and adults. Furniture that can be left uncovered all year and requires the minimum of care and maintenance is the most practical.

Interpreting the style

A family garden is about sharing your space. The dining area is the social hub around which the design revolves, and can be created with a paved or decked terrace that links into a lawn or into more structured play areas with integrated or temporary play equipment. Swimming pools or natural ponds make reflective centrepieces for gardens where older children play.

△△ **Versatile space**
A large-scale chessboard is both a design feature and a challenging family game, making the most of a quiet retreat surrounded by textured foliage planting.

△ **Safe play area**
This built-in sandpit is close enough to the house to be monitored, but planting creates the illusion of another world. A cover will provide protection from the weather.

▷ **Star attraction**
Central to the design of this contemporary garden, the turquoise pool is both functional and decorative. Safety covers or security fences may be introduced if necessary.

▷▷ **Secret hideaway**
In a secret corner of this densely planted garden, a den of willow and brushwood becomes the focus of adventure and discovery, providing an escape from the adult world.

◁◁ **Adventure playground**
A play house that can only be accessed via a footbridge – fun for kids, but perhaps too precarious for adults – allows children to escape, and control who visits.

◁ **Wildlife haven**
This large reflective pond and the reed margins provide a range of wildlife habitats that can be observed from the various vantage points located around the banks.

▽ **Family fun**
As well as exercise, a trampoline offers a perfect outlet for letting off steam, which is beneficial for both children and adults alike.

"Helping to bring families closer together is perhaps the garden's most important role"

△ **Colourful entertaining**
This vibrant area is part of a modern design, and combines cooking, dining, and relaxation, offering a fun area where the whole family can decamp to escape the confines of the house.

▷ **Tree-top retreat**
A tree house takes pride of place here, acting as both a retreat for children and a decorative focal point. It also offers a hideaway for adults when the children are in bed.

FAMILY GARDENS TO VISIT

ALNWICK GARDEN, Northumberland
Created with children in mind, with water features and a gigantic tree house.
www.alnwickgarden.com

CAMLEY STREET NATURAL PARK,
King's Cross, London
Ponds and meadows, and hands-on activities.
www.wildlondon.org.uk

CAMDEN CHILDREN'S GARDEN,
Camden, New Jersey, USA
Four-acre interactive garden for families.
www.camdenchildrensgarden.org

MILLENNIUM PARK, Chicago, USA
Offers a programme of interactive family events and workshops.
www.millenniumpark.org

CASE STUDY

FAMILY VALUES

Family gardens should be places of fun, where children have freedom to explore and play safely. Successful designs cater for both young and older users, providing features to entertain little ones, and areas for adults to relax and enjoy the scenery.

Shady canopies

The white birch stems echo the white blooms, while contrasting with the understorey of green foliage. The trees punctuate the design with their bright vertical trunks, and their canopies also offer essential shade, helping to protect youngsters from sunshine.

Soft to touch

Easy-care plants that are soft to touch are ideal for family spaces. Here, shade-loving perennials, shrubs, and evergreen ferns create a leafy blanket, while star jasmine clads the walls, its tiny blooms scenting the air.

Designers **Nick Buss and Clare Olof**
Show **RHS Hampton Court Palace Flower Show**
Award **Silver Medal**

Colourful journey

The curved path is colourful
and confident, creating a
visually exciting journey and
a focal point through the
duo-tone planting. The small
brick setts also lend detail and
texture, and complement
the tiled box stool.

Bubbling tubes

A great way to introduce water
safely into a family garden is
with these eye-catching "bubble
tubes" filled with clear and
dyed water. The sound and
movement will fascinate
children, while also producing
a soothing, calming effect.

Hide and seek

The hollowed tree trunk and
woven willow playhouse (far
left) bring an element of fairy
tale to the design, to fire the
imagination and provide places
to play and hide. Such naturalist
structures blend tonally with
the planting and wider design.

Family garden plans

Integrating functional spaces for different age groups is the challenge in family gardens. These gardens – the first designed by Ian Kitson, with planting by Julie Toll, and the second designed by Claire Mee – take contrasting but equally successful approaches to the family garden brief. Ian's curved, informal layout blurs the line between adults' and children's areas, while Claire's follows formal lines with a more discreet spot for play.

Gently rolling

In this London family garden, Ian Kitson has created a spacious lawn where the children can play, while the terrace provides a place for family dining and social occasions. The two areas are divided by a snaking dry-stone and log wall, and by soft planting, designed by Julie Toll.

Key ingredients
1 *Acer palmatum* 'Sango-kaku'
2 *Geranium* 'Jolly Bee'
3 *Echinacea purpurea*
4 *Crataegus monogyna*
5 Dry-stone walling
6 *Lavandula angustifolia*
7 *Calamagrostis* x *acutiflora* 'Karl Foerster'
8 *Brunnera macrophylla* 'Jack Frost'

Ian says:
"Julie and I call this the 'snakes and ladders' garden – the layout is curvilinear, but the detailing is sharp and precise. The garden previously featured a sudden drop in level, but the retaining walls, steps, and planting have softened this.

"Lighting is included within the steps and between the logs in the curving dry-stone and log walls, which give the garden an organic quality.

"The terrace is used for outdoor dining, and there's room on the lawn for games. I like the way the grass oozes around the wall, and the fact that it's transformed into a carpet of daffodils in spring."

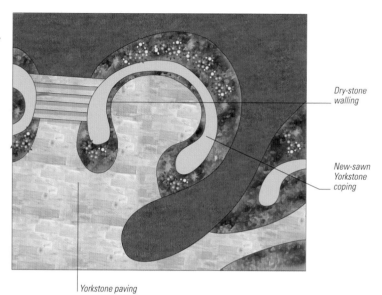

Dry-stone walling

New-sawn Yorkstone coping

Yorkstone paving

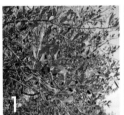

Corner piece

The sophisticated look of this family garden by Claire Mee was achieved with an elegant decked terrace for dining, while the pergola at the end of the plot gives the children a play area, complete with swing. The spaces are divided by a grove of olive trees, which offer privacy and add height. The tree canopies have been lifted to leave clear stems that create dramatic shadows; light also reflects on the silvery foliage.

Key ingredients
1 *Olea europaea*
2 *Buxus sempervirens* 'Latifolia Maculata'
3 *Allium hollandicum* 'Purple Sensation'
4 *Sisyrinchium striatum*
5 Bark chippings
6 *Origanum vulgare* 'Aureum'

Claire says:
"This urban garden occupies a corner plot, so it's an unusual shape. My ideas for the design were developed from the house's architecture, and from the interior design and decor. I'm often influenced by the interiors of hotels, restaurants, and bars, which use different materials so well.

"Wide windows look down the length of the garden, and we used clear-stemmed olives to provide privacy without blocking this view. Elsewhere, I like the contrast between the softer planting and the architectural specimens. The client also wanted a terrace outside the French doors to match the floor-level in the house, and I designed a large timber deck to make this link (legally, a paved surface would have to be lower to avoid the damp course)."

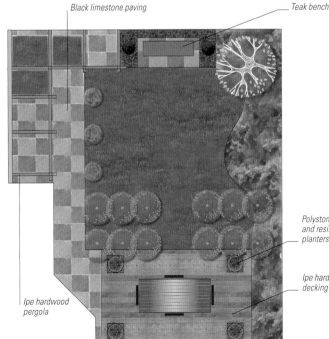

Black limestone paving

Teak bench

Polystone (fibreglass and resin composite) planters

Ipe hardwood decking

Ipe hardwood pergola

Natural gardens

Naturalistic gardens are nothing new, with influential designers from the 18th to the 21st century striving to emulate the natural world in a variety of ways. Today, this style focuses primarily on sustainability, with designers incorporating plants and materials that do not diminish the world's dwindling resources. A natural garden will typically include recycled and renewable materials and a diverse mix of plants that offer food and habitats to wildlife.

Introduced in the late 20th century, the New Perennial Movement – as espoused by plantsmen such as Piet Oudolf – increased interest in naturalistic gardening styles and has influenced many contemporary designers today. This style combines hardy perennials with grasses, matching plants with their sites so that they flourish with little maintenance. More recently, British, Dutch, and German research into sustainable plant communities has also set new design trends.

There is a popular idea that natural gardens must be rustic in character, but this need not be the case, and many modern, elegant designs include local or renewable materials, such as timber from certified plantations, and sophisticated recycled materials.

Recycled materials create key features in this modern design.

Most owners of natural gardens adopt an organic approach to controlling pests and diseases, keeping them at bay through use of biological controls and balanced ecosystems, rather than chemical pesticides. Habitats that support local species and help to increase biodiversity are key to these designs, but natural gardens do not rely exclusively on native species; non-invasive exotic plants that attract beneficial insects and wildlife are also highly useful, offering extra colour and year-round interest.

Extensive prairie and meadow planting is often used in large gardens, but wild flowers and bee-friendly species can easily be included in smaller spaces, too, providing a range of different habitats in tiny gardens.

A sympathetically designed swimming pond will attract wildlife.

What is natural style?

A natural garden should be capable of working as an effective ecosystem, with reduced or minimal levels of intervention – it is this approach that sets it apart from a traditional garden. Ecological principles play an essential role in creating habitats in which planting neighbours thrive, competition between them is balanced, and species are closely matched to the prevailing soil and climatic conditions.

Natural gardens in detail

The materials used in a natural garden need to be assessed against a series of criteria. Recycled products are a good idea as they reduce the exploitation of new resources, but sometimes they have a higher carbon footprint, whereas sourcing new timber from managed, renewable and, preferably, local plantations may be a better option.

Other factors to consider include the permeability or drainage of hard-landscaped surfaces. These should be either porous, in order to top up groundwater, or designed to allow water to run off into a collection unit or water butt, thereby reducing the strain on supplies.

In a sustainable, natural garden, planting is key, and a healthy variety of wildlife habitats essential. Choose plants that thrive in the prevailing conditions and complement each other, which in turn will help to reduce the incidence of pests and diseases, although other forms of biological control may also be needed. Soil improvers should come from your own compost heap and organic manures.

Rosemary Weisse's garden at Westpark in Munich.

DESIGN

The change from purely ornamental planting to the creation of successful plant communities started when William Robinson (1838–1935) advocated the integration of native and exotic species, which he called "wilderness planting". The development of American prairie planting, championed by Jens Jensen in the 1920s and '30s, responded to Robinson's ideas, and was later taken up in Europe by the New Perennial Movement. Large drifts of grasses and perennials, like those seen in the schemes of Rosemary Weisse in Munich, are typical of this approach. In the UK, the Department of Landscape at the University of Sheffield has produced significant research into sustainable prairie and meadow planting.

Key design elements

1 **Green roofs**
Green roof systems manage rainwater run-off and provide insulation. Convert existing roofs using pre-planted sedum mats. New structures can accommodate more elaborate habitats.

2 **Encouraging wildlife**
Increased diversity is achieved by creating effective habitats for wildlife. The more habitats there are, such as old logs, bee hotels, and insect-friendly planting, the greater the diversity.

Wildlife haven

Designed as a naturalist, sustainable garden by Stephen Hall (*left*) this beautiful design shows how precious resources, such as water and wildlife, can be supported and protected. The garden includes a range of diverse habitats, including a pile of decaying logs and tree stumps to provide homes for rare beetles, small mammals, and overwintering amphibians, such as frogs and toads. The traditional-style building is built entirely from sustainably sourced cedar, and features a green roof planted with sedum species. Research shows that green roofs help to insulate buildings and keep them cool when temperatures rise, reducing the need for heating and air-conditioning. They also attract beneficial insects when the plants are in flower.

Eco-friendly building with an insulating green roof

Harmonious design
The gravel path that weaves through Stephen Hall's garden and around the pond allows visitors to enjoy the different plants and features at close hand, and integrates perfectly into this naturalistic setting.

Nectar-rich planting attracts beneficial insects

Gravel, pebbles, and boulders suit the natural style

Wildlife pool attracts insects, birds, and small mammals

3 Rainwater harvesting
However small, water butts are an excellent way to catch and store rainwater. If you need something with a larger capacity, underground storage and pump mechanisms are available.

4 Rustic garden furniture
Wherever possible, support your local economy by commissioning a craftsman close to home to make your furniture. All products should be made from responsibly sourced, natural materials.

5 Recycling features
The recycling of organic waste through composting is vital. Several compost bins may be required in order to maintain and rotate supply. Think carefully about their location, as they need regular access.

6 Naturalistic ponds
Wildlife ponds with sloping sides that allow easy access, and margins planted to provide cover, offer a natural habitat for aquatic creatures, as well as birds and insects, such as dragonflies.

Interpreting the style

A natural garden can follow a formal layout, but most are informal, with relaxed planting drifts and apparently random mixes of grasses and perennials, indigenous trees, and shrubs. You can then organise these into habitats, such as wetland, meadow, or woodland, and use recycled materials, sourced locally or from renewable plantations, and permeable paving.

△△ **Desert oasis**
American designer Steve Martino produces elegant and modernistic gardens in the Arizona desert, using billowing natives and drought-tolerant species, interspersed with key plants such as Agave.

△ **Bird haven**
Feeders and bird tables will help attract wildlife, especially during harsh winters when food sources may be scarce.

△ **Ideal match**
For a successful meadow, it is essential to match planting to the environment. Here, the elegant nodding heads of Fritillaria meleagris suggest damp conditions.

▷ **Sleek combination**
Diffused mixes of meadow or prairie perennials and grasses provide a perfect foil to sharply detailed contemporary architecture, existing happily side by side.

"Natural gardens offer food and habitats for beneficial insects and other wildlife"

GARDENS TO VISIT

THE BETH CHATTO GARDENS, Essex
Famous gardens that have been developed, and are gardened, on ecological principles.
bethchatto.co.uk

LONDON WETLAND CENTRE, London
A network of ponds and wetland meadows, with observation hides and interactive features.
wwt.org.uk/wetland-centres/london

BOTANICAL GARDENS, University of Göttingen, Germany
Ecological and habitat-based gardens.
uni-goettingen.de/en/108651.html

WEIHENSTEPHAN UNIVERSITY GARDEN, Freising, Germany
Where the New Perennial Movement began. hswt.de/en/weihenstephan-gardens

WESTPARK, Munich, Germany
This public park includes the herbaceous drift and steppe planting of Rosemary Weisse.
muenchen.de

△◁ **Waterside planting**
Pond margins provide one of the richest garden habitats, bringing together aquatic, marginal, moisture-loving and dry planting schemes. Keep planting groups large and associations simple for the best results.

◁ **Lasting interest**
Sown prairie planting mixes, typically combining Echinacea and Rudbeckia with grasses such as Panicum, provide an effective display and long season of interest.

△△ **Safe habitat**
A simple timber structure provides dry storage for logs, an important habitat for over-wintering insects.

△ **Mixed species**
Allowing native plants to colonise among meadow grasses aids the conservation of species endangered through urban development or intensive farming.

CASE STUDY

WILD ONE

Blending wild flowers, ornamental plants, a small
woodland and recycled materials, this design shows
how a natural garden can also look sophisticated
and exciting, while offering a range of habitats for
birds, beneficial insects, and other forms of wildlife.

Bug hotels

These dry-stone walls not only
break up the space, injecting
eye-catching sculptural forms,
but they are also designed as
bug hotels, with ready-made
nest holes, cracks, and crevices
for solitary bees and other
beneficial insects to inhabit.

Roosting sites

This small copse of white-
stemmed birch trees (*Betula
utilis* var. *jacquemontii*) offers
a home for birds to roost
and nest, while ferns and
other shade-tolerant plants
below offer further habitats
for small creatures.

Beneficial planting

A mix of wild flowers, nectar-rich ornamentals, such as the orange *Geum* 'Prinses Juliana', and houseleeks (*Sempervivum*) on the tops of the walls, offers plenty of visual interest and food for the bees, and is low-maintenance once established.

Upcycled office

Creating a dramatic statement, this stylish outdoor office is made from an old shipping container. The circular panel decorations are filled with cones, bits of wood, and bamboo canes, which provide further homes for insects.

Flood defence

A series of shallow, linked pools form the reservoirs of a "storm-water chain" drainage system, designed to capture excess rainwater after a heavy downpour to prevent it running off the garden and causing flooding.

Natural garden plans

To keep their varied planting in some order, many natural gardens have quite structured layouts, and despite their abundant and seemingly uncontrolled appearances, the gardens designed by Nigel Dunnett and James Barton, shown here, are held together with well-defined lines and shapes. They also include water, which provides an important habitat for many types of wildlife, and permeable hard-landscaping surfaces.

Practising what you preach

Nigel Dunnett is a Professor at the University of Sheffield and a landscape designer. He is renowned for his research into sustainable planting and urban drainage systems, and this small garden, which sits on a north-facing slope, puts many of his findings into practice.

Key ingredients

1 *Euphorbia palustris*
2 *Geranium sylvaticum*
3 *Lonicera periclymenum* 'Serotina'
4 Green roof
5 *Euonymus alatus* 'Compactus'
6 *Astilbe chinensis* var. *taquetii* 'Purpurlanze'
7 *Caltha palustris*
8 *Acorus calamus*

Nigel says:

"I wanted to create a woodland glade, with closely planted birch forming a light canopy and linking with the surrounding countryside. Clipped hornbeam hedges provide enclosure and structure alongside softer successional planting.

"Perennials form a dense ground cover, almost eliminating the need for weeding. The planting is 50 per cent natives and 50 per cent cultivated garden plants – together they give almost year-round colour. The pond is filled with run-off from the paved surfaces and helps to manage the drainage in the garden, which has been a huge success."

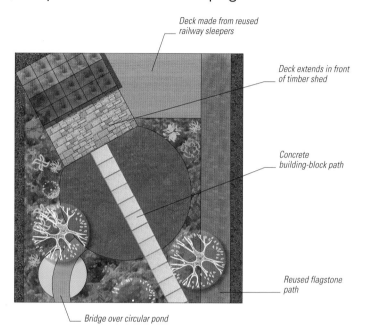

Deck made from reused railway sleepers

Deck extends in front of timber shed

Concrete building-block path

Reused flagstone path

Bridge over circular pond

Compact sustainability

Dr James Barton and his wife developed the design of their sustainable garden in Westphalia, Germany, over a number of years. The garden is modest in size, yet includes a rich range of planting – ornamental and native species, selected for interest and their ability to thrive as good neighbours, are intermingled. A system of pathways provides easy access to them.

Key ingredients
1 *Nymphaea alba*
2 *Iris sibirica*
3 *Fagus sylvatica*
4 *Angelica archangelica*
5 *Carpinus betulus*
6 *Lychnis flos-cuculi*

James says:
"In its early days, this was a family garden, but since our children left home it has evolved into something else.

"We develop areas as we gain new ideas, but the basic layout of the garden, as a series of "rooms", remains the same.

We have structured the spaces with beech and box hedges, or with fences, and we have also created a range of small, informal seating areas to provide different views through the garden. In the main, we use perennials and shrubs, with some annuals added as necessary to provide splashes of colour.

"For inspiration, we have visited many open gardens, primarily in the Netherlands and southern England. However, we were originally inspired by a visit to a small private garden in Germany, the owner of which was the president of a local society, the Gesellschaft der Staudenfreunde, of perennial plant enthusiasts."

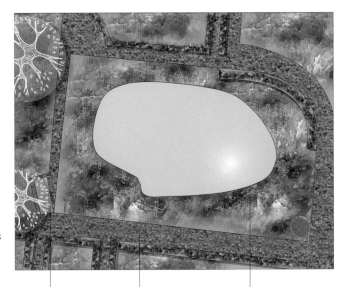

Timber bench, made from recycled wood

Paths made from granite, reclaimed when local streets were re-paved

Dense planting around pool offers habitats for wildlife

Urban gardens

Gardens have always had a presence in cities, but since the late 19th century, when urban populations began to increase dramatically, they have taken on an ever more important role as relaxing oases. City gardens are generally small spaces, and though there are plenty of ways to design them successfully, simplicity usually produces the best results.

Many urban garden designers, keen to use space efficiently, employ plans based on squares and rectangles that fit snugly into small, regular-shaped plots. Other designers organize layouts on the diagonal, which can make an area seem larger. Free forms are also increasingly popular as urban designers become more experimental.

But whatever their size or shape, modern city gardens should be flexible, since they may have to offer areas for play, as well as for outdoor dining, entertaining, and relaxation. A simple palette of hard-landscaping materials creates clean, practical surfaces, while careful planting along the boundaries can increase privacy.

Lighting is an essential addition to these architectural spaces. It can emphasize both the hard landscaping and the planting, as well as extend the garden's use after dark.

In small urban gardens, planting is often restricted to a handful of high-performing plants used to create interest all year round, with vertical planting, in the form of climbers and wall shrubs, softening the

Repetition of forms adds impact.

edges. Owners of city gardens can also try their hand at growing vegetables, fruit, and herbs, using containers and pots to create a mini allotment. Some urban garden designers also choose to minimize open spaces in favour of dense planting and a complex range of plant species, which can increase the feeling of seclusion and privacy. Architectural minimalism, a proliferation of plants, or both? You decide.

A neat mix of materials offers contrasts in colour and texture.

What is urban style?

Today's city gardens have to work hard, providing space for planting, relaxation, play, and entertaining. As the high price of land in urban areas has squeezed the size of gardens, new ideas for small spaces have emerged. Approaches vary, but most urban gardens are treated either as functional spaces or as green oases – both offer a private escape or retreat from hectic city life. In the former, hard surfaces dominate, creating a stage for multiple uses. Architectural treatments to boundary walls, furniture, and water features create elegant "rooms", often lit after dark to create extensions to the home. In the latter, planting dominates, often taking over areas that could have been used for entertainment or play. This intensive planting approach benefits the keen urban gardener, who may even use the space as a productive allotment.

Urban style in detail

The urban garden layout needs a simple, clear geometry. Planting similarly needs careful thought, as space is limited – the trend has been for fewer species that work harder seasonally, providing architectural or sculptural interest. Grasses and large-leaved foliage plants are popular with designers of this style.

In many city gardens, sliding or folding doors create a seamless transition between interior and exterior "rooms", extending the living area. Paved or decked surfaces help to increase functional space; materials are often selected to match interior finishes, further unifying indoors and outdoors. Pergolas or pleached trees offer privacy in overlooked minimalist spaces, while dense planting can achieve the same effect in more naturalistic urban gardens.

Sculpture provides a focal point, often combined with water used in jets or cascades rather than pools. Built-in seating fits architecturally, but can limit the flexibility of the garden. Stylish furniture and identical containers in a row add drama and rhythm.

City garden
Here, garden designer Philip Nixon has created a simple but decorative plan with timber-clad walls complementing the furniture, and folding doors that lead out from the house (right). Planting is a mix of perennials, grasses, and evergreens, with the addition of tall pleached hornbeams, which provide valuable screening.

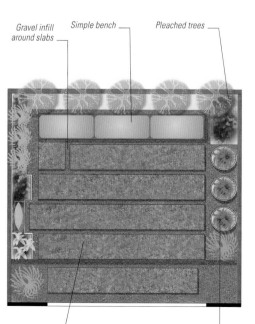

Gravel infill around slabs   Simple bench   Pleached trees

Slate slabs   A row of potted grasses

DESIGN INFLUENCES

Evocative of country gardens, early city designs were often heavily planted and complex in layout. Today, they have become much simpler.

In 1839, JC Loudon – the Scottish botanist, garden designer and garden magazine editor – responded to increasing urbanization and the diminishing size of city gardens in his book *The Suburban Gardener and Villa Companion*. In it, he classified different design approaches to the small urban garden, including low-maintenance designs.

More than a century later, John Brookes published a series of successful books that, like Loudon before him, addressed designs for smaller plots, and explored the idea of the "outdoor room".

More recently, the Japanese have led the way in designing tiny outdoor spaces. In their densely populated cities, balconies or light wells are often the only areas available for planting.

A John Brookes design for a London garden.

Key design elements

1 Dramatic containers
Clay, stone, or steel plant containers are often repeated for effect. Fill them with clipped box or – for a softer, more informal look – a mix of perennials and grasses.

2 Sculptural furniture
Artfully designed furniture – in the shape of bespoke built-in benches, coordinated tables and chairs, or recliners – gives the garden focus and answers a functional need.

3 Lighting
With the introduction of low-voltage and LED systems, lighting has become more sophisticated. Use it to emphasize your garden's contours and plants.

4 Pleached trees
In overlooked city gardens, pleached trees (which look like hedges on stilts) provide privacy while using little floor space. Use lime, hornbeam, or evergreen holly oak.

5 Stylish materials
Designers often employ a mix of materials to maximize texture and interest. Both natural and man-made materials, such as concrete, glass and steel, are popular.

Interpreting the style

If hard surfaces for outdoor living dominate, planting has to work harder to compensate. Choose simple, bold, architectural combinations, which are stylish and easily maintained. Lighting, strategically placed, will flatter the space in the evening. For densely planted areas, keep paving simple, using strong textural foliage and colour as a foil to the built-up environment.

△△ **Soothing retreat**
Vertical or wall planting optimises the restricted space, while retaining a softening effect. A textured panel of basalt provides sound as water trickles over the surface.

△ **Formal welcome**
Here, box hedging defines dense, foliage planting that softens the paved areas. Pleached trees obscure the outline of the surrounding buildings and create privacy.

△△ **Geometrical harmony**
The decorative grid of paving reinforces the soft lawn surface, while the simple, rectilinear geometry of the garden and its planting complement the house.

△ **A place to entertain**
Raised beds also provide informal seats for relaxing around the fireplace. The mix of ornamental grasses and Allium creates a diffuse screen between two areas.

"As space diminishes, the urban garden becomes an increasingly precious resource"

△△ **Hidden gem**
A suspended canopy adds style and privacy to a seating area. Planting is minimal and restricted to containers, tonally linking to the cushions on the benches.

△ **Outside living**
A room outside in which to eat and relax, with extra seating provided by the raised beds. Water spilling from the wall and over the slabs creates a sensuous sound.

GARDENS TO VISIT

RHS CHELSEA FLOWER SHOW, London
The five-day show contains a specific section of gardens designed for urban situations. Held in May of each year.
www.rhs.org.uk/chelsea

KENSINGTON ROOF GARDENS, London
Located on top of a Grade II listed building in the heart of a busy London street.
www.roofgardens.com

THE NGS YELLOW BOOK, UK
A comprehensive list of many privately-owned urban gardens open to the public.
www.ngs.org.uk

THE GARDENS OF APPELTERN, Holland
A whole range of gardens, including urban style.
www.appeltern.nl

PALEY PARK, 53rd Street, New York, US
One of New York's famous pocket-handkerchief spaces offering cooling water and shade.
www.pps.org

CASE STUDY

HIDDEN DEPTHS

This sunken garden maximises the sense of space in a compact urban plot by introducing different height levels, while the edges of the hard landscaping are softened by restful planting that relies on contrasting leaf shapes, colours, and textures.

Eye-level intimacy

The sunken seating area is surrounded by raised beds that bring the planting up to eye level, achieving a sense of intimacy with nature in an urban setting. Different levels also partly conceal the space to create a sense of discovery.

City shades

Grey stone paving creates a contemporary look. It is made from a traditional material, but the colour – which reflects the urban landscape – brings it up to date, while creating a foil for the plants' foliage.

Calm contrasts

The textural planting blends a range of leafy plants, such as *Epimedium* and grass-like *Libertia*, with a sprinkling of floral interest from the likes of *Aquilegia* and *Anthriscus sylvestris* 'Ravenswing'. The restrained colour scheme creates a tranquil effect.

Stylish furnishings

Dressing a garden to suit your taste helps to personalise the space. The seats here are perfect for two people to relax in, away from the noise of the city beyond, while the bold red and grey cushions add a contemporary note.

Secret spaces

Solid screens reinforce a feeling of privacy and help to shield the social space of the garden from neighbouring properties. Tall shrubs and perennials are used to lightly veil other areas of the garden, affording glimpses through to tempt in visitors.

Urban garden plans

Small gardens demand big ideas, and in their designs for these two city plots, Andy Sturgeon and Sam Joyce have certainly delivered. Andy has found a clever solution to the particular problems that a roof garden presents – such as an overall weight limit, and increased exposure to the elements for plants and people. Sam has made the most of a very small plot with a useful, yet uncluttered and colourful design.

Up on the roof

In a restricted city space, this roof-top garden by Andy Sturgeon makes excellent use of the great outdoors. The low-maintenance design creates an extra room in which to entertain, with materials providing the focus and simple planting offering shelter and privacy.

Key ingredients
1 *Fargesia rufa*
2 Iroko bench
3 *Astelia chathamica*
4 Gas-fired flambeaux

Andy says:
"This space suited the client, who was young and enjoyed entertaining friends, but wasn't particularly interested in gardening.

"The water became the focus of the garden. It is very shallow, to reduce the weight on the roof, but highly reflective to excite and entrance; combining it with fire proved a particularly complex detail.

"I normally design larger spaces that are not so minimal, but my approach to this project suited the client and the roof-top location, and I enjoyed responding to the challenge. More specifically, the client wanted to be able to sit outside in all weather and seasons, hence the canopy and the water, fire, and bench combination.

"I call upon a wide range of inspirations, from shop-window treatments to contemporary art, and find this input particularly useful in urban situations."

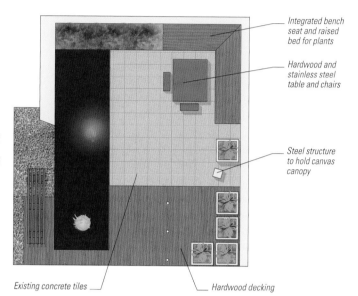

Integrated bench seat and raised bed for plants

Hardwood and stainless steel table and chairs

Steel structure to hold canvas canopy

Existing concrete tiles

Hardwood decking

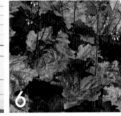

Tiny retreat

When designing a small area, you have to make a very short wish list of uses and then prioritize: what is essential and what can you do without? In this suburban back yard, Sam Joyce's choices were limited, but she responded to her client's main request for room to entertain and relax with a fitted seating area, a simple deck that offers space for extra chairs to be brought out from the house, and statement plants to soften the lines without cluttering up the garden.

Key ingredients

1 Trachelospermum jasminoides
2 Miscanthus sinensis
3 Musa basjoo
4 Buxus sempervirens
5 Electric wall light
6 Heuchera 'Plum Pudding'

Sam says:

"This is a very small yard attached to a Victorian terraced house. The client is a single professional with grown-up children who live away from home, and the space was to be used primarily for relaxing and entertaining – there is a strong sense of community in this area, and neighbours regularly socialise in each other's gardens.

"The bench provides seating for several guests, and doubles as a sun lounger. It also helps to disguise the various utilities in the garden, and creates a colourful contrast for the planters filled with box balls behind it. The white wall comes alive with the silhouettes of the plants in front of it when the garden is lit at night, and this architectural planting adds impact to the design during the day."

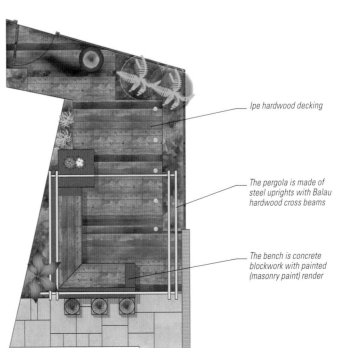

Ipe hardwood decking

The pergola is made of steel uprights with Balau hardwood cross beams

The bench is concrete blockwork with painted (masonry paint) render

Country gardens

For centuries, people living in cities have been tantalized by a romantic vision of a garden in the country. Today, improved transport links have made it possible to work in town but return to the country on a daily or weekly basis, and the dream of a country garden has become a reality for many.

In the 18th century, the Landscape Movement turned from classical formality to more natural designs, and the country garden idyll was born. The style was later developed by the Arts and Crafts designers during the Edwardian period, and it now refers generally to large, heavily planted gardens, often split into a range of smaller spaces. Areas for pleasure may include swimming pools, tennis courts, lawns, and terraces for entertaining. Orchards, woodland, meadows, or lakes provide habitats for wildlife, as well as vantage points to view the landscape beyond.

For designers, it is the scale of these gardens that presents a challenge. The most effective layouts tend to favour rectilinear formality close to the house, with increasing informality and curvilinear geometry as you move through the garden.

Planting is also generally formal around the house, terraces and main lawns, but becomes more naturalistic towards the boundaries. Natural stone or brick are typical paving materials, although concrete may be appropriate for the modern country house. Further from the house, gravel is often used, with grass paths taking over in the wider landscape.

Views and vistas are enhanced and exploited, while features, such as stone seats, pergolas, ornamental pools, and sculpture, are important as destinations and help to define the garden landscape. Hedges contain garden spaces or define views, and tree-lined avenues are also effective if space allows.

Taking inspiration from the past and infusing it with innovative contemporary ideas, country garden style continues to evolve.

Golden spikes of *Kniphofia* provide colour and structure.

Stone steps and a narrow waterfall lead to a tranquil pool with *Pontederia*.

What is country style?

The advantage of a country garden is the ample space it affords, which can accommodate a range of areas and planting schemes. The designer's challenge is to bring together the various elements in a coherent composition. Formal symmetry often dominates the styling close to the house, with more informal spaces in the outlying garden, moving from man-made features to natural landscapes. These different areas are linked with paths and visitors are led by focal points, viewing areas, and resting places.

Country style in detail

Many large country gardens are exposed to wind, which can seriously limit or damage plant growth, sometimes even preventing the plants' establishment altogether. Consequently, windbreak planting is frequently the first element to be introduced, but this can obscure surrounding views. Compromises have to be made, often producing limited or narrow vistas, yet this restriction forces designers to evaluate views and different perspectives carefully, which can increase the drama.

Hedges provide soft structure within the garden, creating rooms in traditional-style gardens or independent screens in more contemporary schemes. Hard paving materials are generally used near the main property, with routes through the garden in more economical gravel. Planting schemes have to be appropriate for the large scale.

Lawns often cover the greatest area, but meadows with mown paths or prairie planting provide more texture and seasonal colour. Woodland and lakes also offer a variety of experiences and habitats.

William Robinson's natural style.

Curved concrete wall — *Boulders provide structure*

Decked walkway adds texture — *Lawn links planting and paths*

Country contrast

Here, Andy Sturgeon uses concrete, decking, and bound gravel paths to create a fluid transition between level changes in this contemporary country garden in south-east England (right). Large boulders stand out against the soft sweeps of planting beyond, and grasses provide movement and light in the deep planting beds. The wide boundary hedges screen views of neighbouring properties.

Bound gravel path meanders through the garden

Key design elements

1 Luxuriant planting
Extensive borders provide the opportunity for dynamic planting, using colour and texture in drifts or *en masse*. Meadow-style planting is also used for its potential wildlife value and decorative aspects.

2 Large pools and streams
Natural springs may provide the basis for ponds and streams, but they can be introduced artificially to create reflective surfaces and wildlife habitats, or for new planting opportunities.

3 Views into the landscape
The garden experience can be dramatically enriched by linking it to the landscape. Long, narrow views, which open up to a wide natural panorama beyond, produce spectacular effects.

4 Sweeping lawns
Lawns are used both as a functional surface and as a decorative foil to more textured or colourful planting. Lawns and grass pathways should be as wide and open as possible, as the surface can wear with heavy use.

5 Hedging and screens
Hedges define space and control views. Yew produces a dark, dense backdrop that is perfect for colourful borders. Low box hedges are ideal for parterres, and mixed hedges work well on a larger scale.

6 Natural materials
Local stone that weathers to produce varied surface textures, such as York-stone, is often seen in traditional country gardens. A more contemporary quality is achieved with concrete and decking.

Interpreting the style

There has recently been a move away from complex mixed border schemes to a more limited planting palette, such as the architectural hedges and monocultures typical of Jacques Wirtz's designs, or the large drifts of colour evident in the work of Piet Oudolf. Both designers rely on the movement and light-capturing qualities of grasses, which provide a long season of interest.

△△ **Graphic design**
Rows of clipped hedges and billowing grasses are interspersed with the white trunks of closely planted birches, creating strong shadow patterns, rhythm, and movement. The simple palette of green foliage plants emphasizes line and texture.

△ **Painting with flowers**
Christopher Lloyd experimented with vivid colour in his garden at Great Dixter, shown above. He combined clashing pinks and reds, flouting conventional colour theory.

▷ **Autumn glory**
The mahogany seedheads of Phlomis stand out against the green, silver, and bronze mounds of grasses and perennials in these stunning deep borders.

◁ **Exuberant border**
Splashes of colour illuminate this haze of planting and emerge skywards, adding vertical interest. Transparent veils of grasses and perennials create the romance.

▷ **Catching the light**
These graceful borders, planted with a mix of golden feathery grasses and eye-catching red Sedum, encircle this sunny seating area with movement and light.

▽ **Mirror image**
The glassy surface of the pond is the main feature in this garden. Marginal planting is restrained to maximize the reflective surface, and the terrace provides space for outdoor entertaining.

▽▷ **Virtuoso planting**
In his own garden, Piet Oudolf mixes broad masses of colour with drifts of grasses to create a soft meadow effect. The wave-clipped yew hedges provide a contrast in architectural form.

"The luxury of space and abundant planting create the magic"

GARDENS TO VISIT

BORDE HILL, West Sussex
Combines many different garden and planting styles, including water gardens.
bordehill.co.uk

GREAT DIXTER, East Sussex
Inspiring garden that uses colour creatively.
greatdixter.co.uk

HESTERCOMBE, Somerset
A garden by Edwin Lutyens and Gertrude Jekyll, plus an 18th-century landscape garden.
hestercombe.com

KIFTSGATE COURT, Gloucestershire
An outstanding 20th-century garden.
kiftsgate.co.uk

ROUSHAM PARK HOUSE, Oxfordshire
William Kent's early 18th-century masterpiece. rousham.org

SCAMPSTON HALL, Yorkshire
Includes Piet Oudolf's dazzling walled garden.
scampston.co.uk

CASE STUDY

UPDATED COUNTRY

For some, a country garden is traditional, formal, and large; this garden proves otherwise, mixing many of the style's key ingredients – such as burgeoning flower borders, lawns, and sculpture – into a clean, contemporary design in a relatively small space.

Elegant borders

Despite its modern design, the planting is firmly rooted in the past. Roses, such as the red 'Chianti', provide old-fashioned scent, alongside pink foxgloves (*Digitalis purpurea*), geraniums, and blue salvias, to form a traditional country border.

Sculptural focus

The modern piece of figurative sculpture brings a contemporary note to the traditional setting, a focal point that helps lead the eye along the path and presents a destination to draw in the visitor.

Verdant lawn

A lawn is an essential element of country garden style and the soft grass is married here with a tidal rill, designed to emulate a sparkling stream. While a lawn suggests formality, its oval shape and rill edging provides a link with the natural landscape.

New perspectives

The curved pathway through the garden is echoed by the false-perspective bench, which is wider at one end to create the illusion of greater length. The curves contrast with bronze upright fins along the boundary, which add drama.

Hidden secrets

Few country gardens reveal all the interest and features they possess in one go. This garden uses the same trick, offering glimpses through trees and borders of areas yet to be discovered, as the visitor journeys through the space.

Country garden plans

The expansive nature of country gardens gives designers room to luxuriate in planting. The first of these two examples is open to the public and was designed by Piet Oudolf – the influential Dutch designer, nurseryman, and author, who is also a leading figure of the "New Perennial" movement. The second, by Fiona Lawrenson, is a private space where the plants, although just as abundant, feel a little more contained.

Garden meadows

Piet Oudolf's garden for Sir Charles and Lady Legard at Scampston Hall in Yorkshire is one of his most arresting. It mixes formal elements with drifts of informal grasses and shapely perennial flowers – Piet's signature planting, which injects dramatic seasonal impact.

Key ingredients
1 *Achillea* 'Summerwine'
2 *Rudbeckia occidentalis*
3 *Monarda* 'Scorpion'
4 *Phlomis russeliana*
5 *Echinacea pallida*
6 *Stachys officinalis* 'Hummelo'
7 *Panicum virgatum* 'Rehbraun'
8 *Salvia* x *sylvestris* 'Mainacht'

Piet says:
"The garden at Scampston covers about four acres and sits within protective walls. It used to be a working garden, but my clients wanted to create a contemporary space rather than a reconstruction.

"I worked with the large scale of the garden to create something of interest to the visiting public, so not all of the planting is typical of what I do. I aimed to link the past with the present by using formal elements, such as hedges and clipped specimens, between more relaxed perennials.

"I am influenced by contemporary architecture, art, and nature; and I think that, at Scampston, there is interest in both the planting and the strong design."

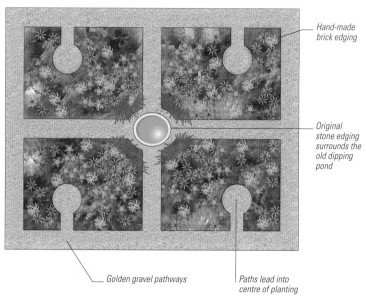

Hand-made brick edging

Original stone edging surrounds the old dipping pond

Golden gravel pathways

Paths lead into centre of planting

Stepping out

In this large garden designed by Fiona Lawrenson, stepped, circular lawns provide an elegant transition from the terrace outside the house to the main garden. Planting surrounds these circles, softening their geometry – a key quality in the country garden.

Key ingredients

1 *Rosa* 'Rambling Rector'
2 *Salvia nemorosa* EAST FRIESLAND
3 *Sambucus racemosa* 'Plumosa Aurea'
4 *Campanula poscharskyana*
5 *Centranthus ruber*
6 *Acanthus spinosus*

Fiona says:

"This Hampshire property has an old-fashioned country pedigree – Jane Austen used to live nearby and visited regularly to collect milk. Its garden stands on a south-facing hillside with views across a valley, and I wanted to create a gentle descent into it from the house, with the wide circular steps gradually turning to take advantage of the view. Originally there was a narrow path and a vertical drop down into the main garden, so the new terrace and steps created space and a link into the main garden.

"The owners were a young family who needed usable space and wanted a spot from which they could enjoy views of the setting sun, hence the 'gin' terrace.

"I like to link a house with its surrounding landscape through its garden, and I am strongly influenced by the architecture I work with. But plants are my first love, so they take centre stage. This garden's bedrock is chalk with heavy clay soil on top, and its planting suits these conditions."

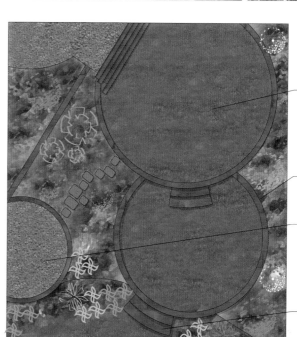

Two large, circular lawns provide a soft, lush link between the house and main garden

Flint panels clad the brick retaining walls, and echo details on the house

Gravel viewing terrace

The bricks used for the steps match the colour of those on the house

Cutting-edge gardens

Influenced by art as much as horticulture, cutting-edge gardens break design conventions and free up designers to make their own set of rules. Conceptual gardens, which are often based on an idea or theme, fit into this category and examples can be seen at various festivals around the world, including RHS Hampton Court Palace Flower Show in the UK, Chaumont-sur-Loire in France, and Reford Gardens in Métis, Canada. Cutting-edge design can also be used to describe any contemporary garden that does not fit neatly into a more conventional style.

Many cutting-edge designs celebrate new technologies and employ man-made materials, such as concrete, steel, rubber, fabric, glass, and Perspex, to create impact and visual interest. Lighting is also used to great effect in many of these gardens.

Planting is not intrinsic to a successful cutting-edge garden, but can support the overall message conveyed by the design. When used, planting is often included for its sculptural qualities, and may also emphasize colour, texture, and movement. For some designers, ideas are inspired by ecology or the environment, and their gardens may feature plants that showcase a particular place or habitat.

Architectural foliage and flowers provide focal points.

Design concepts can be applied on a whim, but the best results are achieved where there is a relationship between the garden, its location, and the personality of its owner, or its history and cultural significance.

Key figures in cutting-edge design include the landscape architects Martha Schwartz and Kathryn Gustafson, who have both created ground-breaking gardens. Land art has also been influential in the evolution of this style. Examples include the works of Richard Long and Andy Goldsworthy; both designers are renowned for their natural sculptures, which form part of the landscape and intensify visitors' experience of a place.

Manufactured materials are mixed with natural elements.

What is cutting-edge style?

This style is a mix, sometimes accidental, but often deliberate, drawing from a wide range of genres. Short-lived and more experimental, show gardens offer a platform for these eclectic creations and allow designers the freedom to innovate. Colour, sculpture, and garden art provide focal points and interest, while planting often focuses on architectural specimens and lighting adds to the drama.

Cutting-edge gardens in detail

Rendered walls are typical of this style, as they provide backdrops or surfaces on which art and sculpture can be displayed. Colour, usually intense and bold, is also important, creating a vibrant atmosphere. A wide range of materials are associated with the style, and in some gardens the combinations can be quite complex. Designers often use a mixture of man-made and natural surfaces, such as concrete and timber, or stone and steel, and by keeping the overall plan simple, these textural contrasts are more clearly appreciated.

Furniture is frequently used to express particular architectural or stylistic references, or it may also introduce colour. Sculptural plants add scale and drama, and are sometimes repeated to amplify ideas. In addition, colourful and textural planting is a common feature, with containers used to reinforce stylistic concepts.

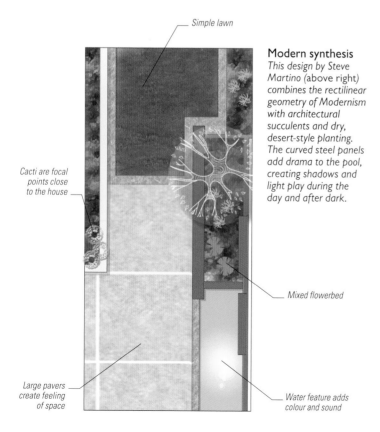

Simple lawn

Cacti are focal points close to the house

Large pavers create feeling of space

Mixed flowerbed

Water feature adds colour and sound

Modern synthesis
This design by Steve Martino (above right) combines the rectilinear geometry of Modernism with architectural succulents and dry, desert-style planting. The curved steel panels add drama to the pool, creating shadows and light play during the day and after dark.

DESIGN INFLUENCES

This style borrows from a range of ideas with energy and confidence. Travel, a shrinking world, and the Internet have opened up access to a wide range of plants, materials, and influences – from jungle planting to Japanese gravel, Modernism to Mediterranean, and formal to conceptual style. This gazebo by Michael Schultz and Will Goodman uses Japanese elements with Art Deco and Post-Modernist overtones. The personality of the resulting designs may not please the purists, but cutting-edge style is all about breaking the rules.

The Hurst garden by Schultz Goodman.

Key design elements

1 Modern materials
Cutting-edge designs often include materials that are not traditionally associated with gardens, such as glass, steel, and Perspex, with planting softening the lines.

2 Sculptural plants
Although a wide variety of plants are used in cutting-edge gardens, many have sculptural qualities – grasses, *Yucca or Astelia* are typical, and palms are used for height.

3 Water cascades and fountains
Cascades, fountains, and water blades – controlled by a smart phone to produce complex displays – provide movement, atmosphere, and sound.

4 Lighting
Light effects are key style devices, picking up architectural details, specimen plants, and decorative topiary. The development of lighting technology and LEDs produces spectacular results and can also inject additional colour.

5 Eclectic floor plan
The mixing of styles can produce interesting and complex layouts, with Modernist designs mixed with drought gravel planting, or formality combined with the asymmetry of Japanese gardens.

6 Vibrant colours
Bold colours are often used in surface finishes to make connections between plants and hard materials. Rendered walls, ceramics, paving and lighting can all contribute colour and drama while creating an exciting ambience.

Interpreting the style

Designing cutting-edge gardens is a liberating and fun experience, where rules can be rewritten. Colour can be a controlling element, with rich or strident tones making clear connections between materials and planting. Also try using irregular shapes and mix solid materials with transparent glass or Perspex to create a bold, unique design.

△ Bedrock of design
Like a geological phenomenon, these angled layers of red sandstone rise out of a pond and are juxtaposed with dry, Mediterranean planting combinations with glaucous foliage.

◁ Solid seating
A touch of the interior design is brought to this outdoor terrace, with its concrete seating and coffee table. Planting softens the effect in places, and cushions would make the furniture more comfortable.

▷ Playing with the elements
In his garden of sculptural Jura limestone, Peter Latz uses fog as a device with which to create a sense of mystery. Its wisps veil and reveal the stone forms in turn.

▷▷ Golden brown
This gravelled courtyard space is unified by striking colour and strong shadows. The simple grove of Mexican fan palms (Washingtonia robusta) creates a brilliant connection with the modern Mexican architecture, too.

"Cutting-edge designs mix up conventional ideas and bend the rules of garden-making"

△**Water and earth**
Water gently cascades over this ledge, cantilevered from a rendered wall, and into the trough below, creating an oasis in this desert garden. The warm earth-tones echo the sandy soil and glow in the sun.

◁ **Blocks and undulations**
White concrete cubes are counterpoints to the turf that ripples across this garden. They create a sculpted quality that offsets the stark walls of the house.

GARDENS TO VISIT

RHS HAMPTON COURT PALACE FLOWER SHOW, UK
Show with a section of conceptual gardens.
rhs.org.uk/hamptoncourt

GARDEN OF AUSTRALIAN DREAMS,
Canberra, Australia
Richard Weller and Vladimir Sitta's garden.
nma.gov.au

FESTIVAL OF GARDENS
Chaumont-sur-Loire, France
domaine-chaumont.fr

CORNERSTONE, Sonoma, California, USA
Regularly changing showcase of innovative design. cornerstonesonoma.com/gardens

CASE STUDY

DESIGN FUSION

Fusing a range of styles, from Mediterranean to Modernist, this cutting-edge garden weaves Jurassic period inspirations into a harmonious design, with large metal structures – inspired by the bony back plates of a stegosaurus – defining the space.

Natural structure

Evergreen trees, including the Mediterranean oak (*Quercus ilex*) and strawberry tree (*Arbutus unedo*), lend structure and a sense of permanence to the garden, while other hardy exotic trees with finely-cut foliage soften the look.

Prehistoric slabs

The seemingly random floor plan and irregular-shaped paving stones throw out the design rule book. They help to evoke a rugged landscape that references the earth's tectonic plates as they collide to form new geological features.

Bridging the gap

The pathway in the garden steps up to form a bridge across the water, giving the impression that the water has been here for a long time and the paving is a new addition. In other areas, stone slab-like benches suggest ancient rock formations.

Steel screens

Bronze-coated steel slabs stand proud, cutting dramatic shapes that resemble a dinosaur's back plates, and providing a focus along the perimeter of the garden. They also present a foil to the firepit and create hidden areas that heighten the intrigue.

Artful planting

The planting seems informal, even "shaggy" in parts, but this belies a considered approach. Sculptural plants, such as *Corokia × virgata* with its tangle of black stems, jostle with colourful perennials, including the fiery orange kangaroo paw (*Anigozanthos*).

Cutting-edge garden plans

Gardens that use a range of stylistic references can easily become confused, yet these spaces by Paul Cooper and Tony Heywood manage to maintain clarity of vision. One, a north-facing plot, has been lightened with reflective surfaces and enlivened by its various influences; the other represents the power of nature, with swirling landforms and contrasting textures.

Two become one

This private garden is an amalgamation of two of Paul Cooper's designs – Hanging Gardens, and the Cool and Sexy Garden – both for the RHS Chelsea Flower Show. Paul embraced an exciting mix of modern materials to create a garden where height and structure dominate.

Key ingredients
1 *Hedera helix* 'Kolibri'
2 Stainless steel water feature
3 *Phyllostachys nigra* f. *henonis*
4 *Wisteria floribunda* 'Macrobotrys'
5 *Santolina chamaecyparissus*
6 Painted wrought-iron railings
7 *Lavandula pedunculata* subsp. *lusitanica*
8 *Mahonia* x *media* 'Buckland'

Paul says:
"My client on this project was great. He was forward-thinking and didn't want a conventional garden. The plot is north-facing, cool, and gloomy, so I emphasized verticals to create the feeling of escaping these restrictions. And, with its theatrical lighting and reflective surfaces, this garden really performs at night.

"I'd say the design is typical of my work. I originally trained and worked as a sculptor, and I can definitely see a three-dimensional character here. Contemporary architecture was, and is, a big influence, but there are some Japanese elements in there, too."

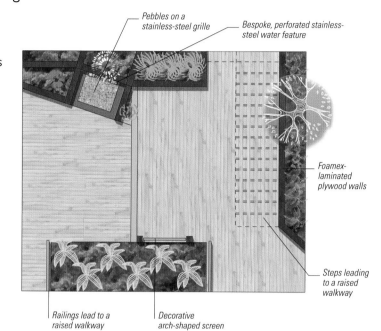

Pebbles on a stainless-steel grille

Bespoke, perforated stainless-steel water feature

Foamex-laminated plywood walls

Steps leading to a raised walkway

Railings lead to a raised walkway

Decorative arch-shaped screen

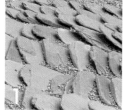

Helter skelter

Artist, horticulturist, and garden designer, Tony Heywood, created this public garden at a junction of two busy streets in the centre of London. The design revolves around a vortex of bubbling water, with the other elements spiralling towards it, alluding to the energy and speed of the traffic and people passing by. Contrasting textures heighten this sense of excitment, while the disparate elements are bound together by a simple colour palette of blues and greens.

Key ingredients
1 Slate fish scales
2 Tipping box topiary
3 Mirrored steel sculpture
4 Minimalist planting palette
5 Man-made materials
6 Lead and slate wall feature

Tony says:
"I wanted the garden to look like it had been created by a powerful natural force that was pulling the land like a twisted carpet into a central vortex. The railings were bent, yew topiaries are tipping over, and jagged vertical slates erupt from the ground. Inspired by Japanese gardens, I used stones to represent rivers and pushed this idea further, with slates up the wall made to look like a rock face."

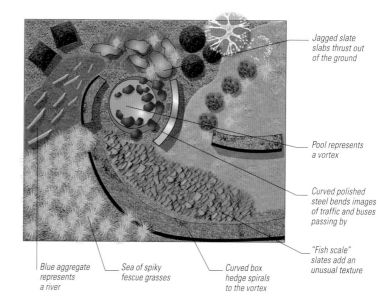

Jagged slate slabs thrust out of the ground

Pool represents a vortex

Curved polished steel bends images of traffic and buses passing by

"Fish scale" slates add an unusual texture

Blue aggregate represents a river

Sea of spiky fescue grasses

Curved box hedge spirals to the vortex

Building preparations

Creating a new garden from scratch, or tackling a major hard landscaping project, is a serious undertaking. If you decide to do the work, but only have weekends free, or do all the ground preparations by hand, it could take months to finish. The upside, however, is the immense satisfaction of having done it yourself, and the savings on labour. Detailed preparation is paramount, and it is essential that you calculate the cost of all materials, hire equipment, and any professional fees in your budget.

DIY vs employing professionals

Depending on your experience, you may feel confident about tackling a simple paving project, erecting trellis, or building a deck. In fact, many modern building materials and garden features are specifically designed for ease of construction. When taking on any work yourself, ensure you are equipped with the appropriate safety equipment, such as

Laying surfaces DIY style
If you have the necessary strength, skill, and experience (such as in using specialist cutting equipment), you may consider building your own new patio or wooden deck.

eye protection when sawing timber, and steel-toed boots for any construction work. Jobs involving heavy materials or a high level of skill are often best left to professionals. Natural stone, for instance, often comes in large pieces that require skill to cut and lay. Similarly, in a modern garden, crisp design demands a very high-quality finish to avoid it standing out for all the wrong reasons. Experience and expertise count, especially when it comes to safety. Wet soil, for example, can be very heavy, so employ a professional to construct retaining walls.

If in any doubt about your ability to take on a project, seek expert advice from garden designers or civil engineers; source them via their professional organisations (see p.368). Also remember when hiring a contractor that they are responsible for taking out insurance, and ensuring that work complies with all safety standards and building codes.

Laying paving in difficult places
Building stepping-stones that appear to float on the surface of a pool is not easy, as water shows up the slightest discrepancy in levels. Since the steps are to be walked on, they must also be rock solid to avoid accidents.

Sequencing workflow

The value of an experienced contractor is that they know how long it takes to perform various tasks, such as digging and laying foundations, or constructing brick walls. They should also be able to pull together the necessary skilled workforce, just as the next phase is about to commence.

Any project can be dogged with unforeseen difficulties, such as bad weather or delayed deliveries, which hamper the work. As established contractors often have several projects running simultaneously, delays in these other gardens can also have a knock-on effect on yours. Project managers must maintain good communications with all parties, anticipate problems and find ways to maintain a free-flowing operation. Sit down with your contractors, and go through the details of construction together. Then draw up an agreed schedule and refer to it regularly.

Consider lighting before landscaping
Integrated light effects need to be planned well in advance of construction so that fixtures can be built in and cables suitably camouflaged.

KEEPING TO A BUDGET

If you hire a contractor to run a project from start to finish, and have a contract drawn up detailing completion deadlines, material selection and costs, you shouldn't run into difficulties over the budget. Problems commonly arise when you make changes to the plan mid-way through the build, or alter the specifications of the materials used. Good organization is vital if you run the project yourself, especially when hiring a workforce. Workers standing idle, waiting for materials to be delivered, still have to be paid.

Special effects
Some lighting and water features need expert installation, and many materials also require specialist preparation. Always check that your contractors have the relevant experience.

Pre-construction checklist

Once you have completed a site survey, and prepared your design, it is time to work out when the construction and planting should take place, and who will do the work. You may decide to do some of the preparations yourself and bring in specialist contractors only for specific jobs. Either way, try to visualize the project from start to finish to make it run as efficiently as possible.

STAGE	PROJECT NAME	DETAILED INFORMATION
1	PERMISSION	Major building work, such as the construction of a conservatory or changing access, may need planning permission. Check if in any doubt, and talk to neighbours to explain plans and settle concerns.
2	HIRING CONTRACTORS	One or more contractors may oversee the project, bringing in specialists as needed. If project-managing the job yourself, you will need to find and hire bricklayers, pavers, joiners, electricians, etc.
3	SELECTING MATERIALS	Ask contractors to provide samples of landscaping materials, or visit stone and builder's merchants, and timber yards yourself. Personally select feature items and commission bespoke pieces.
4	MATERIALS ORDER/DELIVERY	Double-check amounts to avoid under- or overbuying. Arrange deliveries to coincide with different construction stages. This avoids materials getting in the way and having to be relocated later.
5	SITE CLEARANCE	Peg out area and hire a skip. Remove unwanted hard landscaping materials and features. If it is to be re-laid, lift current lawn with a turf-cutting machine. Also lift and move existing plants for reuse.
6	TOPSOIL REMOVAL	Save quality topsoil for reuse and do not mix with subsoil. Remove it manually or with a mini digger. Locate topsoil away from the construction site and pile it up on the future planting areas.
7	MACHINERY HIRE/ACCESS	If your plan requires a lot of heavy digging, trenching and re-levelling, hire a mini digger and operator. Ensure suitable access, clearing pathways and removing fence panels, as required.
8	FOUNDATIONS AND DRAINAGE	Establish different site levels and excavate accordingly. Organize the digging of foundations and drainage channels, then pour foundations and lay drainage pipes. If needed, move existing drains.
9	LIGHTING AND POWER	Bring in an electrician or lighting engineer to install the cabling grid for all garden lighting and powered features. Some of these shouldn't be wired up until the garden has been completed.
10	BUILDING AND SURFACES	Build all hard landscaping features, including all walls, steps, terraces, pathways, water features, and raised beds. Construct timber decks, pergolas and screens. Prepare new lawn areas.
11	BOUNDARY CONSTRUCTION	Once the contractors, builders or landscapers no longer require access across the boundary for their machinery, vehicles, and materials, walls and fences can be completed and/or repaired.
12	TOPSOIL AND PLANTING	Some basic planting may have to be done during the dormant season, while construction continues. Replace or buy in topsoil to make up levels, then carry out remaining planting.

Building garden structures

Permanent features and hard surfaces, such as footpaths, patio areas, fences, raised beds, ponds and pergolas, provide the structural framework for your garden design, underlining and enhancing softer areas of lawn and planting.

Many garden structures are easy to construct, and there are several simple projects that gardeners with few building skills – or none at all – can tackle safely, and achieve satisfying results in just a day or two. For example, pergola kits are widely available and quite simple to assemble, and you can buy pressure-treated timber pre-cut to length for features such as raised beds or decking.

When executing your design, start with the hard surfaces, but, before you begin, take time to measure your garden carefully. Check that you have sufficient space for a path that will be easy to negotiate, and that the area for a proposed patio or terrace will accommodate your chosen furniture. It may even be worth selecting furniture before you finalize your design plans; it's surprising how much room you need for a dining table and chairs, allowing for the chairs to be moved back comfortably with space to walk around them. Paths for main routes should be at least 1.2m (4ft) wide, and preferably paved or laid with gravel. These will be easier to navigate than narrow, winding routes or a course of stepping stones. Wide paths also provide space for mature plants to spill over the edges without impeding free movement.

Stepping stones are easier to lay than a paved pathway.

Building patios and some paths can be major DIY projects, and if you intend to pave or deck big areas it may be worth considering professional help, especially if your plans include heavy materials, such as stone or composite slabs. Small setts or bricks laid in intricate designs also require expertise. A gravel surface requires less skill to lay, and is ideal for an area around planting, or a path.

Informal ponds are beautiful features and quite easy to construct, although for a large site, a digger would be helpful.

Pergola kits make construction relatively easy and the results can be stunning.

Laying a path

Small paving units, such as blocks, bricks, and cobbles, offer flexibility when designing a path. For this project we used carpet stones (blocks set on a flexible mat), which are quick and easy to lay. If you use recycled bricks, check they are frostproof and hardwearing; ordinary house bricks are not suitable.

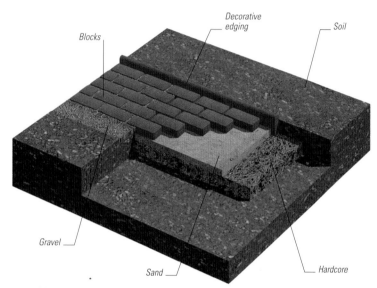

Decorative edging
Soil
Blocks
Gravel
Sand
Hardcore

You will need

- Tape measure
- Long pegs and string
- Hammer
- Spade
- Spirit level
- Nails
- Shuttering boards
- Hardcore
- Hand rammer or plate compactor
- Sharp sand
- Carpet stones
- Sledge hammer
- Post-hole concrete
- Edging stones
- Rubber mallet
- Broom
- Sharp knife, trowel
- Compost, herbs
- Gravel
- ⏱ 1 day

Marking out a path

1 Measure the path and mark with string and long wood pegs, spaced every 1.5m (5ft). Don't forget to allow for shuttering boards (*Step 4*) and decorative edging. Hammer in the pegs gently so they are firm.

2 Dig out the soil between the string to a depth sufficient to accommodate layers of hardcore and sand, as well as the thickness of the blocks. Check levels along the course of the path using a spirit level.

Laying the path

5 Spread an 8–10cm (3–4in) layer of hardcore along the length of the path. You could use excavated soil if the path will only get light use. Use a hand rammer or hired plate compactor to tamp it down.

6 Spread a layer of sharp sand over the hardcore. Level the surface by pulling a length of timber across the path towards you – use the shuttering along the sides as a guide. Fill any hollows with extra sand.

Adding edging and finishing off

9 Carefully knock the shuttering and pegs away and remove the string. Use a spade to create a "vertical face" to the edge – dig down as far as the hardcore base on both sides of the path.

10 Spread a strip of hardcore along each side of the path and tamp firm with a sledge hammer. If you're using heavy edging stones, lay a foundation strip of post-hole concrete mix on top.

11 Position edging stones and bed them in place by tapping them gently with a rubber mallet. Set stones flush with the path, or leave proud to stop soil migrating on to the path's surface. Backfill with soil.

12 Brush sharp sand into the joints (unlike mortar it allows rain to drain away). Remove the occasional block from the edge of the path to form a planting pocket. Carpet stones must be cut from the backing mat.

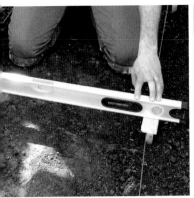

3 To prevent puddles on the surface, the path must slope gently to one side to drain into soil or a soakaway. Angle it away from the house or garden walls to avoid damp problems. Check levels again.

4 Carefully nail the shuttering boards to the pegs to enclose the area of the path. Check the levels once more with the spirit level; any necessary adjustments can be made by easing the pegs up and down.

CUTTING BLOCKS

When you are laying a path you may need to cut blocks or bricks to fit the pattern or to run around an obstacle, such as a manhole cover or the edge of a wall.

To make a neat cut, place the block on a firm, flat surface. Then, using a bolster chisel, score a line across the block where you want to cut it. Position the bolster on the score line and hit it sharply with a club hammer. Use the chisel to neaten up any rough areas. Remember to wear goggles to protect your eyes while working.

Cutting a block to size.

7 Tamp down the sand (see *Step 5*), ensuring the surface remains level. Begin laying whole blocks. Carpet stones come prespaced, as do most blocks, but if laying bricks, you will need to use spacers.

8 Once you have finished laying whole blocks, fill any gaps with blocks cut to fit (see *top right*). Bed the blocks into the sand with a hand rammer on a flat piece of wood, or a plate compactor.

13 Use a trowel to remove sand and hardcore from the planting pocket and replace it with loam-based potting compost. Plant a clump-forming aromatic herb, such as thyme. Water well.

14 Brush gravel into the joints between the blocks. If, as here, you have left a strip of soil along one side of the path to act as a soakaway, apply a topping of gravel to keep it looking neat and tidy.

Up the garden path

A well-laid path not only provides safe access through a garden, but is also a feature in itself. For period charm, use Victorian-style rope tiles as an edging.

Laying a patio

Pavers are available in a wide range of shapes, sizes, and materials, including concrete and natural or reconstituted stone, and make a practical hardwearing surface for paths and patios. Laying large pavers, while heavy work, is quick and easy; preparing the foundations is the hardest part of the job.

You will need

- Pegs and string
- Builder's square
- Spade
- Turf cutter (optional)
- Hand rammer or plate compactor
- Spirit level

- Hardcore, sharp sand
- Rake
- Pavers
- Bricklayer's trowel
- Ready-mix mortar

- Club hammer
- Wood spacers
- Stiff brush
- Pointing tool
- Masking tape

⏱ 2–3 days

Pavers — *Compacted sharp sand*

Lawn — *Soil*

Compacted hardcore

Marking out the patio

1 For a rectangular or square patio, mark out the paved area with pegs set at the height of the finished surface and joined with taut string. Use a builder's square to check the corner angles are 90 degrees.

2 Skim off turf using a spade or hire a turf cutter. (Reuse turf elsewhere, or stack rootside up for a year to make compost.) Dig out the soil to a depth of 15cm (6in) plus the thickness of the paving.

Laying the paving slabs

5 Top the hardcore with a levelled and compacted 5cm (2in) layer of sharp sand. Lay the first line of pavers along the perimeter string, bedding each one on five spaced trowelfuls of ready-mixed mortar.

6 Tamp down each paver with the handle of a club hammer. Maintain even spacing by inserting wood spacers in the joints. Check and keep checking that the pavers are sitting level.

The finishing touches

7 Wait about two days before removing the wood spacers. Then, either brush dry ready-mix mortar (or one part cement to three parts builder's sand) into the joints, or, for a neater, more durable finish, you could use a wet mortar mix (see Steps 8 and 9).

8 In dry weather, pre-wet the joints to improve adhesion of the mortar. For wet mortar, add water to the ready-mix mortar and push it into the cracks between the pavers using a bricklayer's trowel.

9 Firm the mortar in place with a pointing tool (above). Wet mortar may stain some pavers, but masking tape along the joints will protect them when pointing. Brush off excess mortar before it sets.

3 Use a hand rammer or plate compactor to tamp down the area. Set pegs at the height of the finished surface, allowing for the patio to have a slight slope so rain drains away. Check with a spirit level.

4 Spread a 10cm (4in) deep layer of hardcore over the area, rake level (ensuring you retain the slight slope), then tamp firm with a hand rammer or a plate compactor (*above*).

Keep it clean
If you don't stand patio pots on saucers, water and mud from them may stain the pavers. Where this occurs, clean the patio with a pressure washer.

Cutting a curve into a slab

Although pavers are available in a wide range of shapes, you may have to cut them to size, or to accommodate a curve in your design. Pavers, which are usually made from stone or concrete, are surprisingly brittle; to prevent them cracking when they are being cut, lay them flat on a fairly deep, level bed of sand.

1 Protect yourself with goggles, ear defenders, anti-vibration gloves, and a dust mask. Mark the curve on the paver with chalk, then, using an angle grinder fitted with a stone-cutting disc, slowly cut part-way through the paver, going over the line several times.

2 Mark out parallel lines on the waste area with chalk. Cut along the lines part-way through the paver, again going over each one slowly several times. Make sure you don't cross or damage your neatly cut curved line.

3 Starting on one side of the paver and working across to the other, tap firmly along the length of each cut strip with a rubber-headed mallet. Make sure that the paver is well supported.

4 Grip each strip firmly and snap it sharply along the cut. Remove all the strips in this way. Trim off any roughness along the curved edge with the angle grinder.

Cutting corners
A few shapely curves can completely transform a rectangular patio. Here, the corners have been opened up to form a planting pocket and to give a gentle sweeping curve to the adjacent area of lawn.

Laying decking

Decking is adaptable and blends with most garden styles. It can be built from hard- or softwood, or, more popularly, recycled plastic. If using timber, ensure supplies come from responsibly managed sources, and check building regulations and planning requirements for large or above-ground structures.

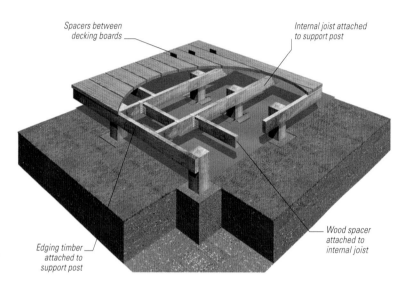

Spacers between decking boards

Internal joist attached to support post

Wood spacer attached to internal joist

Edging timber attached to support post

You will need

- Pegs and string
- Builder's square
- Geotextile membrane
- Tape measure
- Spade
- Hardcore
- Metal pole

- 75 x 75mm (3 x 3in) support posts
- Post-hole concrete
- Spirit level
- 100 x 50mm (4 x 2in) timber lengths
- Drill and router

- Galvanized bolts, washers, screws and nails
- Saw and hammer
- Decking boards
- Chisel, wood spacers

🕐 2 days

Putting up support posts

1 Mark out a square or rectangular deck with pegs and string. Check the corners are at a 90-degree angle using a builder's square. Mow grass very short, or skim off turf to use elsewhere in the garden.

2 Lay a geotextile membrane over the area, overlapping joins by 45cm (18in). As well as the four corner posts, you will need extra support posts on each side; mark these with pegs about 1.2m (4ft) apart.

Making the deck frame

5 Leave the concrete to set for 24 hours before building the deck frame. Cut edging timbers to length – note that joins should coincide with a post. Predrill bolt holes, countersinking them with a router.

6 Hold the first edging timber in place against the frame (you may need help); mark and drill a bolt hole on the post. Insert a washer and bolt and tighten up, but not too tight; leave a little room for movement.

Building the internal frame and laying the decking

9 Internal joists strengthen the deck. Run them across the shortest span set 40cm (16in) apart. Support joists with extra posts (cut the membrane when you concrete them in) aligned with those on the outer frame.

10 Once all the joists are bolted to the support posts, insert short lengths of wood set 1.2m (4ft) apart to hold them rigid. Nail or screw the joists and spacers in place or use joist hangers (see *top right*).

11 Lay a decking board on the frame (at right angles to the joists) and cut to length, leaving a slight overhang at each end to fit a fascia board (optional). Centre any joins in the board over a joist.

12 Predrill holes in the board, then attach it to each joist using two corrosion-resistant decking nails or countersunk screws. Cut the remaining boards to size and screw them to the joists.

3 Dig out post-holes about 30cm (12in) square and 38cm (15in) deep, and fill the bottom 8cm (3in) with hardcore. Tamp firm with a metal pole, insert the post and pack upright with more rammed hardcore.

4 Fill the hole with water to dampen the hardcore and allow to drain. Pour in post-hole concrete mixed to a pouring consistency. Use a spirit level to check the post is vertical; adjust as necessary.

JOIST HANGERS

If your deck is at ground level, you can screw or nail the frame together. More robust alternatives are advisable for raised decks or where the joists butt against a wall. Timber-to-timber joist hangers, made from galvanized mild steel, are nailed or bolted on to the joists and then attached to the edging timbers. Stronger steel joist hangers can be mortared into the wall. You may find it easier to bolt a length of timber to the wall first, and then hang the joists from it with timber-to-timber joist hangers.

Timber-to-timber joist hanger.

Joist hanger mortared into a wall.

7 Lift the edging timber into position, use a spirit level to check it's horizontal, and mark and drill the timber where it coincides with a post. Insert a bolt and washer as Step 6.

8 Attach all the edging timbers in the same way, butting the corner joints neatly. Bolt the timbers to all intermediate posts to complete the frame. Cut the tops off the posts flush with the frame.

13 Use a chisel to lever the boards into place, spacing them 5mm (¼in) apart with thin strips of plastic or wood. Spacing allows the decking boards to expand in the heat and to let rainwater drain away.

14 Fascia boards fixed around the edge of your deck make for a neat finish. Overlap them precisely where they meet at the corners. If your decking is built on a slope, fascia boards will hide any ugly gaps.

Wood treatments
Pretreated decking timber can be left natural, or you can choose from a huge variety of coloured stains or treatments. This deck is a contemporary grey-brown.

Putting up fence posts

The strength of a fence lies in its supporting posts. Choose 75 x 75mm (3 x 3in) posts made from a rot-resistant timber, such as cedar or pressure-treated softwood, and set them in concrete or metal post supports. Treat the timber with wood preservative every three to four years to prevent it rotting, and replace old posts when you spot signs of deterioration.

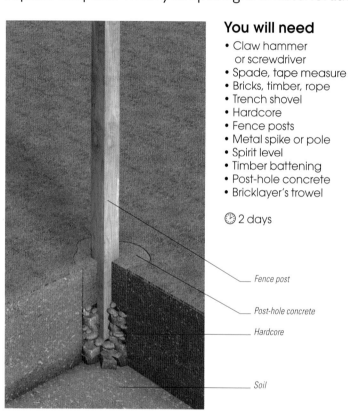

You will need

- Claw hammer or screwdriver
- Spade, tape measure
- Bricks, timber, rope
- Trench shovel
- Hardcore
- Fence posts
- Metal spike or pole
- Spirit level
- Timber battening
- Post-hole concrete
- Bricklayer's trowel

⏱ 2 days

— Fence post

— Post-hole concrete

— Hardcore

— Soil

Replacing old fence posts

1 Use a claw hammer or screwdriver to free one end of the panel. Remove metal clips and fixings. Clear soil away from the base of the panel, then free the other end. Leave the top fixing brackets until last for support.

2 Before putting in a new post, first remove the old concrete footing. Once you have removed the fence panels, dig out the soil from round the base of each post to expose the concrete block.

Concreting the posts

7 To test that the post is vertical, hold a spirit level against each of its four sides. Make any adjustments as necessary, and check that the post is the right height for the fence panel.

8 To hold the post upright while you're concreting it in place, tack a temporary wood brace, fixed to a peg driven firmly into the ground, to the post. Don't attach it to the side that you'll be hanging the panels on.

Fixing bolt-down supports

When erecting posts on a solid level surface, such as paving, use bolt-down, galvanized metal plates. These can be fixed in place relatively easily and will help to prolong the life of the timber posts by holding them off the ground.

1 Measure and mark the exact position of the post, as there will be no opportunity to change it later. Position the base plate, marking the position of each of the corner bolt holes with a pencil.

2 Use a percussion or hammer drill fitted with a masonry bit to drill the bolt holes. Keep the drill upright and make sure you penetrate right through the paving into the hardcore underneath.

3 Fill the drilled holes with mortar injection resin and insert Rawl bolts. After the recommended setting time, tighten the bolts using a spanner – the bolts will expand to fill the hole.

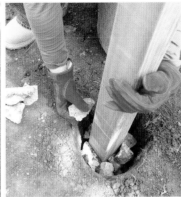

3 Using the post as a lever, loosen the block in the hole. Tie a length of timber to the post, balance it on a pile of bricks (*as shown*) and use this simple lever to help minimize any strain as you lift out the post.

4 If a new post is to go in the same spot, refill the hole and compact the soil before digging a new post hole using a trench shovel. Make it about 60cm (2ft) deep and 30cm (12in) across.

5 Fill the bottom of the hole with a 10cm (4in) layer of hardcore. Stand the post on the base, check it's level with the original fence line, and pack the hardcore around the sides.

6 Use a metal spike or pole to ram the hardcore in place, working the post gently back and forth to help settle the material. Aim to fill the hole to about half its depth with compacted hardcore.

9 Fill the post hole to the top with water and leave it to drain. This will help settle the hardcore and improve adhesion of the concrete. Make up post-hole concrete to a pouring consistency.

10 Pour concrete into the hole, stirring gently to remove air bubbles. Shape it around the post, using a trowel, so rain runs off. Rehang panels after 48 hours. Remove bracing after three weeks.

METAL SPIKE SUPPORTS

If you have firm, undisturbed ground, use metal spike supports. Position the spike in place and insert a "dolly", a special post-driver, into the square cup. Hit the dolly with a mallet to drive the spike into the ground. Check the angle of the spike with a spirit level to ensure that it is going in straight – twist the dolly handles to correct any misalignment. When the spike is in the ground, remove the dolly. Clamp the square cup around the post and tighten up.

New posts, new panels
A new fence makes a beautiful natural backdrop to planting, or, for a more contemporary look, try staining the wood matt black or dark blue.

Laying a gravel border

Gravel isn't just for driveways and paths – when used as a decorative mulch in the border it sets plants off to great effect. If you spread a thick layer of gravel on top of a geotextile membrane, it will also suppress weeds and help retain moisture in the soil.

You will need
- Scissors or sharp knife
- Geotextile membrane
- Metal pins
- Pea gravel
- Tape measure

⏱ 1 day

Gravel

Brick edging

Geotextile membrane

Soil

Laying the membrane

1 Cut a piece of geotextile membrane to fit your bed or border. For large areas, you may need to join several strips together – in which case, leave a wide overlap along each edge and pin in place.

2 Presoak container-grown plants in a bucket of water for about half an hour. Position plants, still in their pots, on top of the membrane. Check the labels to make sure that each plant has enough room to spread – once the gravel is down, moving them isn't easy.

3 Use scissors or a sharp knife to cut a large cross in the membrane under each plant. Fold back the flaps. Make the opening big enough to allow you to dig a good-sized planting hole.

Planting up the border

4 Remove the plants from their pots and lower each one into its allocated planting hole. Plants should sit in the ground at the same depth as when in the pot. Fill in around the root ball with soil.

5 Firm in the root ball with your hands, then tuck the flaps back around the base of the plant. If necessary, trim the membrane to fit neatly around the plant's stems. Water thoroughly.

6 Cover the membrane with a thick, even layer of gravel. A depth of 5–8cm (2–3in) should prevent any bald patches appearing. Should you need to move plants in the future, pin a piece of membrane over the old planting area to stop weeds popping up through the cut.

AGGREGATE OPTIONS

You can lay most aggregates over a geotextile membrane in the same way as gravel. Other decorative options for a planting area include slate chips (*shown right*), small pebbles, ground recycled glass, crushed shells, and coloured gravels. (*See pp.354–355 for more information on these materials.*)

Permeable paths

The main advantage of using permeable surfacing in a garden is that it allows rain water to drain through to the soil. But when you discover that the materials are durable, easy to lay, and cost-effective, they're definitely an attractive alternative to paving.

Loose gravel
Look carefully and you'll see that this gravel has been poured into a honeycomb grid. This cleverly designed plastic matting, which you lay like a carpet, prevents gravel migrating all over the garden or driveway.

Self-binding gravel
Gravels are usually washed clean of soil and stones, but self-binding gravels, such as Breedon gravel, are not. When compacted, these fine particles bind the material together to form a strong, weed-free, permeable surface.

Shredded bark
Bark is pleasantly springy underfoot. Lay it over a geotextile membrane, or straight on to compacted soil. Whichever you decide to do, the bark will start to break down after a couple of years and should be replaced with fresh bark.

Keep it neat
A gravel surface works best when it's contained by a solid edge. If the gravel border is next to a lawn, consider laying a brick mowing strip (see p.275).

Building a pergola

A pergola is essentially a series of arches linked together to form a covered walkway. The framework provides the perfect support for climbing plants, such as fragrant honeysuckle and roses. Although often made from timbers or metal components, many designers choose to use a wood frame kit, as shown here, the instructions for which are pretty universal.

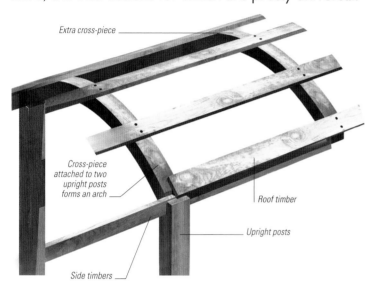

Extra cross-piece

Cross-piece attached to two upright posts forms an arch

Roof timber

Upright posts

Side timbers

You will need

- Pergola kit
- Pegs and string
- Builder's square
- Vice
- Drill
- Screwdriver
- Galvanized screws or bolts
- Tape measure
- Hammer
- Wooden battening
- Spray paint
- Spade
- Hardcore
- Metal spike
- Spirit level
- Ready-mix concrete

🕐 2 days

Making the arches

1 Unpack and identify all the pieces. Mark the layout of the pergola on the ground with pegs and string – use a builder's square to check right angles. If the area is to be paved, lift and reuse the turf elsewhere.

2 Arrange the pieces flat on the ground to check the fit of the joints. Make adjustments as necessary. If the wood isn't predrilled, clamp the timber in a vice and drill holes for the screws and bolts.

Erecting the arches

5 Mark the two upright post positions for the first arch using spray paint. Dig out the holes making them about 60cm (2ft) deep and 30cm (12in) across. Fill with 10cm (4in) of hardcore (see Step 5, p.269).

6 Ram the hardcore firmly in place with a metal spike or pole. Place the upright posts in the holes and test that each one is vertical by holding a spirit level against each of its four sides.

Constructing the roof

9 Dig two holes for the uprights on the second arch (see Steps 5 and 6). Do a final check on the relative position of the two arches by positioning the side timbers on top of their respective uprights.

10 Using a spirit level, check that the side timbers are horizontal and the uprights are vertical before concreting them in position. Repeat Steps 5–10 until all the arches are concreted in place.

11 Leave the concrete to set for 48 hours, then screw or bolt all the side timbers in place, butting the joints tightly together. To avoid splitting the wood it's best to predrill the holes.

12 Most pergolas have extra cross-pieces to strengthen the roof (these do not sit on uprights so are unsupported). Mark their position midway between the uprights. Predrill screw holes in each piece.

3 To make an arch, attach each end of a cross-piece to the top of an upright post using galvanized screws or bolts. Support the wood on a board to help steady and align the pieces as you work.

4 Measure the distance between the upright posts at the top and bottom of each arch, adjust the posts until the spacing is the same, and then nail wooden battening across to stop them splaying.

7 To hold the upright posts vertical while you're concreting them in, tack a temporary wooden brace to them (see *Step 8, p.268*). Concrete the posts in place (*see Steps 9 and 10, p.269*).

8 To position the second arch, lay a side timber on the ground to work out the spacing. Mark the position of the post holes with paint. Allow for a slight overlap where the side timbers will rest on the uprights.

A shady retreat
Walking under a shady, plant-covered pergola is a real treat on a hot summer's day. It would also be the perfect spot for outdoor entertaining.

13 Screw or bolt the cross-pieces in place – you will need someone to hold them steady to stop them twisting when you're drilling. Check that all the fixings on the pergola frame are tight.

14 Position the roof timbers on top of the cross-pieces. Mark and predrill holes, and then screw in place. Leave the bracing on the uprights for three weeks until the concrete has completely set.

WIRING FOR CLIMBERS

A system of wires attached to the upright posts of your pergola will give plants the support they need to start climbing. Fix screw eyes at 30cm (1ft) intervals around the four sides of an upright. Attach galvanized wire to the lowest screw eye, run it through all the eyes on the same side of the upright, and secure it firmly to the top one. Repeat on the other three sides of the upright. Guide shoots of twining plants on to the wires; tie in shoots of stiffer-stemmed climbers.

Set up a system of wires for climbers.

Making a raised bed

Creating a square or rectangular timber-framed raised bed is easy, especially if the pieces are pre-cut to length. Buy pressure-treated wood, which will last for many years, or treat it with preservative before you start. If the bed is to sit next to a lawn, make a brick mowing strip by following the steps opposite.

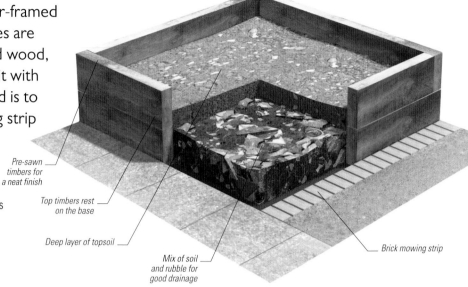

Pre-sawn timbers for a neat finish

Top timbers rest on the base

Deep layer of topsoil

Mix of soil and rubble for good drainage

Brick mowing strip

You will need

- Spade
- Pre-cut wooden sleepers
- Spirit level
- Tape measure
- Rubber mallet

- Drill, screwdriver
- Heavy-duty coach screws
- Rubble and topsoil
- Bark

🕐 1 day

Measuring up the base

1 Dig out strips of turf wide enough to accommodate the timbers. Pressure-treated wood is an economic alternative to rot-resistant hardwoods, such as oak. Or consider buying reclaimed hardwood.

2 Lay out the timbers *in situ* and check that they are level with a builder's spirit level (use a plank of wood to support a shorter spirit level). Check levels diagonally, as well as along the length of the timbers.

3 Make sure the base is square by checking that the diagonals are equal in length. For a perfect square or rectangular bed, it is a good idea to have the timbers pre-cut to size at a local timber yard.

Building the bed

4 Using a rubber mallet, gently tap the wood so that it butts up against the adjacent piece; it should stand perfectly level and upright according to the readings on your spirit level. Remove soil as needed.

5 Predrill holes through the end timbers into the adjacent pieces at both the top and bottom to accommodate a couple of long, heavy-duty coach screws. Secure the timbers with the screws.

6 Arrange the next set of timbers on top of the base; make sure they overlap the joints below to give the structure added strength. Check with a spirit level before screwing together (see *Step 5*).

7 For extra drainage, partially fill the base with rubble. Then add topsoil that is free of perennial weeds. Fill the bed up to about 8cm (3in) from the top with soil, plant up, then mulch with bark or gravel.

RAISED VEGETABLE BED

Raised beds are ideal for growing vegetables, fruit and herbs. They provide better drainage on heavier soils and a deeper root run for crops like carrots and potatoes. Raised beds also lift up trailing plants, such as strawberries, which helps to prevent rotting. If you buy in fresh topsoil that's guaranteed weed- and disease-free, your crops will have a better chance of doing well.

Raise your profile
As well as providing an eye-catching feature, a raised bed gives you a better view of your plants and, by lifting them up, less strain on your back when tending them.

Laying a mowing strip

Grass doesn't grow well too close to a raised bed, since the soil tends to be dry and any overhanging plants create shade. A strip of bricks, sunk slightly lower than the level of the turf, creates a clean edge to allow for easy mowing.

1 Using a spare brick to measure the appropriate width for your mowing edge, set up a line of string to act as a guide. Dig out a strip of soil deep enough to accommodate the bricks, plus 2.5cm (1in) of mortar.

2 Lay a level mortar mix in the bottom of the trench as a foundation for the bricks. Set them on top, leaving a small gap between each one. (This design is straight, but mowing edges can be set around curves.)

3 With a spirit level, check that the bricks are aligned and slightly below the surface of the lawn (when set in place, you should be able to mow straight over them). Use a rubber mallet to gently tap them into position.

4 Finally, use a dry mix to mortar the joints between the bricks, working the mixture in with a trowel. Clean off the excess with a stiff brush.

A clean cut
The mowing strip makes a decorative feature and allows you to manoeuvre the mower more easily.

Making a pond

Designing a pond with a flexible butyl rubber or PVC liner, rather than a rigid preformed type, allows you to create a feature of almost any size and shape. To work out how much liner you need, add twice the depth of the proposed pond to its maximum length plus the width. Choose somewhere sheltered and sunny for your water feature, avoiding heavy shade under trees.

You will need

- Hosepipe
- Spade
- Pickaxe
- Spirit level/plank
- Sand
- Pond or carpet underlay
- Flexible pond liner
- Waterproof mortar, bucket, trowel
- Sharp knife
- Decorative stone

🕐 2 days

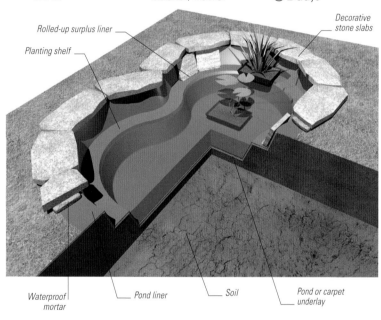

Rolled-up surplus liner

Planting shelf

Decorative stone slabs

Waterproof mortar

Pond liner

Soil

Pond or carpet underlay

Digging an informal pond

1 Use a hose to mark the outline of the pond. Aim for a curved, natural shape without any sharp corners. To prevent it freezing solid in winter, a section of the pond must be at least 45cm (18in) deep.

2 Before you start digging, skim off any turf for reuse elsewhere. Keep the fertile topsoil (which you can also reuse) separate from the subsoil. Loosen compacted subsoil with a pickaxe.

Lining and edging

5 To protect the liner, line the sides and base of the pond with pond underlay. If using old carpet underlay, beware stray tacks. On stony soils, spread a 5cm (2in) layer of sand over the base first.

6 Centre the liner over the hole, letting it slide down under its own weight into the base. Leaving plenty of surplus around the rim, pleat the liner to help fit it to the shape of the pond. Fill with water.

Making a rill

A rill or channel of water adds light and movement to a garden. Employ a qualified electrician to install a power supply for you.

You will need

- Pegs and string
- Spade
- Sand
- Spirit level
- Plastic reservoir
- Plastic liner
- Sharp knife
- Bricks
- Waterproof mortar
- Submersible pump, flexible pipe, filter
- Gravel, cobbles
- Metal grille
- Geotextile fabric

🕐 1 day

1 Clear and level the site. Mark out the length and width of the rill with pegs and string. Dig out the area to a depth of 15–20cm (6–8in). Cut a shallow shelf all around the rill for the brick edging.

2 Line the rill with sand, compacting it with a piece of wood. Use a spirit level to check the base is flat. Dig a hole at one end and insert the reservoir – check the rim is level with the base of the rill.

3 Line the rill with the plastic liner, smoothing out any creases. Use a sharp knife to trim the liner at the reservoir end so that it drapes over the rim. Leave 20cm (8in) surplus material along the other three sides.

3 Dig out the pond to a depth of 45cm (18in). Make the sides gently sloping. Leave a shelf 30–45cm (12–18in) wide around the edge, then dig out the centre to a further depth of 45cm (18in).

4 Use a spirit level placed on a straight piece of wood to check that the ground around the top of the pond is level. Remove any loose soil and all large or sharp stones from the sides and bottom of the pond.

7 When the pond is full, trim the surplus liner leaving 45cm (18in) around the rim. Pleat the excess liner so it lies flat and bury the edges in the ground. Lay a bed of waterproof mortar for the edging stones.

8 Bed the edging stones into the mortar, overhanging them by 5cm (2in) to hide the liner. When positioning vertical stones, stand them on a piece of rolled-up surplus liner to protect the liner from being torn.

Planting up
Wait a week for the mortar to set before placing water lilies on the bottom of the pond and marginals on the shelf (see pp.98–99 for more on aquatic plants).

4 Edge the rill with bricks on three sides (not the reservoir end). Bed bricks on a 2cm (1in) layer of waterproof mortar, making sure that it doesn't fall into the rill. Mortar between the bricks.

5 Place the pump in the reservoir. Push the pipe on to the pump outlet, run the pipe along the length of the rill, and cut it to fit at the far end. Fit a filter on the free end of the pipe to prevent blockages.

6 Cover over the pipe in the rill with a level bed of gravel. Place a metal grille over the reservoir and top with cobbles. If you sit them on a sheet of geotextile fabric it will stop debris falling into the water.

Finishing touches
Fill the reservoir with water, prime the pump, and adjust the flow, according to manufacturer's instructions. Slate chips make an attractive edging material.

Planting techniques

Having designed a beautiful garden, assessed your soil and aspect, and worked out what plants to buy, it is now time to bring them home, get them into the ground and put your ideas into practice. Take your time when planting; tackling the task in a measured way will help to ensure your treasures thrive.

Choose a dry, fine spell when the soil is not frozen or too wet. Before starting, gather all necessary tools together – fork, spade, fertilizer and watering can – so you have everything to hand. Also make sure the soil is free of weeds, especially any pernicious perennials, before forking in fertilizer and digging holes. The new plants will need a thorough soaking prior to planting, and the best way to do this is to emerse them in water while they are still in their pots, leave until the bubbles disperse, then remove and allow to drain. Bare-rooted trees, roses or shrubs should be planted between autumn and early spring; container-grown plants can go in the ground at any time, but hardy plants are best planted in autumn when the soil is still warm and moist. Leave more tender types until spring, as young plants may not survive a cold, wet winter.

Add fertilizer to the soil for a fine display of lupins.

Allow space for shrubs and trees to spread – the area needed should be indicated on the plant label. Bare patches can always be filled in with seasonal flowers, or screened by containers or an easily moved ornament, such as a bird bath or light-weight sculpture.

Early spring or early autumn are the best times to establish a lawn, whether you are using turves or sowing seed, and avoid walking on new grass for a few months, if possible. Water it frequently in the early stages and in dry spells.

Giving your new purchases a good start will repay dividends for years to come in the form of strong, healthy plants that continually give a good show, season after season.

Leave space for trees like this *Acer palmatum* to spread.

How to plant trees

A well-planted tree will reward you with years of healthy growth. Container-grown trees can be planted at most times of the year, but the best time is in autumn, when the leaves are starting to fall. Bare-root plants are a cheaper option and are available in autumn and winter. Unless it's very frosty or there's been a long dry spell, you should plant them as soon as you get them home.

You will need
- Bucket
- Spade and border fork
- Well-rotted organic matter
- Bamboo cane
- Tree stake
- Mallet and nails
- Tree tie with spacer
- Chipped bark mulch
- 🕐 up to 2 hours

Planting a container-grown tree

1 Soak the tree thoroughly and leave it to drain. Meanwhile, clear the planting area of weeds and debris. Place the tree, still in its pot, in its planting position, making sure that it won't be crowded by other plants.

2 Loosen the soil over a wide area, to the same depth as the tree's root ball. Add organic matter to heavy clay or sandy soils. Dig a large hole no deeper than the tree's pot but ideally three times the root ball's diameter.

Planting and staking

5 With a container-grown tree, you may find that the roots are packed together tightly. If this is the case, gently tease out any encircling roots, as these could prevent it from establishing well.

6 With a helper holding the tree upright, backfill the hole with the excavated soil. Make sure there are no air pockets by working the soil in between the roots and around the root ball with your fingers.

7 Once you are satisfied that there are no gaps or air pockets around the roots, continue to hold the tree upright and firm it in using your foot with your toes pointing towards the trunk.

8 Small trees do not require staking but top-heavy or larger specimens should be staked. Drive into the soil a wooden tree stake at an angle of 45 degrees. Ensure you do not damage the root ball.

Planting a hedge

An informal mixed hedge of native species will provide a rich habitat for wildlife, as well as attractive flowers, fruits, and nuts. The best time to plant a bare-root hedge is autumn, when plants first become available.

You will need
- Spade
- Rake
- Tape measure
- String and canes
- Secateurs

🕐 up to 3 hours

1 A few weeks before planting, remove weeds and dig the area over, working in organic matter (as Step 2 above). At planting time, weed the area again, tread the ground until firm, and rake level.

2 Mark the planting line with pegs and string. If you have space, put in a double row of plants for extra screening. It's also less likely to suffer gaps if plants die. Set the rows 40cm (16in) apart.

3 Set the plants 80cm (32in) apart. Spacing is critical for hedging, so use a tape measure or marked canes rather than guessing. Dig holes large enough to accommodate the roots comfortably.

3 Puncture and scuff up the walls and base of the hole to allow for easy root penetration; the result will be a stronger tree. Don't loosen the base too much as the tree may sink after planting.

4 Remove the tree from its pot. Lower it into the hole and check that the first flare of roots will be level with the surface after planting – try scraping off the top layer of compost if you can't see the flare.

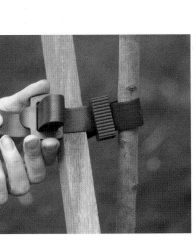

9 The stake should be a third of the height of the tree, and the end should face into the prevailing wind. Fit a tree tie with a spacer to the stake and trunk. This can be adjusted as the tree grows.

10 Knock a nail through the tree tie into the stake to prevent it slipping down. Water the tree thoroughly and apply a mulching mat around the trunk, which will keep the area around the tree free of weeds.

Spring blossom in a woodland border
In small- to medium-sized gardens choose compact trees with an attractive overall habit. This hawthorn (*Crataegus*) is ideal, with pretty, pink blossom in spring, followed by ornamental fruits.

4 Plant the bare-root hedging plants at the same depth as they were growing in the field; you'll see a dark soil stain on the stem. Plant roses slightly deeper for stability. Firm plants in with your hands.

5 Stagger the plants on the second row to maximize coverage. Position the first plant 40cm (16in) in from the edge of the front row. Keep bare-root plants wrapped until planted to stop their roots drying out.

6 Check that the soil around the plants is firmed in, and then water each plant thoroughly. Prune back the tips of any tall or leggy shrubs to encourage new, bushy growth from the base of the plant.

Wildlife-friendly screen
A mixed hedge will attract wildlife all year. Don't clip too hard if you want summer flowers and fruit in autumn, and take care not to disturb nesting birds in spring.

How to plant shrubs

Shrubs form the backbone of a planting scheme, providing structure as well as flowers and foliage. Plants grown in containers can be planted all year round if you avoid days when the ground is frozen, or excessively wet or dry. Before planting, always check the label for the shrub's preferred site and soil.

You will need
- Spade and fork
- Organic matter
- Bucket
- Mulching material

🕐 1 hour

1 Dig over the soil, removing any weeds and working in plenty of well-rotted manure or garden compost. Make the planting hole twice the diameter of the container and a little deeper.

2 Stand the plant in its container in a bucket of water and leave it to soak. Remove the plant from its pot and tease out any thick, encircling roots. Plant at the same depth that it was in its pot. Backfill with soil.

3 Firm in gently, ensuring the shrub is upright and that it is sitting in a shallow depression to assist watering. Water in well, then spread a mulch of organic matter (see right), keeping it away from the stems.

How to plant perennials

Unlike annuals and tender patio plants, herbaceous perennials come up year after year. Many modern varieties need little maintenance other than deadheading and cutting back in spring. Give them a good start by improving the soil at planting time and minimize competition for water and nutrients by controlling weeds.

You will need
- Spade and fork
- Organic matter
- General fertilizer
- Bucket
- Mulching material

🕐 up to 1 hour

1 Prepare the planting area, removing perennial weeds and large stones. On dry ground or heavy clay, work in organic matter (as Step 1, above). On sandy soil, also apply a general fertilizer.

2 Dig a hole a bit deeper and wider than the pot. After soaking the plant, remove the pot. Add soil to the hole so that the top of the root ball is level with the soil surface. Backfill and firm in lightly with your hands.

3 Water in well. Apply a thick mulch to conserve moisture, suppress weeds and protect roots from penetrating frosts. Take precautions against slugs and snails, and watch for aphids on shoot tips.

Mulch options

Mulches conserve water, which is why they are always applied after planting when the ground is moist. Some improve soil structure and most discourage weeds, which compete with garden plants for water and nutrients. Gravel mulches look attractive while others, such as leafmould, offer a habitat for beneficial creatures such as ground beetles.

Garden compost
Well-rotted compost and manure lock moisture and nutrients into the soil. As the mulch rots down it releases plant food and improves the soil structure. Apply a layer 10cm (4in) deep in late winter to minimize weed growth.

Leafmould
Although low in nutrients, leafmould is excellent for improving soil and retaining moisture, and it looks good around woodland-style plantings. To make it, fill perforated bin bags with autumn leaves, seal up and leave for about 18 months.

Chipped bark
A popular mulch, bark comes in various grades, the finest being the most ornamental. It rots down slowly and is a good weed suppressor and moisture conserver, but doesn't add many nutrients. Top up worn areas annually.

Gravel mulch
Gravel laid over landscape fabric creates a decorative weed-suppressant foil for alpines and Mediterranean-style plantings. Plant through the fabric by cutting a cross and folding back the flaps before replacing the gravel (see also pp.270–271).

Seasonal colour and interest
A mixture of shrubs and perennials provides a rich tapestry of colour, form, and texture that changes in mood as the seasons progress. If space allows, plant the perennials in drifts for greater impact.

How to plant climbers

Walls, fences, and trellis offer planting space for a wide range of climbers and wall shrubs. Using plants vertically is especially important in courtyard gardens, where space is at a premium. Flowers and foliage soften bare walls and boundary screens, as well as creating potential nesting sites for birds. Avoid over-vigorous climbers that could overwhelm their situation.

You will need
- Vine eyes
- Galvanized wire or trellis
- Border fork and spade
- Bulky organic matter
- Granular fertilizer
- Bamboo canes
- Garden twine
- Trowel or hand fork
- Chipped bark mulch
- 🕐 1 to 2 hours

Preparation for planting

1 Before soil preparation, attach a support of vine eyes and horizontal wires, or a trellis, to the wall or fence. Set the lowest wire about 50cm (20in) above soil level, and space the wires 30–45cm (12–18in) apart.

2 Dig over a large area around the planting site. Work in plenty of bulky organic matter, such as well-rotted manure or garden compost, to combat dryness at the base of the fence.

Planting and aftercare

5 Arrange a fan of bamboo canes behind the planting hole, leaning them back towards the fence. The canes will lead the climber's stems up to the horizontal wires and spread them over a wider area.

6 Plant the climber, backfilling the hole with enriched soil. Untie the stems from their original support and untangle them carefully. Cut off any weak shoots and spread them out ready to attach.

7 Tie the stems to the canes using soft garden twine and a loose figure-of-eight knot. Train the outer stems on to the lower wires and train the central stems upwards to cover the higher wires.

8 Firm the climber in using your fists and then take a trowel or hand fork to fluff up the soil where it has been compacted. Next, create a shallow water reservoir (with a raised rim) around the base of the plant.

Support for climbers

Climbers and wall shrubs scale vertical surfaces in a variety of ways, and the support you provide depends on their vigour and method of climbing. Some, such as jasmine, honeysuckle, and wisteria, are twiners; clematis have coiling leaf stalks; and sweet peas, passionflowers, and vines cling with tendrils.

Horizontal wires
These offer the most adaptable support for climbers, wall-trained shrubs, and fruit trees. Training stems horizontally increases flower and fruit production.

Trees and other host plants
To encourage a rambler rose to clamber up into a fruit tree, plant it 1m (3ft) away from the trunk and give it a rope to climb (peg it to the ground and run it to the lowest branch).

Trellis
Wooden trellis can be used against a wall or as a screen. Climbing roses, honeysuckle, clematis, and passionflower may secure themselves, but tying them in also helps.

3 On poor soils, apply a dressing of all-purpose granular fertilizer (follow manufacturer's instructions). Water the climber well a few hours before planting, or plunge the pot into a bucket of water.

4 Dig a planting hole 45cm (18in) from the fence, and twice the diameter of the root ball. Check the depth is the same as the original compost level, though clematis should be planted 10cm (4in) deeper.

9 Water well, then apply a mulch of chipped bark to help combat weeds, conserve moisture, and keep the roots of plants such as clematis cool. Ensure the mulch doesn't touch the stems.

Fragrant cover
The honeysuckle in this planting sequence will eventually produce a mass of evening-scented flowers, loved by bees and moths. Good ground preparation will ensure that the plant won't run short of water, which can lead to powdery mildew.

Obelisks
These provide ideal support for large-flowered clematis, jasmine, and climbing roses, and annual climbers, such as sweet peas, morning glory, and runner beans.

No support needed
Plants such as Boston ivy have tendrils that adhere to walls without support. Ivy and climbing hydrangea have self-clinging roots on their stems. Some initial support is useful.

PLANTING IN POTS

Large containers, especially glazed ceramic pots or oak half barrels, create the opportunity for covering walls, fences, and screens, even without a bed or border. Some pots and troughs come with integral, freestanding trellis support, but you can also add a trellis fan as shown here. Try small- to medium-sized species and cultivars, such as *Clematis alpina* and *C. macropetala*, as well as annual climbers like *Eccremocarpus scaber* (Chilean glory vine) and morning glory (*Ipomoea*).

Laying a lawn

The best time to lay, or seed, a new lawn is early autumn or spring. Dig the area, adding a margin of 15cm (6in), and improve the drainage of heavy clay and wet soils by working lots of grit into the topsoil. For free-draining soils, dig in an 8–10cm (3–4in) layer of bulky organic matter to conserve moisture and fertility.

You will need
- Spade or fork
- Rake and hoe
- All-purpose granular fertilizer
- Sieved topsoil mixed with horticultural sand
- Wooden plank
- Broom
- Hosepipe
- Edging iron

🕐 1 day

Preparing the ground

1 Dig over the lawn area, removing big stones and perennial weeds, and break up the surface into a fine crumb structure. Rake level, then, keeping your weight on your heels, walk over the length of your plot, and then across the width.

2 Rake the ground level to remove any depressions left after walking. Leave for five weeks to allow weed seeds to germinate, then hoe lightly to remove them. Rake level and apply a dressing of all-purpose granular fertilizer.

Laying the turf

3 Arrange turf delivery a few days after applying fertilizer. Carefully unroll the turf, laying whole pieces and working out from an edge. Stand on a plank to distribute your weight. Tamp down turfs with a rake.

4 To ensure that the grass knits together, butt the edges of the turfs, lifting them so that they are almost overlapping when pushed down. This helps to combat any shrinkage. Firm again with a rake.

Finishing and shaping

5 Continue to lay the next row of turf, ensuring that the joins are staggered like wall bricks. This produces a much stronger structure. Use an old knife for cutting, and avoid using small pieces at the edges.

6 To help adjacent pieces of turf to grow together and root firmly, brush in a blend of sieved topsoil and horticultural sand. Use a stiff broom to work in the top dressing and raise flattened grass.

7 Water well during dry spells to prevent shrinkage. Shape lawn edges when the turf has rooted (try gently lifting an edge). Lay out curves with a hosepipe and cut using an edging iron or border spade.

SPOT WEEDING

During lawn establishment, perennial weeds often take root, especially rosette-forming dandelions and thistles, which can smother the turf. Use an old kitchen knife, forked daisy grubber, or long-handled, lawn-weeding tool to extract them. Try to remove all the taproot. Do not use lawn weedkillers for at least six months.

A green carpet
The velvet green of a well-maintained lawn is the perfect foil for border flowers. Lawns create a sense of space in the garden and provide colour, even in the depths of winter.

Seeding a lawn

For large areas of lawn, seeding is the cheapest option and, although it will be about a year before the grass can take heavy use, it should start to green up and look good in under a month. Worn patches in existing lawns can also be repaired by reseeding with an appropriate grass mix.

1 Select a seed mix that suits your conditions and lawn use, eg, hard-wearing family or fine, ornamental lawn. Weigh out seed for 1sq m (1sq yd) following pack directions. Pour into a paper cup; mark where the seed reaches.

2 You should have dug, firmed, levelled, and raked the lawn bed at least five weeks previously (see opposite). A few days before sowing, remove any weeds and add a top dressing of fertilizer. Rake level, removing any stones.

3 Sow in early autumn when the soil is warm and moist, or in spring when plants start to grow actively. Mark out 1sq m (1sq yd) sections using canes, and measure out the grass seed using the marked paper cup.

4 Scatter half the seed in one direction, and then go over at right angles with the remainder, keeping within the template. Move the template along and repeat the process. As a guide, one handful of seed weighs roughly 30g (1oz).

5 Work over the seeded lawn lightly with a rake until the seed is just covered with soil. Protect from birds using netting. Seedlings should appear within 14 days. Once the grass has reached 5cm (2in), cut with the mower blades set high.

Aftercare and maintenance

Making a garden is a process that doesn't end when the construction and planting stages are complete. Even in low-maintenance plots, gardens only thrive when the plants are tended and the soil replenished. Some jobs are regular weekly tasks, but many others are only annual or twice yearly.

When and how to water

Environmentally conscious gardeners and people living in drought-prone areas are increasingly aware of the need to save water. Containers, together with some types of vegetable crops and bedding plants, may need regular summer irrigation. Shrubs, trees and perennials need watering only at planting time and during dry spells in the first year or two, or until they are well established. No matter how brown the grass may turn, established lawns never actually need watering and will eventually recover from drought.

If you need to water, do so in the cool of the morning or evening to minimize evaporation, and water close to the soil rather than overhead, targeting specific plants. Mulches, such as bark and spent mushroom compost, help seal in moisture and reduce competition from weeds. It is better to water heavily, with extended intervals between (allowing moisture to penetrate well into the soil and encourage deep rooting) than to water lightly but more frequently.

Preventing erosion
With shallow-rooting plants like this box, frequent watering can wear away the protective coating of compost. Reduce the problem by directing water on to a large crock or tile so that flow is gently dissipated.

Making watering easy

Although watering can be an enjoyable task, if you are pressed for time or have a large plot, some shortcuts are welcome. Automatic irrigation can be very efficient and, if properly managed, helps to save water. It also makes sense to collect rainwater at sites around the garden and to make use of recycled or grey water, eg, from the bath or washing-up (but only if no strong or heavily perfumed products have been used).

Leaky hose
A perforated hosepipe (leaky hose) connected to an outdoor tap or water butt will channel water directly to where it is needed; eg, snaking through leafy crops, or through a newly planted border.

Timed watering
If you are often away from the garden for more than a couple of days or are too busy to water all your patio containers regularly, consider installing an automatic irrigation system with a timer.

Water butts
Raised up high enough so that you can comfortably fit a watering can under the tap, water butts are a convenient way to reduce dependence on the mains supply. Consider fitting extension kits to increase capacity.

Deadheading promotes new flower growth.

BENEFITS OF DEADHEADING

The aim of the plant is to set seed and reproduce: to achieve this it makes flowers and diverts most of its resources to develop a seedhead. To encourage more flowers you need to remove faded blooms before they have a chance to form seed. This is especially important for annuals which can stop flowering altogether and even die if you don't deadhead regularly. But perennials, including so-called patio plants, can also be encouraged to flower for much longer if they are deadheaded. Removing old, blemished heads also improves the appearance of plants and reduces the risk of disease.

The benefits of pruning

It is not essential to prune any plant, but thinning and cutting back to varying degrees or selectively removing whole branches can produce many useful effects. It can rejuvenate an old, congested specimen, giving it a new lease of life; help short-lived shrubs to live longer; increase the supply of flowering or fruiting wood; improve the shape and appearance of a plant; and reduce the incidence of disease.

The right cut
Cut back to just above a strong bud or pair of buds. Cutting halfway between buds causes die-back, which can introduce disease.

Alternate buds
Where buds form alternately along a stem, make a slanting cut, as shown, so that rain water drains away from the bud.

Removing branches

As a tree matures, it may become too large for its site, or send out branches in inconvenient directions, and require pruning. Damaged or diseased branches and crossing limbs also need to be taken out to maintain the health of the tree. Hire a qualified tree surgeon to tackle very large branches, or those higher than head height. When pruning, take off a branch in sections – if you remove it with one cut close to the trunk, it will be pulled down by its own weight and may tear the bark on the trunk, leaving the tree vulnerable to infection.

1 To cut back branches, make two incisions: one, half way through, from beneath the branch; the second from the top to meet the undercut.

2 Remove the remaining branch stub, starting from the upper surface of the branch, just beyond the crease in the bark where the branch meets the trunk. Angle the cut away from the trunk.

3 This pruning method produces a clean cut, leaving the plant's healing tissue intact. The tree will soon produce bark to cover the exposed area.

Feeding and weeding

Clay loams are naturally fertile, while sandy soils tend to be nutrient poor. Adding bulky organic matter, such as well-rotted manure, improves the quality and structure of both types of soil as well as providing nutrients. During the growing season, wherever you garden intensively, you'll need to add extra fertilizer. Control weeds by digging them out or hoeing, or with a glyphosate weedkiller, except on turf which will require a lawn weedkiller.

Containers
Flowering container plants, in particular, require extra fertilizer. Try a convenient, slow-release formula if you can't manage weekly feeds.

Soluble food
Liquid feeds are fast acting and ideal for bedding and patio plants in containers, as well as greenhouse crops such as tomatoes.

Weedkillers
For convenience and for treating pernicious weeds, use a synthetic- or natural-based weedkiller, which is absorbed through the leaves to kill the roots.

Weeding by hand
Among existing plants, remove weed seedlings by hand. Use a hoe on dry days, severing the stems where they meet the roots just beneath the soil, or dig them out with a fork.

PLANT AND MATERIALS GUIDE

Plant guide

Selecting the right plant for the right place is an essential skill for any garden designer, and this directory will help you to make those critical decisions.

RHS hardiness ratings

The table below shows the corresponding lowest temperature range for each of the ratings in the RHS system of hardiness, used in this directory alongside the more general ratings. Please see p366 for more information.

H1a	warmer than 15°C (59°F)
H1b	10–15°C (50–59°F)
H1c	5–10°C (41–50°F)
H2	1–5°C (34–41°F)
H3	-5–1°C (23–34°F)
H4	-10–-5°C (14–23°F)
H5	-15–-10°C (5–14°F)
H6	-20–-15°C (-4–5°F)
H7	colder than -20°C (-4°F)

Large trees

Acacia dealbata

Mimosa is an evergreen tree with fern-like, silvery grey-green divided leaves. Orange in bud before turning yellow, the fragrant clusters of flowers add colour and scent from winter to spring. Susceptible to frost, so plant in a sheltered site in full sun.

↕15–30m (50–100ft) ↔6–10m (20–30ft) ❋ ❋ H3 ☀ ◊ ◠

Acer campestre

The lobed leaves of the deciduous field maple are red when young, green by late summer, then yellow and red in autumn. The green flowers in spring produce the helicopter fruits that children love to play with. *A. campestre* 'Schwerinii' makes an excellent hedge, or can be grown in a large container.

↕8–25m (25–80ft) ↔4m (12ft) ❋ ❋ ❋ H6 ☀ ☀ ◊ ◠

Acer platanoides 'Crimson King'

The Norway maple is a vigorous, spreading, deciduous tree. 'Crimson King' has large, lobed, dark red-purple leaves that turn orange in autumn. The red-tinged yellow flowers are borne in mid-spring. Fast-growing, it makes a useful screen, but is at its best centre stage as an ornamental specimen.

↕25m (80ft) ↔15m (50ft) ❋ ❋ ❋ H7 ☀ ☀ ◊ ◠

Acer rubrum 'October Glory'

By autumn, the lobed, glossy, dark green foliage of the red maple has turned bright red; erect clusters of tiny red flowers are produced in spring. 'October Glory' is a reliable cultivar, though for best colour, grow it in acid soil. To fully appreciate its beauty give this large deciduous tree plenty of space.

↕20m (70ft) ↔10m (30ft) ❋ ❋ ❋ H6 ☀ ☀ ◊ ◠

Alnus glutinosa 'Laciniata'

The deciduous common alder produces yellow-brown male catkins in late winter or early spring, and small, egg-shaped fruit in summer. Most types have long, rounded, dark green leaves, but those of 'Laciniata' are mid-green with triangular lobes. Will do well in a coastal garden; useful as a screen.

↕25m (80ft) ↔10m (30ft) ❋ ❋ ❋ H7 ☀ ◊ ◠

Betula nigra

Red-brown when young, becoming blackish or grey-white as it ages, the peeling bark of the black birch is its main attraction. Yellow-brown catkins appear in early spring, and its glossy, diamond-shaped leaves turn buttery yellow in autumn. If space allows, plant in a group for maximum impact.

↕18m (60ft) ↔12m (40ft) ❋ ❋ ❋ H7 ☀ ◊ ◊ ◊ ◠

❀ ❀ ❀ H7–H5 fully hardy   ❀ ❀ H4–H3 hardy in mild regions/sheltered sites   ❀ H2 protect from frost over winter   ❀ H1c–H1a no tolerance to frost

☀ full sun   ☼ partial sun   ☀ full shade   ◊ well-drained soil   ◖ moist soil   ◆ wet soil   ♤♡♧◊♤♧♕ tree shape

Betula utilis var. jacquemontii
The smooth, peeling white bark of this Himalayan birch comes into its own in a winter garden. Oval, tapered dark green leaves turn yellow in autumn, and yellow-brown catkins appear in early spring. The reliable cultivar 'Silver Shadow' has an eye-catching pure white trunk.

↕18m (60ft) ↔10m (30ft) ❀ ❀ ❀ H7 ☀ ◊ ♤

Cedrus atlantica f. glauca
Glaucous blue-green foliage, erect, cylindrical cones in autumn and a silvery-grey bark are the attractions of this coniferous tree. The blue Atlas cedar does well growing on chalk and is striking as a specimen in a sunny lawn, but its eventual size makes it unsuitable for all but the largest of gardens.

↕40m (130ft) ↔10m (30ft) ❀ ❀ ❀ H6 ☀ ◊ ♤–♤

Cercidiphyllum japonicum
The leaves of this fast-growing deciduous tree are bronze when young, turning mid-green, then yellow, orange and red in autumn. Acid soil produces the best colour. Fallen leaves smell of burnt sugar when crushed. The Katsura tree is best used as a specimen in a woodland setting.

↕20m (70ft) ↔15m (50ft) ❀ ❀ ❀ H5 ☀ ◊ ♤

Eucalyptus gunnii
To encourage the rounded, bluish young leaves of the cider gum, which are more attractive than the grey-green adult foliage, cut back hard in early spring. The whitish-green bark is shed in late summer, revealing pink- or orange-tinted new bark, as clusters of small white or cream flowers appear.

↕25m (80ft) ↔15m(30ft) ❀ ❀ ❀ H5 ☀ ◖ ◊ ♡

Fagus sylvatica 'Riversii'
The beauty of this cultivar of the common beech lies in its deep purple leaves, which need full sun for best colour. A spreading, deciduous tree, it can be used for hedging, in a woodland garden, or as a focal point. For dramatic effect, plant next to a golden-leaved tree, such as *Catalpa bignonioides* 'Aurea'.

↕25m (80ft) ↔15m (50ft) ❀ ❀ ❀ H6 ☀ ◊ ♡

Pinus wallichiana
The Bhutan pine is a graceful, broadly conical, evergreen tree with long, drooping, blue-green leaves and smooth, grey bark, which is grey-green when young but later becomes darker, scaly and fissured. It produces fresh green foliage in spring, and decorative pine cones that ripen to brown in autumn.

↕20–35m (70ft) ↔6–12m (20–40ft) ❀ ❀ ❀ H6 ☀ ◊ ♡

Quercus ilex
A majestic, round-headed evergreen tree, the holm oak has glossy, dark green leaves, which are silvery-grey when young. Striking yellow catkins are followed in autumn by small acorns. It makes a good screen and hedge, and thrives on exposed coastal sites. It also does well on shallow chalk.

↕25m (80ft) ↔20m (70ft) ❀ ❀ H4 ☀ ☼ ◊ ♡

Taxus baccata
A slow-growing evergreen conifer with distinctive dark green, needle-like leaves, the common yew is a familiar sight in churchyards. When closely-clipped it is excellent for hedging and topiary. The golden-leaved cultivar 'Standishii' is ideal for brightening a shady area. All parts of the plant are poisonous.

↕20m (70ft) ↔10m (30ft) ❀ ❀ ❀ H7 ☀ ☼ ☀ ◊ ♤

TREES FOR EVERGREEN INTEREST

- *Acacia dealbata* p.292
- *Arbutus unedo* p.296
- *Cedrus atlantica* f. *glauca* p.293
- *Chamaecyparis pisifera* 'Filifera Aurea' p.294
- *Cupressus macrocarpa* 'Goldcrest' p.297
- *Eucalyptus gunnii* p.293
- *Laurus nobilis* p.298
- *Olea europaea* p.298
- *Picea breweriana* p.295
- *Picea pungens* 'Koster' p.295
- *Pinus sylvestris* 'Aurea' p.295
- *Pinus wallichiana* p.293
- *Quercus ilex* p.293
- *Taxus baccata* p.293
- *Taxus baccata* 'Fastigiata' p.299
- *Tsuga canadensis* 'Aurea' p.299

Medium-sized trees

Acer negundo 'Variegatum'
There are maples for spring flowers, summer foliage or autumn colour. A fast-growing, deciduous tree, *A. negundo* is known as the ash-leaved maple because of its divided leaves; those of the cultivar 'Variegatum' are splashed white at the margins. It looks good planted near dark-leaved plants.

‡15m (50ft) ↔ 10m (30ft) ❄ ❄ ❄ H6 ☼ ◑ ◊ ◐ ♤

Carpinus betulus 'Fastigiata'
The dependable, deciduous, spring-flowering common hornbeam has glowing coppery autumn colour and is great for hedging. It is an excellent substitute for beech on drier soils. The narrow, upright cultivar 'Fastigiata' opens up as it matures, making a striking specimen tree.

‡15m (50ft) ↔ 12m (40ft) ❄ ❄ ❄ H7 ☼ ◑ ◊ ◐

Catalpa bignonioides 'Aurea'
The beautiful, spreading, deciduous Indian bean tree is popular for its large, dramatic heart-shaped leaves, clusters of tubular flowers, and long bean-like seed pods. It makes a striking specimen tree, but can also be grown in a border. The leaves of 'Aurea' are bronze when young, maturing to yellow.

‡12m (40ft) ↔ 12m (40ft) ❄ ❄ ❄ H6 ☼ ◊ ◐ ♤

Chamaecyparis pisifera 'Filifera Aurea'
This hardy evergreen tree tolerates most soils other than waterlogged sites, and can be grown as a specimen or as hedging. *C. pisifera* 'Filifera' has slender, whip-like shoots and dark green leaves; 'Filifera Aurea' is similar, but has golden yellow leaves and is slower to reach maturity.

‡12m (40ft) ↔ 5m (15ft) ❄ ❄ ❄ H7 ◊ ◐ ♤

Davidia involucrata
The elegant handkerchief tree is so known because of the conspicuous white bracts that surround the small flowerheads in spring. It is deciduous, with sharp-pointed, red-stalked leaves and smooth grey bark. Ridged fruits hang from long stalks in autumn. A fine specimen tree.

‡15m (50ft) ↔ 10m (30ft) ❄ ❄ ❄ H5 ☼ ◑ ◊ ◐ ♤

Fraxinus excelsior 'Pendula'
The common ash is a vigorous, deciduous tree, grown for its rounded habit and attractive foliage. In autumn it produces bunches of winged fruits, and in winter conspicuous black buds appear. 'Pendula' is a graceful, weeping form with long branches that droop, often as far as the ground.

‡15m (50ft) ↔ 10m (30ft) ❄ ❄ ❄ H6 ☼ ◊ ◐ ♧

Gleditsia triacanthos 'Sunburst'
Also known as honey locust, this striking deciduous tree has delicate, fern-like foliage, spines on the trunk and branches, and long, curved seed pods in autumn. The cultivar 'Sunburst' is fast-growing and thornless, with golden yellow foliage in spring and autumn. Best as a specimen tree.

‡12m (40ft) ↔ 10m (30ft) ❄ ❄ ❄ H6 ☼ ◊ ♤

Morus nigra
The black mulberry forms a rounded, deciduous tree with heart-shaped leaves that have rough upper surfaces and toothed margins. The fruit is green, turning red and then purple-black, becoming edible only when fully ripe. Beware of planting next to pale paving as the fruit will stain it when it falls.

‡12m (40ft) ↔ 15m (50ft) ❄ ❄ ❄ H6 ☼ ◊ ◐ ♤

Nyssa sinensis
Grown for its pretty foliage and brilliant autumn colour, the Chinese tupelo forms a broadly conical, deciduous tree. The slender, tapered leaves turn bright shades of orange, red, and yellow in autumn, making it a valuable ornamental. Grow as a specimen tree; it looks very effective alongside water.

‡12m (40ft) ↔ 10m (40ft) ❄ ❄ ❄ H5 ☼ ◑ ◊ ◐ ♤

※ ※ ※ H7–H5 fully hardy　※ ※ H4–H3 hardy in mild regions/sheltered sites　※ H2 protect from frost over winter　❀ H1c–H1a no tolerance to frost

☼ full sun　☀ partial sun　☀ full shade　◊ well-drained soil　◑ moist soil　◆ wet soil　◗◖◗◊◭◊◗♈ tree shape

Paulownia tomentosa

This fast-growing, deciduous tree is grown for its graceful habit, attractive large leaves, and showy, foxglove-like flowers. The fragrant, pinkish-lilac flowers, marked yellow and purple inside, open in late spring before the leaves appear. The tree can be pollarded, which will result in very large leaves.

↕12m (40ft) ↔10m (30ft) ※ ※ ※ H5 ☼ ◊ ♈

Picea breweriana

The popular Brewer's weeping spruce is a hardy, slow-growing, blue-green conifer with horizontal branches and long, slim, pendent branchlets that give it a distinctive appearance. Purple cones decorate the branches in autumn. It can be grown as an effective windbreak or as a specimen tree.

↕15m (50ft) ↔4m (12ft) ※ ※ ※ H6 ☼ ◊ ◑ ◗

Picea pungens 'Koster'

A hardy evergreen tree with scaly, grey bark and sharp, stout, bluish-green leaves. Cultivars of the Colorado spruce make wonderful ornamentals where space permits; 'Koster' has needle-like, silvery-blue leaves that fade to green with age and cylindrical light brown cones with papery scales.

↕15m (50ft) ↔5m (15ft) ※ ※ ※ H7 ☼ ◊ ◑ ◊–♈

Pinus sylvestris 'Aurea'

The Scots pine is widely grown for its timber, but its cultivars make excellent garden trees, either planted singly or in groups. Upright conifers, they have whorled branches when young, and develop a rounded crown with age. 'Aurea' has striking golden yellow leaves in winter.

↕15m (50ft) ↔9m (28ft) ※ ※ ※ H7 ☼ ◊ ◑ ♈

Prunus padus 'Watereri'

A deciduous, spreading tree, the bird cherry produces slender, pendent spikes of fragrant, star-shaped white flowers in mid-spring, followed by small black fruits. The leaves turn red or yellow in autumn. The conspicuous long flower spikes of the cultivar 'Watereri' create a spectacular spring display.

↕15m (50ft) ↔10m (30ft) ※ ※ ※ H6 ☼ ◊ ◑ ♈

Robinia pseudoacacia 'Frisia'

Deciduous and fast-growing, false acacia has elegant dark green leaves and coarsely fissured bark. Pea-like flowers are borne in early summer, followed by dark brown seed pods. The pretty cultivar 'Frisia', with golden yellow foliage that turns orange in autumn, makes a superb focal point.

↕15cm (50ft) ↔8m (25ft) ※ ※ ※ H6 ☼ ◊ ◑ ♈

Salix alba var. *sericea*

The silver willow is a fast-growing, deciduous, spreading tree, conical in shape when young. The leaves are long, narrow and an intense silver-grey, and emerge at the same time as the yellow catkins in early spring. The foliage sparkles in the breeze, and it makes an elegant specimen tree.

↕15m (50ft) ↔8m (25ft) ※ ※ ※ H6 ☼ ◊ ◑ ♈

Salix x *sepulcralis* 'Chrysocoma'

A wide-spreading, deciduous tree with supple yellow stems that reach the ground, the golden weeping willow is grown for its beautiful cascading habit. Slender yellow or green catkins are borne with the narrow yellow-green leaves in spring. It looks particularly striking when planted by water.

↕15m (50ft) ↔15m (50ft) ※ ※ ※ H5 ☼ ◊ ◑ ♈

TREES AS FOCAL POINTS

Small trees

Acer griseum
The chief attraction of this deciduous maple is its unusual bark, which is orange to mahogany-red and peels laterally in papery rolls. The dark green leaves turn bright crimson and scarlet in autumn, and the ornamental bark gives this spectacular tree a valued winter role in small gardens.

↕10m (30ft) ↔10m (30ft) ❄ ❄ ❄ H5 ☼ ☼ ◊ ◊ ♀

Acer japonicum 'Vitifolium'
A pretty, deciduous tree with broad, fan-shaped leaves that turn scarlet, gold, and purple in autumn. The leaves are similar to those of a grapevine, hence the cultivar name. In mid-spring it bears clusters of small, delicate, reddish-purple flowers. Can be grown as a bushy tree or large shrub.

↕10m (30ft) ↔10m (30ft) ❄ ❄ ❄ H6 ☼ ☼ ◊ ◊ ♀

Acer palmatum 'Bloodgood'
Japanese maples make lovely ornamental trees. 'Bloodgood' forms a deciduous, bushy-headed shrub or small tree and is grown for its deeply cut, dark reddish-purple leaves, which turn bright red in autumn. Small purple flowers are borne in mid-spring, followed by attractive red-winged fruits.

↕5m (15ft) ↔5m (15ft) ❄ ❄ ❄ H6 ☼ ☼ ◊ ♀

Acer palmatum 'Osakazuki'
A stunning Japanese maple for autumn colour. The mid-green leaves are larger than average and turn a brilliant scarlet before falling. Dainty red-winged fruits appear in late summer. It can be grown in a large container but must not be allowed to dry out, and needs shelter from cold winds.

↕6m (20ft) ↔6m (20ft) ❄ ❄ ❄ H6 ☼ ☼ ◊ ♀

Acer palmatum 'Sango-kaku'
For colour interest all year, this delicate Japanese maple is a perfect choice. The divided leaves are orange-yellow in spring, maturing to green, then turning yellow in autumn before they fall. In winter, the new shoots, borne on ascending branches, turn coral-pink, deepening in colour as winter advances.

↕6m (20ft) ↔5m (15ft) ❄ ❄ ❄ H6 ☼ ◊ ♀

Amelanchier lamarckii
With abundant white flowers in spring and brilliant red leaf colour in autumn, this deciduous hardy shrub or small tree provides plenty of seasonal interest. The young oval leaves unfold bronze before the star-shaped flowers emerge, and the small red fruits that follow are attractive to birds.

↕10m (30ft) ↔12m (40ft) ❄ ❄ ❄ H7 ☼ ☼ ◊ ◊ ♀

Arbutus unedo
This handsome evergreen with flaky, red-brown bark and attractive, glossy green leaves forms a large shrub or small tree in sheltered gardens. Lily-of-the-valley-like blooms appear in early winter and the rounded fruits, ripening to red in autumn, give rise to the common name, strawberry tree.

↕8m (25ft) ↔8m (25ft) ❄ ❄ ❄ H5 ◊ ♀

Cercis canadensis f. alba 'Forest Pansy'
A pretty, multi-stemmed tree or shrub with vivid, reddish-purple, heart-shaped leaves that are velvety to the touch. Magenta buds open to pale pink, pea-like flowers in mid-spring before the characteristic leaves appear. Impressive as a single specimen but also useful for the back of the border.

↕10m (30ft) ↔10m (30ft) ❄ ❄ ❄ H5 ☼ ☼ ◊ ◊ ♀

Cercis siliquastrum
The Judas tree is an eye-catching, spreading, bushy tree, with bright purple-rose spring flowers and long, purple-tinted pods that appear in late summer. Its heart-shaped leaves are bronze when young, turning yellow in autumn. Although hardy it originates from the Mediterranean, so avoid very cold sites.

↕10m (30ft) ↔10m (30ft) ❄ ❄ ❄ H5 ☼ ☼ ◊ ◊ ♀

Cornus controversa 'Variegata'

This elegant deciduous tree with horizontally-tiered branches creates a distinctive architectural profile. Flat heads of star-shaped white flowers appear in summer, followed by blue-black fruit. 'Variegata' has bright green leaves with creamy white margins, and makes a beautiful focal point.

↕8m (25ft) ↔8m (25ft) ❋ ❋ ❋ H5 ☼ ◇ ♤

Cornus kousa var. chinensis 'China Girl'

A broadly conical deciduous tree, this dogwood has tiny green flowerheads in summer surrounded by decorative petal-like white bracts. Fleshy red fruits develop later, followed by rich, purple-red autumn leaves. 'China Girl', free-flowering even when young, has large creamy-white bracts that age to pink.

↕7m (22ft) ↔5m (15ft) ❋ ❋ ❋ H6 ☼ ☼ ◐ ◇

Corylus avellana 'Contorta'

The corkscrew hazel is a slow-growing, small deciduous tree or shrub with unusual twisted shoots, which are seen at their best in winter when the long yellow catkins appear. Ideal as a focal point in a winter garden, the stems can also be cut for striking indoor displays.

↕5m (15ft) ↔5m (15ft) ❋ ❋ ❋ H6 ☼ ☼ ◐ ◐ ♤

Crataegus orientalis

Hawthorns are widely used for hedges and as ornamentals. Many are thorny but *C. orientalis* is almost thornless. It is an attractive, compact, deciduous tree with deeply cut, dark green leaves. White flowers appear in profusion in late spring, followed by yellow-tinged red fruit.

↕6m (20ft) ↔6m (20ft) ❋ ❋ H6 ☼ ☼ ◇ ◐ ♤

Crataegus persimilis 'Prunifolia'

An excellent small deciduous tree, with rich brown bark and long, dramatic thorns. It is grown mainly for its polished, deep green leaves that turn brilliant orange and red in autumn. Dense heads of white flowers are produced in early summer followed by clusters of long-lasting, bright red berries.

↕8m (25ft) ↔10m (30ft) ❋ ❋ H7 ☼ ☼ ◇ ◐ ♤

Cupressus macrocarpa 'Goldcrest'

The Monterey cypress is a coastal tree in the wild and will tolerate dry growing conditions, which makes it useful as a hedge or windbreak in exposed sites. 'Goldcrest' is a handsome, narrowly conical tree with lemon-scented golden foliage. It looks stunning grown against a dark background.

↕5m (16ft) ↔2.5m (8ft) ❋ ❋ H4 ☼ ◇ ◊

Dicksonia antarctica

A spectacular and hardy tree fern, *D. antarctica* brings drama into the garden. In spring its arching pale green fronds unfurl from the top of a mass of fibrous roots that form the trunk. It is evergreen in mild climates, but in cold winters protect the crown by covering it with straw.

↕6m (20ft) ↔4m (12ft) ❋ ❋ H3 ☼ ☀ ◇ ◐ ♈

Ficus carica 'Brown Turkey'

A popular variety of fig that thrives in cool climates, 'Brown Turkey' has large lobed leaves and pear-shaped edible fruits, green at first, maturing to purple-brown. Grow as a fan against a sunny wall or as a freestanding tree; in cold areas keep in a pot and move under cover in winter.

↕3m (10ft) ↔4m (12ft) ❋ ❋ H4 ☼ ◇ ◐ ♤

TREES FOR SPRING INTEREST

Small trees

Laburnum × watereri 'Vossii'
This elegant, spreading, deciduous tree has glossy green leaves, cut into oval leaflets, and bears magnificent long golden chains of pea-like flowers in late spring. It makes an impressive specimen tree in a small garden, but can also be trained over a pergola. The leaves and seeds are poisonous.

↕8m (25ft) ↔8m (25ft) ❋ ❋ ❋ H5 ☀ ◊ ♀

Larix kaempferi 'Pendula'
Unusually among the conifers, larches are deciduous. A small grafted weeping cultivar, 'Pendula' has fine green linear leaves that turn bright yellow in autumn. It needs to be trained; the height of the stake will determine how tall the plant is. Its compact, waterfall-like habit makes it ideal for a small garden.

↕to 5m (15ft) ↔to 3m (10ft) ❋ ❋ ❋ H7 ☀ ◊ ♀

Laurus nobilis
Bay laurel is a conical evergreen tree grown for its aromatic, leathery, dark green leaves, which are used as flavouring in cooking. Clusters of small, greenish-yellow flowers appear in spring, followed by black berries in autumn. It can be grown in a pot, and looks attractive when trimmed into formal shapes.

↕to 10m (30ft) ↔to 8m (25ft) ❋ ❋ H4 ☀ ☀ ◊ ◊ △

Malus 'Evereste'
This crab apple is an excellent choice for a small garden as it forms a neat, conical shape. A profusion of white, shallow, cup-shaped flowers open from pink buds in late spring, followed by small, red-flushed, orange-yellow fruit. The green leaves turn yellow and orange in autumn before falling.

↕7m (22ft) ↔6m (20ft) ❋ ❋ ❋ H6 ☀ ☀ ◊ ◊ △

Malus 'Royalty'
This pretty crab apple is smothered in deep pink to bright purple flowers, which open from dark red buds in spring. The glossy leaves are dark red-purple and maintain their colour well through the season, turning red in autumn. Inedible small purple fruits follow the flowers. A fine specimen tree.

↕8m (25ft) ↔8m (25ft) ❋ ❋ ❋ H6 ☀ ☀ ◊ ◊ ♀

Olea europaea
An elegant, slow-growing evergreen, the olive tree has grey-green leaves and tiny, fragrant, creamy-white flowers in summer. The green olives only ripen to black in hot, dry conditions. It makes a stunning feature in a sunny, sheltered spot, or grow in a large pot and move under cover in winter.

↕10m (30ft) ↔10m (30ft) ❋ ❋ H4 ☀ ◊ ♀

Prunus 'Mount Fuji'
Ornamental cherries make very attractive specimen trees for small gardens. This beautiful deciduous tree has pale green young leaves, darkening to deep green, then turning orange and red in autumn before they fall. Clusters of fragrant, white, cup-shaped flowers are borne in mid-spring.

↕6m (20ft) ↔8m (25ft) ❋ ❋ ❋ H6 ☀ ◊ ◊ △ ♀

Prunus serrula
A dramatic choice for winter interest, this deciduous tree is prized for its glossy mahogany bark with pale horizontal lines. Small white flowers are produced at the same time as the new leaves in late spring, followed by small inedible cherries on long stalks. The leaves turn yellow in autumn.

↕10m (30ft) ↔10m (30ft) ❋ ❋ ❋ H6 ☀ ◊ ◊ ♀

Prunus 'Spire'
Attractive over a long season, the leaves of this upright, deciduous cherry are bronze when young, green in summer, then orange and red in autumn. In spring, bowl-shaped, soft pink flowers emerge in clusters against the new leaves. Makes a beautiful feature in a small garden.

↕10m (30ft) ↔6m (20ft) ❋ ❋ ❋ H6 ☀ ◊ ◊ ♀

Prunus × subhirtella 'Autumnalis Rosea'
A popular tree for its early-flowering nature, this delicate spreading cherry is perfect for a small garden. Clusters of tiny, double, pale pink flowers appear in winter during mild spells. The green leaves are narrow and bronze when young, turning golden-yellow in autumn.

‡8m (25ft) ↔ 8m (25ft) ※ ※ ※ H6 ☼ ◊ ◑ ♀

Pyrus salicifolia 'Pendula'
This delightful ornamental pear tree has an elegant weeping habit and silvery-grey, willow-like leaves. An abundant show of creamy-white flowers in spring is followed by small, hard, inedible pears in late summer. Grow as specimen tree on a lawn, where its graceful habit can be seen to advantage.

‡5m (15ft) ↔ 4m (12ft) ※ ※ ※ H6 ☼ ◊ ♀

Rhus typhina
Known as the stag's horn sumach because of its red velvety shoots, this distinctive deciduous tree is particularly fine in autumn when its deeply divided leaves turn shades of orange and red. The fruits are formed in dense, hairy, crimson-red clusters on female plants. Plant singly or in a shrub border.

‡5m (15ft) ↔ 6m (20ft) ※ ※ ※ H6 ☼ ◊ ◑ ♀

Sorbus aria 'Lutescens'
A pretty deciduous tree, this eye-catching whitebeam has striking silvery-grey young foliage that gradually turns grey-green. White flowers in late spring are followed by orange berries in autumn. A freestanding tree of great beauty, it can also be used for mass planting or screening.

‡10m (30ft) ↔ 8m (25ft) ※ ※ ※ H6 ☼ ☀ ♀ ♀

Sorbus commixta
Sorbus are excellent ornamental trees for city gardens as they tolerate atmospheric pollution. *S. commixta* bears large white flowerheads in spring and has elegant foliage, which turns shades of yellow, red, and purple in autumn. 'Embley' has bright red leaves in late autumn, and plenty of crimson fruit.

‡10m (30ft) ↔ 7m (22ft) ※ ※ ※ H6 ☼ ☀ ◊ ◑ ♀

Stewartia sinensis
A good choice for autumn foliage colour, this small deciduous tree is also prized for its unusual peeling red-brown bark and showy, white fragrant flowers that appear in midsummer. Autumn brings an impressive display of red, orange, and yellow leaves. It prefers acid soil.

‡6m (20ft) ↔ 3m (10ft) ※ ※ ※ H5 ☼ ☀ ◊ ◑ ♀

Taxus baccata 'Fastigiata'
Irish yew has a narrow, upright habit, eventually forming a distinguished, columnar shape. This makes it useful as a focal point or accent plant in a border. Small red berries appear in summer. 'Fastigiata Aurea' is similar but has variegated yellow-green leaves. All parts are poisonous.

‡10m (30ft) ↔ 2m (6ft) ※ ※ ※ H6 ☼ ☀ ◊ ◑ ●

Tsuga canadensis 'Aurea'
A graceful species of conifer, there are many varieties of Eastern hemlock available. 'Aurea' is an elegant, compact, and fairly slow-growing tree with golden yellow juvenile foliage, which darkens to green with age. It is useful for evergreen interest in partially shaded areas.

‡8m (25ft) ↔ 4m (12ft) ※ ※ ※ H6 ☼ ☀ ◊ ◊ ♀

TREES FOR AUTUMN COLOUR

Large shrubs

Aralia elata 'Variegata'

The Japanese angelica tree, *A. elata*, is an elegant, deciduous shrub with striking grey-green leaves that turn many shades of yellow, orange, or purple in autumn. Large heads of small white flowers appear in late summer. The leaves of 'Variegata' have creamy-white margins that shine out in a shady border.

↕5m (15ft) ↔5m (15ft) ❋ ❋ ❋ H5 ☼ ◊ ◗

Azara microphylla

An attractive evergreen shrub or small tree with large sprays of small, glossy, dark green leaves. Small clusters of vanilla-scented, deep yellow flowers are borne in late winter and early spring, making it a useful shrub for winter interest. It will tolerate part-shade and grows well against a wall.

↕7m (22ft) ↔4m (12ft) ❋ ❋ ❋ (boderline) H4 ☼ ◐ ◗

Buddleja alternifolia 'Argentea'

The slender, arching branches of this robust deciduous shrub have narrow grey-green leaves and carry dense clusters of very fragrant lilac flowers in summer. Its weeping habit makes it suitable for training as a standard. Prune after flowering to prevent branches from becoming tangled.

↕4m (12ft) ↔4m (12ft) ❋ ❋ ❋ H6 ☼ ◐ ◊ ◗

Buddleja globosa

This eye-catching upright shrub has handsome, semi-evergreen, dark green leaves. Small, bright, orange-yellow balls of fragrant flowers appear in early summer, and will brighten up a border. It prefers a sunny position and tolerates chalky soil, but does not respond well to hard pruning.

↕5m (15ft) ↔5m (15ft) ❋ ❋ ❋ H5 ☼ ◊

Camellia reticulata 'Leonard Messel'

Camellias are invaluable evergreen spring-flowering shrubs for acid soils in sheltered sites. 'Leonard Messel' produces a profusion of large, semi-double, pink flowers in spring that stand out vividly against a background of matt, dark green leaves. It is ideal as a specimen or in a woodland setting.

↕4m (12ft) ↔3m (10ft) ❋ ❋ ❋ H5 ☼ ◊ ◗

Chimonanthus praecox 'Grandiflorus'

Known as wintersweet, this deciduous shrub produces pale yellow flowers that hang from its bare stems throughout winter, perfuming the air with intoxicating scent. Grow it as a specimen shrub, as part of a border planting, or train it on a sunny wall. The stems can be cut for indoor displays.

↕4m (12ft) ↔3m (10ft) ❋ ❋ ❋ H5 ☼ ◊

Clerodendrum trichotomum var. fargesii

This spectacular deciduous shrub has an upright habit and attractive bronze young leaves. Fragrant, white, star-shaped flowers with green sepals open from pink and greenish-white buds in late summer. Jewel-like, bright blue berries, surrounded by pronounced maroon calyxes, follow the flowers.

↕6m (20ft) ↔6m (20ft) ❋ ❋ ❋ H5 ☼ ◊

Cordyline australis 'Red Star'

The New Zealand cabbage palm is a popular evergreen shrub grown for its striking foliage. In warm regions, it makes an eye-catching architectural plant for a sheltered courtyard garden; in frost-prone areas, keep it in a pot in a cool greenhouse during winter. 'Red Star' has rich red-bronze, sword-like leaves.

↕3–10m (10–30ft) ↔1–4m (3–12ft) ❋ ❋ ❋ H3 ☼ ◐ ◊

Cornus mas

Shrubs that flower in winter, such as this Cornelian cherry, are a valuable asset to the designer. It bears little clusters of tiny yellow flowers on bare branches in late winter, before the leaves appear. Bright red fruits are produced in late summer, and the leaves turn red-purple in autumn.

↕5m (15ft) ↔5m (15ft) ❋ ❋ ❋ H6 ☼ ◐ ◊

Corylus maxima 'Purpurea'

The intense colour of this deciduous, deep purple-leaved hazel makes an immediate impact in a garden. Attractive purple-tinged catkins appear in late winter, and edible nuts ripen in autumn. Grow as a specimen plant or as a focal point in a shrub border. The best colour is produced in full sun.

↕6m (20ft) ↔5m (15ft) ❀ ❀ ❀ H6 ☼ ☼ ◊

Cotinus coggygria Rubrifolius Group

This bushy, deciduous shrub is known as the smoke bush because its fluffy plumes of pale pink summer flowers produce a smoky effect. The dark purple leaves colour best in full sun, and turn scarlet and orange in autumn. A fine structural shrub to plant on its own, it is also useful at the back of a border.

↕5m (15ft) ↔5m (15ft) ❀ ❀ ❀ H5 ☼ ☼ ● ◐

Cotinus 'Grace'

A vigorous smoke bush cultivar that can be grown as a small bushy tree or as a tall multi-stemmed shrub. Large, dark pink flower clusters appear above the foliage in summer, and the soft purple-red leaves turn a brilliant orange-red before falling. An excellent choice for autumn colour.

↕6m (20ft) ↔5m (15ft) ❀ ❀ ❀ H5 ☼ ☼ ◊ ◐

Cotoneaster frigidus 'Cornubia'

A large, arching, semi-evergreen shrub, this cotoneaster has narrow green leaves that are tinted bronze in autumn. Creamy-white, early summer flowers are produced in profusion, followed by heavy clusters of bright red fruit that are attractive to birds. It can be trained as a single-stemmed tree.

↕10m (30ft) ↔10m (30ft) ❀ ❀ ❀ H6 ☼ ◊

Cotoneaster lacteus

This dense, evergreen shrub sports distinctive, dark green, leathery leaves. Cup-shaped, milky-white flowers appear in summer, followed by clusters of dark red fruit that persist well into winter. It makes an attractive hedge or screen, and it can also be grown as a small tree.

↕4m (12ft) ↔4m (12ft) ❀ ❀ ❀ H6 ☼ ☼ ◊

Cytisus battandieri

An elegant, deciduous shrub, the pineapple broom gained its common name from the scent of its yellow pea-like flowers, which emerge in summer. Its attractive, silvery-green leaves are covered in soft, silky hairs. Ideal as a freestanding shrub, but grow it against a sunny wall in colder areas.

↕5m (15ft) ↔5m (15ft) ❀ ❀ ❀ H5 ☼ ◊

Dipelta floribunda

This handsome deciduous shrub offers interest through the seasons. Masses of fragrant pale pink flowers with yellow markings appear in late spring, its light green leaves turn yellow in autumn, and it has attractive peeling bark in winter. Grow it as a specimen plant or in a shrub border.

↕4m (12ft) ↔4m (12ft) ❀ ❀ ❀ H5 ☼ ◊

Elaeagnus × ebbingei 'Gilt Edge'

A hardy, evergreen, dense shrub, 'Gilt Edge' has brown scaly stems and glossy leaves with green centres and golden yellow margins. Small, lightly-scented flowers are produced from mid- to late autumn. The plant's hardiness makes it a good choice for a shelter belt or hedge, especially in coastal areas.

↕4m (12ft) ↔4m (12ft) ❀ ❀ ❀ H5 ☼ ◊

SHRUBS FOR FOCAL POINTS

Large shrubs

Elaeagnus 'Quicksilver'

With silvery shoots and narrow, silver-grey leaves, this fast-growing shrub makes a great foil for dark-leaved plants. Although bushy, with a loose, spreading crown, it can be trained as a small tree. Star-shaped, fragrant, creamy-yellow flowers open from silvery buds in late spring or summer.

‡4m (12ft) ↔4m (12ft) ❀ ❀ ❀ H5 ☼ ◊

Hamamelis × intermedia 'Pallida'

Witch hazel is a handsome shrub that produces spider-like scented flowers on bare branches in winter. There are many cultivars. 'Jelena' has large, coppery-orange flowers and orange and red autumn foliage. 'Pallida' bears large, fragrant, yellow flowers and has golden autumn leaves.

‡4m (12ft) ↔4m (12ft) ❀ ❀ ❀ H5 ☼ ◐ ◊ ◗

Hippophae rhamnoides

Sea buckthorn thrives in harsh conditions and makes an excellent screening plant for a coastal garden. It has a bushy habit, but can be trained to make a small tree, and has thorny stems with narrow, silver-grey leaves. Grow male and female plants together to produce brilliantly orange-coloured berries.

‡6m (20ft) ↔6m (20ft) ❀ ❀ ❀ H7 ☼ ◊ ◗

Hydrangea paniculata 'Unique'

Hydrangeas are mainly grown for their showy flowerheads, but some have pretty bark and others develop good autumn colour. *H. paniculata* 'Unique' bears large, creamy-white flowerheads from midsummer to early autumn, and its leaves turn yellow before falling. It's best planted singly or in a shrub border.

‡3–7m (10–22ft) ↔2.5m (8ft) ❀ ❀ ❀ H5 ☼ ◐ ◊ ◗

Ilex aquifolium 'Silver Queen'

Common holly has dark green leaves, but there are many cultivars with white, cream, or yellow variegation. 'Silver Queen' is a male variety (it does not bear berries); it forms an upright evergreen, with purple stems and striking leaves with broad, creamy-white margins. It is ideal for hedges and screens.

‡10m (30ft) ↔4m (12ft) ❀ ❀ ❀ H6 ☼ ◐ ◊ ◗

Itea ilicifolia

This spectacular evergreen shrub has holly-like, shiny, dark green leaves. Long catkins made up of small, greenish-white flowers appear in late summer, and a honey-like scent is discernible on warm evenings. A fine freestanding shrub for mild areas, but plant it against a wall in more exposed sites.

‡3–5m (20–15ft) ↔3m (10ft) ❀ ❀ ❀ H5 ☼ ◊

Juniperus communis 'Hibernica'

Junipers tolerate a wide range of soils and growing conditions, are tough enough for hot, sunny sites, and need little pruning. 'Hibernica', also known as the Irish juniper, forms a slender column of crowded, needle-like leaves, each with a silver line, and makes an excellent structural plant for formal schemes.

‡3–5m (10–15ft) ↔30cm (12in) ❀ ❀ ❀ H7 ☼ ◊

Ligustrum ovalifolium 'Aureum'

A vigorous, semi-evergreen shrub, golden privet has variegated leaves with bright yellow margins and bears dense clusters of white flowers in midsummer, followed by black berries. It clips easily and is ideal for hedging and topiary. Shade tolerant, it can be planted to brighten a shady corner of the garden.

‡4m (12ft) ↔4m (12ft) ❀ ❀ ❀ H5 ☼ ◐ ◊

Mahonia × media 'Charity'

With their attractive foliage, bright yellow flowers, and decorative fruits, mahonias make magnificent architectural features in a winter garden. 'Charity' is fast-growing and has spiny holly-like leaves. Bright yellow to lemon yellow flowers are produced in spikes from late autumn to late winter.

‡5m (15ft) ↔4m (12ft) ❀ ❀ ❀ H5 ☼ ◊ ◗

Olearia macrodonta
New Zealand holly is a vigorous evergreen shrub with sharply-toothed, sage green leaves, which provide mellow colour all year. Fragrant, white, daisy-like flowers are borne in early summer. A handsome freestanding shrub in mild areas, it also makes an excellent screen for exposed coastal gardens.

↕6m (20ft) ↔5m (15ft) ❀ ❀ ❀ (borderline) H4 ☼ ◌

Photinia × fraseri 'Red Robin'
This hardy evergreen shrub is grown for its conspicuous, deep red young foliage, which is produced in spring on the tips of the branches. It looks good in a woodland garden or in a shrub border, and can also be used for hedging. 'Red Robin' is a compact cultivar, with especially bright red young leaves.

↕5m (15ft) ↔5m (15ft) ❀ ❀ ❀ H5 ☼ ☀ ◌ ◖

Pittosporum tenuifolium
A charming, upright evergreen shrub with pale grass-green, wavy leaves and attractive black stems. The small dark purple flowers, produced in abundance in spring, are honey-scented at dusk. In mild regions, it can be grown as a specimen plant on a lawn, or used for simple topiary.

↕4–10m (12–30ft) ↔ to 5m (15ft) ❀ ❀ H4 ☼ ☀ ◌ ◖

Rhamnus alaternus 'Argenteovariegata'
This handsome, bushy, evergreen shrub bears glossy grey-green leaves with creamy-white margins. Small yellow-green flowers appear in early summer, followed by spherical red berries in a warm summer, which ripen to black. It does well in coastal and city gardens, but needs shelter in colder areas.

↕5m (15ft) ↔4m (12ft) ❀ ❀ ❀ H5 ☼ ◌

Rhododendron luteum
An elegant deciduous azalea, R. luteum bears rounded clusters of funnel-shaped yellow flowers in late spring, which have a delightful scent. The rich green leaves turn shades of crimson, purple, and orange in autumn, making it a valuable garden plant over a long season. It requires acid soil.

↕4m (12ft) ↔4m (12ft) ❀ ❀ ❀ H5 ☼ ◌ ◖

Syringa vulgaris 'Mrs Edward Harding'
Lilacs form spreading deciduous shrubs with pretty heart-shaped leaves, and make useful screening plants. Sweetly-scented flowerheads appear from spring to early summer. There are over 500 cultivars of common lilac to choose from; 'Mrs Edward Harding' has double, purple-red flowers.

↕to 7m (22ft) ↔ to 7m (22ft) ❀ ❀ ❀ H6 ☼ ◌

Tamarix ramosissima 'Pink Cascade'
Tamarisks are excellent shrubs for exposed coastal gardens where they can make an effective screen. They have attractive, feathery foliage, formed of needle-like leaves. T. ramosissima is deciduous, with arching branches and upright plumes of small, pink flowers; 'Pink Cascade' has rich pink flowers.

↕5m (15ft) ↔5m (15ft) ❀ ❀ ❀ H5 ☼ ◌

Viburnum opulus
The guelder rose is a good choice for a wildlife garden as birds love the translucent red berries; as a bonus, the leaves also turn a rich red in autumn. The late spring blooms are attractive, too, forming lacecap-like heads of white flowers. This deciduous plant is vigorous and is commonly seen in hedgerows.

↕5m (15ft) ↔4m (12ft) ❀ ❀ ❀ H6 ☼ ☀ ◌ ◖

SHRUBS FOR HOT, DRY SITES

- *Artemisia arborescens* p.310
- *Buddleja globosa* p.300
- *Ceanothus thyrsiflorus* var. *repens* p.310
- *Choisya* × *dewitteana* p.305
- *Cistus* cultivars p.311
- *Convolvulus cneorum* p.311
- *Cytisus battandieri* p.301
- *Escallonia* 'Apple Blossom' p.305
- *Helianthemum* 'Wisley Primrose' p.313
- *Lavandula angustifolia* 'Munstead' p.314
- *Lavatera* × *clementii* 'Barnsley' p.307
- *Lonicera nitida* 'Baggesen's Gold' p.314
- *Origanum* 'Kent Beauty' p.314
- *Pinus mugo* 'Mops' p.315
- *Potentilla fruticosa* cultivars p.315
- *Ribes sanguineum* p.308
- *Rosmarinus officinalis* p.316
- *Salvia officinalis* cultivars pp.316–7
- *Santolina pinnata* p.317

Medium-sized shrubs

Abelia x grandiflora

A vigorous, semi-evergreen shrub with glossy dark green foliage and an abundance of fragrant, pink-flushed white flowers from midsummer to mid-autumn. Plant either as a freestanding shrub, or as an informal hedge. It is best fan-trained against a sunny wall in colder areas.

‡3m (10ft) ↔4m (12ft) ❁ ❁ ❁ H5 ☼ ◊

Acer palmatum Dissectum Atropurpureum Group

Most Japanese maples are low-growing and shrubby, and look their best at the front of a border; many have beautiful foliage and fiery autumn colour. *A. palmatum* var. *dissectum* forms a mound of narrow, very finely-toothed leaves, and Dissectum Atropurpureum Group has red-purple leaves.

‡2m (6ft) ↔3m (10ft) ❁ ❁ ❁ H6 ☼ ◐ ◊ ◖

Aucuba japonica 'Crotonifolia'

Hardy evergreen shrubs, spotted laurels are easy to grow and tolerant of a wide range of growing conditions – shade, dry sites, and even areas with polluted air. 'Crotonifolia' has large, glossy green leaves speckled with yellow marks. In mid-spring, small red-purple flowers appear, followed by red berries.

‡3m (10ft) ↔3m (10ft) ❁ ❁ ❁ H5 ☼ ◐ ● ◊ ◖

Berberis darwinii

This vigorous, dense, mounded evergreen shrub has glossy dark green foliage on prickly stems. During spring, it bears drooping clusters of bright orange flowers, which are followed by round blue-black fruit. It makes an attractive informal hedge, and tolerates heavy clay soils.

‡3m (10ft) ↔3m (10ft) ❁ ❁ ❁ H5 ☼ ◐ ◊

Berberis julianae

A handsome evergreen shrub with spiny-margined, glossy deep green leaves, this plant is often used as a screen. From spring to early summer, clusters of scented yellow or red-tinged flowers are produced, followed by egg-shaped, blue-black fruits. It is best planted where its scent will be appreciated.

‡3m (10ft) ↔3m (10ft) ❁ ❁ ❁ H5 ☼ ◐ ◊

Buddleja crispa

Perfect for planting in the shelter of a sunny wall or fence, this deciduous shrub has striking leaves covered in soft, greyish-white down and woolly white young shoots. Small, fragrant, lilac-pink flowers appear in long, dense clusters from mid- to late summer. Attractive to bees and butterflies.

‡3m (10ft) ↔3m (10ft) ❁ ❁ H4 ☼ ◊

Buddleja davidii 'Dartmoor'

An outstanding butterfly bush cultivar, 'Dartmoor' has arching stems and soft green leaves that are white beneath. In late summer and autumn, it bears broad, open-branched plumes of highly scented, pinkish purple flowers. Loved by butterflies and ideally suited to wildlife gardens.

‡2.5m (8ft) ↔2.5m (8ft) ❁ ❁ ❁ H6 ☼ ◐ ◊

Camellia japonica 'Bob's Tinsie'

Camellias make elegant evergreen flowering plants for gardens with acid soil. New variations of *C. japonica* appear every year and there is a huge range of cultivars to choose from. 'Bob's Tinsie' has an upright habit, and bears small, clear red flowers from early to late spring. Shelter from cold, drying winds.

‡2m (6ft) ↔1m (3ft) ❁ ❁ ❁ H5 ☼ ◊ ◖

Ceanothus 'Concha'

Ceanothus are cultivated for their flowers, which may be blue, white, or pink. *C.* 'Concha' is a good choice for a warm, sunny wall or fence. It forms a dense evergreen shrub with finely toothed, dark green leaves and produces masses of reddish-purple buds in late spring that open up to dark blue flowers.

‡3m (10ft) ↔3m (10ft) ❁ ❁ H4 ☼ ◊

Chaenomeles speciosa 'Moerloosei'
Ornamental quinces make reliable garden shrubs, and can even be trained against a shaded wall or fence. This variety (also sold as 'Apple Blossom') bears large clusters of white flowers, flushed dark pink, in spring and early summer, followed by aromatic fruits. Prune after flowering.

↕2.5m (8ft) ↔5m (15ft) ❀ ❀ ❀ H6 ☀ ☼ ◊

Choisya x dewitteana 'Aztec Pearl'
A compact, elegant example of Mexican orange blossom, this pretty evergreen shrub with slim dark green leaves is suitable for a small garden or container. Fragrant clusters of white star-shaped flowers emerge from pink buds in late spring, and appear again in smaller numbers in late summer and autumn.

↕2.5m (8ft) ↔2.5m (8ft) ❀ ❀ H4 ☀ ☼ ◊

Cornus alba 'Aurea'
This golden-leaved, vigorous dogwood offers a combination of summer and winter interest. Throughout summer it forms a mound of broad greenish-yellow leaves and, after these fall in late autumn, the dark red stems create a stunning display. Cut down a third of the stems in spring to rejuvenate the plant.

↕3m (10ft) ↔3m (10ft) ❀ ❀ ❀ H7 ☀ ☼ ◊

Cornus alba 'Sibirica'
A deciduous, upright dogwood, 'Sibirica' forms a dense thicket of young scarlet stems. These are seen at their best in sunshine, and set a dull winter garden ablaze with their fiery colours. It is one of the best cultivars for autumn colour, its dark green leaves turning red before falling.

↕3m (10ft) ↔3m (10ft) ❀ ❀ ❀ H7 ☀ ☼ ◊

Cornus sericea 'Flaviramea'
The winter shoots of this vigorous dogwood display their most vivid colour when grown in a sunny site. The plant bears white flowers from late spring to early summer, and the dark green leaves turn red and orange in autumn. The form 'Flaviramea' produces bright yellow-green winter stems.

↕2m (6ft) ↔4m (12ft) ❀ ❀ ❀ H7 ☀ ☼ ◊

Daphne bholua 'Jacqueline Postill'
A shrub for a border or rock garden, D. bholua is best planted in a sheltered position where the richly fragrant flowers will be appreciated. This cultivar is vigorous, evergreen, and bears clusters of deep purple-pink flowers, white inside, over a long flowering season in late winter. Mulch to retain moisture.

↕2m (6ft) ↔1.5m (5ft) ❀ ❀ H4 ☀ ☼ ◊ ◐

Erica arborea var. alpina
This tree heath makes a dense, compact, upright shrub, crowded with needle-shaped, bright green evergreen leaves. Masses of tiny, fragrant, bell-shaped white flowers appear in spring. Grow it in acid soil for the best results, and prune hard after flowering to keep it in shape and encourage new growth.

↕2m (6ft) ↔90cm (36in) ❀ ❀ H4 ☀ ◊

Escallonia 'Apple Blossom'
Tolerant of maritime conditions, this attractive evergreen shrub with glossy dark green leaves is a good choice for a coastal garden. It is compact and bushy, and produces clusters of pink and white flowers, similar to apple blossom, from early to midsummer. Grow as a hedge or windbreak.

↕2.5m (8ft) ↔2.5m (8ft) ❀ ❀ H4 ☀ ◊

SHRUBS FOR SHADE

- Aucuba japonica 'Crotonifolia' p.304 (dry shade)
- Azara microphylla p.300 (dry shade)
- Buxus sempervirens 'Suffruticosa' p.310 (dry shade)
- Chaenomeles speciosa 'Moerloosei' p.305 (dry shade)
- Cornus alba 'Aurea' p.305 (damp conditions)
- Cornus sericea 'Flaviramea' p.305 (damp conditions)
- Mahonia japonica p.307 (dry shade)
- Rhododendron 'Kure-no-yuki' p.315 (dry shade)
- Sarcococca hookeriana var. digyna p.317 (dry shade)
- Viburnum opulus p.303 (damp conditions)

Medium-sized shrubs

Exochorda x *macrantha* 'The Bride'
Pure white, showy, saucer-shaped flowers on arching branches cover this spreading evergreen shrub in late spring, making a beautiful display. Mound-forming and wider than it is tall, it is suitable for growing as a specimen plant, although it can also be grown in a shrub border.

↕2m (6ft) ↔3m (10ft) ❄ ❄ ❄ H6 ☼ ◐ ◌ ◓

Fatsia japonica
The castor oil plant is valued for its bold evergreen foliage and architectural habit. Its long-stalked, palmate, shiny dark green leaves give a subtropical effect, while striking branched clusters of creamy-white flowers emerge in autumn, followed by small black berries. It is tolerant of coastal exposure.

↕↔1.5–4m (5–12ft) ❄ ❄ ❄ H6 ☼ ◐ ◌ ◓

Fuchsia magellanica
In frost-free regions, this deciduous shrub, the hardiest of the fuchsia species, can be grown on its own or as informal hedging. It carries small, lantern-like flowers with red tubes, long red sepals and purple petals, from midsummer through into autumn. The flowers are followed by black fruits.

↕to 3m (10ft) ↔ to 3m (10ft) ❄ ❄ H4 ☼ ◐ ◌ ◓

Hebe 'Midsummer Beauty'
Hebes are adaptable evergreen shrubs that suit a wide range of growing conditions, including containers. 'Midsummer Beauty', an upright, rounded shrub with purplish-brown stems and bright green leaves, bears tapering plumes of medium-sized, lilac-purple flowers from midsummer to late autumn.

↕2m (6ft) ↔1.5m (5ft) ❄ ❄ H4 ☼ ◌ ◓

Hibiscus syriacus 'Diana'
Large showy flowers are the main allure of hibiscus cultivars. They thrive in a sunny border and flower over a long period. 'Diana' is an erect, deciduous shrub with toothed, dark green leaves that produces trumpet-shaped, white flowers with wavy-margined petals, from late summer to mid-autumn.

↕3m (10ft) ↔2m (6ft) ❄ ❄ ❄ H5 ☼ ◌ ◓

Hydrangea arborescens 'Annabelle'
Excellent as specimen plants or in groups, in a mixed border or in containers, hydrangeas are versatile garden shrubs. 'Annabelle', one of the most elegant cultivars, is deciduous and, from summer to early autumn, bears large, spherical flowerheads, crowded with creamy-white flowers.

↕2.5m (8ft) ↔2.5m (8ft) ❄ ❄ ❄ H6 ☼ ◐ ◌ ◓

Hydrangea aspera Villosa Group
An impressive deciduous shrub with lance-shaped, downy dark green leaves that form an attractive background for the flattened lacecap flowerheads. Produced from late summer to autumn, the lacecaps have large, purple-blue central clusters with a ring of lilac-white flowers on the outer edge.

↕1–3m (4–10ft) ↔1–3m (4–10ft) ❄ ❄ ❄ H5 ☼ ◐ ◌ ◓

Hydrangea macrophylla 'Mariesii Lilacina'
This rounded, deciduous shrub is grown for its mauve-pink to blue, showy lacecap flowers, which appear from mid- to late summer. It makes a fine freestanding shrub, and is also useful for mass planting in shady areas. Leave the flowerheads on over winter to protect the plant from frost damage.

↕2m (6ft) ↔2.5m (8ft) ❄ ❄ ❄ H5 ☼ ◐ ◌ ◓

Hydrangea quercifolia SNOW QUEEN
The oak-leaved hydrangea is grown chiefly for its deeply lobed, dark green leaves, which turn magnificent tints of bronze and purple in autumn before falling. From midsummer to autumn, SNOW QUEEN, also known as 'Flemygea', produces large, white, conical flowerheads, which fade to pink as they age.

↕2m (6ft) ↔2.5m (8ft) ❄ ❄ ❄ H5 ☼ ◐ ◌ ◓

Indigofera heterantha
Elegant, fern-like, grey-green leaves clothe the arching branches of this spreading, multi-stemmed, deciduous shrub. From early summer through to autumn, small, purple-pink, pea-like flowers are carried in dense spikes. It thrives when fan-trained against a sunny wall, especially in colder areas.

↕2–3m (6–10ft) ↔2–3m (6–10ft) ❀ ❀ ❀ H5 ☼ ◊ ◔

Jasminum nudiflorum
Winter jasmine has long, slender, arching, leafless shoots bearing bright yellow flowers from winter to early spring. Oval, dark green leaves emerge after flowering. It is ideal for training on a low wall or trellis. Prune once flowering has finished to maintain a neat shape.

↕3m (10ft) ↔3m (10ft) ❀ ❀ ❀ H5 ☼ ☀ ◊

Kolkwitzia amabilis 'Pink Cloud'
A hardy, deciduous shrub, the beauty bush forms a dense twiggy shape. Bell-shaped pink flowers, with yellow-flushed throats, are borne in profusion from late spring to early summer. Pale, bristly seed clusters follow. It makes a fine freestanding shrub, but can be planted as an informal hedge.

↕3m (10ft) ↔4m (12ft) ❀ ❀ ❀ H6 ☼ ◊

Lavatera × clementii 'Barnsley'
Throughout the summer, this semi-evergreen mallow bears very pale, blush-pink, red-eyed flowers. The lobed leaves are grey-green and downy. The cultivar 'Bredon Springs' has a similar habit and flowering period, but the flowers are mauve-flushed and dusky pink. Both suit sandy soils.

↕2m (6ft) ↔2m (6ft) ❀ ❀ ❀ H5 ☼ ◊

Magnolia liliiflora 'Nigra'
One of the most reliable of all magnolias, this cultivar produces beautiful large, dark purple-red upright flowers in early summer and intermittently into the autumn. It is compact and deciduous, with glossy dark green leaves that provide a foil to the flowers. Grow as a specimen plant for the best effect.

↕3m (10ft) ↔2.5m (8ft) ❀ ❀ ❀ H6 ☼ ☀ ◊ ◔

Magnolia stellata
This graceful, deciduous shrub is slow-growing but well worth the wait. The star magnolia bears pure white, sometimes pink-flushed, star-shaped flowers in early spring, before the leaves emerge. A compact shrub, it is initially bushy and then spreading. Spring frosts may damage early blooms.

↕3m (10ft) ↔4m (12ft) ❀ ❀ ❀ H6 ☼ ☀ ◊ ◔

Mahonia japonica
Invaluable in a winter garden, this handsome evergreen shrub thrives in shady spots. Its spectacular, sharply-toothed, dark green leaves glow with rich red tints in winter. Arching spikes of fragrant, pale yellow flowers appear from late autumn to early spring, followed by blue-purple berries.

↕2m (6ft) ↔3m (10ft) ❀ ❀ ❀ H5 ☼ ☀ ◊ ◔

Myrtus communis 'Flore Pleno'
Myrtle is a sun-loving, evergreen Mediterranean shrub with aromatic foliage. Masses of pretty, fragrant white flowers appear in late summer. The double blooms of 'Flore Pleno' look like small pompons. It thrives in a sunny border and can also be planted in a container, but needs shelter in cold areas.

↕3m (10ft) ↔3m (10ft) ❀ ❀ H4 ☼ ◔ ●

SHRUBS FOR FOLIAGE INTEREST

Medium-sized shrubs

Nandina domestica

The fruit, flowers, and foliage of this evergreen shrub give it a long season of interest. The leaves have warm red tints in spring and autumn, and small star-shaped white flowers emerge in midsummer, followed by bright red berries. The cultivar 'Fire Power' is a compact form with bright red leaves.

‡2m (6ft) ↔1.5m (5ft) ❄ ❄ ❄ H5 ☼ ◐ ◊ ◗

Osmanthus × burkwoodii

This hardy evergreen shrub is grown for its glossy dark green leaves, and clusters of tiny, creamy-white trumpet-shaped flowers, which are sweetly scented and appear from mid-to late spring. Its dense habit makes it useful for hedging and topiary. Trim into shape after flowering.

‡3m (10ft) ↔3m (10ft) ❄ ❄ ❄ H5 ☼ ◐ ◊

Paeonia delavayi

In early summer, this magnificent tree peony produces single, cup-shaped, dark crimson flowers on long lax stems. The handsome, deeply cut, dark green leaves are tinged burgundy in spring. A stunning deciduous shrub for a mixed border; it does not tolerate being moved.

‡2m (6ft) ↔1.2m (4ft) ❄ ❄ ❄ H5 ☼ ◐ ◊ ◗

Philadelphus 'Belle Étoile'

Mock oranges are grown for their beautiful flowers, which are often scented and usually white. 'Belle Étoile' makes an arching, deciduous shrub with tapering leaves. Its fragrant white flowers are single with a maroon flush at the centre, and are freely produced from late spring to early summer.

‡1.2m (4ft) ↔2.5m (8ft) ❄ ❄ ❄ H6 ☼ ◐ ◊

Physocarpus opulifolius 'Diabolo'

Grown chiefly for its attractive purple foliage and upright red stems, this spreading deciduous shrub also produces clusters of small pinkish-white flowers in late spring, followed by maroon fruit. The peeling bark gives additional winter interest. Cut down to the ground in spring to rejuvenate.

‡2m (6ft) ↔2.5m (8ft) ❄ ❄ ❄ H7 ☼ ◐ ◊

Pieris japonica 'Blush'

A versatile evergreen shrub for acid soils, *P. japonica* has narrow, glossy leaves, which are an attractive coppery-red when young. Tassels of white flowers appear from early to mid-spring. The compact cultivar 'Blush' has dark green leaves and its pink-flushed white flowers open from dark pink buds.

‡2m (6ft) ↔2m (6ft) ❄ ❄ ❄ H5 ☼ ◐ ◊ ◗

Pyracantha SAPHYR JAUNE

Firethorns can be grown as a freestanding feature, against a wall, or for hedging. This cultivar, also known as 'Cadaune', is an upright, evergreen shrub with spiny branches, dark green leaves, and small, white late-spring flowers. Its bright yellow autumn fruits provide a flash of colour as winter approaches.

‡4m (12ft) ↔3m (10ft) ❄ ❄ ❄ H6 ☼ ◐ ◊

Ribes sanguineum 'Pulborough Scarlet'

Fairly upright when young, this flowering currant becomes spreading with maturity. It is a vigorous, deciduous shrub, with aromatic leaves and clusters of dark red, white-centred tubular flowers in spring, followed by round, blue-black berries. It is ideal for the back of a mixed border.

‡3m (10ft) ↔2.5m (8ft) ❄ ❄ ❄ H6 ☼ ◊

Rosa 'Geranium'

This spectacular shrub rose has arching branches and small, dark green leaves. A profusion of open, scarlet flowers, with prominent cream stamens is produced in summer, followed by blazing orange-red, bottle-shaped hips in autumn, which extend the season of visual interest.

‡2.5m (8ft) ↔1.5m (5ft) ❄ ❄ ❄ H6 ☼ ◐ ◊ ◗

Rubus 'Benenden'
An ornamental, deciduous member of the bramble family with arching, thornless stems, this shrub is grown for its large, pure white, rose-like flowers, which appear in abundance from late spring to early summer. It is suitable for a shady shrub border, and is very attractive to butterflies.

↕3m (10ft) ↔3m (10ft) ❋ ❋ ❋ H5 ☼ ☼ ○

Sambucus nigra 'Eva'
This graceful elder is attractive for most of the year. The dark purple lacy foliage provides colour contrast in a mixed border. Showy pale pink, lemon-scented, flattened flowerheads appear in midsummer, followed by dark red elderberries. Full sun is best for foliage colour. It is also sold as 'Black Lace'.

↕3m (10ft) ↔2m (6ft) ❋ ❋ ❋ H6 ☼ ☼ ○ ◐

Sambucus racemosa 'Plumosa Aurea'
A bushy plant with arching shoots; the deeply cut leaves, which are bronze in youth and mature to golden yellow, provide a bright splash of colour in a border. Small, creamy yellow flowers appear in mid-spring, followed by round, glossy red fruits in summer. The foliage may scorch in hot sun.

↕3m (10ft) ↔3m (10ft) ❋ ❋ ❋ H7 ☼ ☼ ○ ◐

Skimmia x confusa 'Kew Green'
In spring, this compact, mounded, evergreen shrub produces dense, conical heads of fragrant, creamy-white flowers above deep green, pointed, aromatic leaves. Suitable for a shady border or woodland garden, it also looks attractive in a container. An adaptable shrub, it can cope with polluted air.

↕3m (10ft) ↔1.5m (5ft) ❋ ❋ ❋ H5 ☼ ☀ ○ ◐

Spiraea nipponica 'Snowmound'
At its peak in early summer, this spiraea presents a marvellous display, with clusters of bowl-shaped white flowers carried all along the upper sides of the arching stems. Deciduous, fast-growing, and densely leaved, it forms a spreading shape and is perfect for growing near the back of a sunny mixed border.

↕2.5m (8ft) ↔2.5m (8ft) ❋ ❋ ❋ H6 ☼ ○ ◐

Viburnum x bodnantense
Useful for providing winter interest in a garden, this shrub produces clusters of scented, tubular, rose-tinted flowers on bare stems over a long season, from late autumn to spring. It is upright and deciduous, with toothed, dark green leaves. A range of cultivars is available; 'Deben' has white flowers.

↕3m (10ft) ↔2m (6ft) ❋ ❋ ❋ H6 ☼ ☼ ○ ◐

Viburnum carlesii 'Aurora'
Suitable for a border or woodland garden, this deciduous shrub is densely bushy with irregularly toothed, dark green leaves. 'Aurora' is mainly grown for its clusters of perfumed flowers, which emerge in mid-spring. The buds are initially red and then open up to the pink tubular blooms.

↕2m (6ft) ↔2m (6ft) ❋ ❋ ❋ H6 ☼ ☼ ○ ◐

Viburnum plicatum f. tomentosum 'Mariesii'
This viburnum has distinctive extended horizontal branches that create a striking architectural effect, which is best appreciated when the shrub is grown as a specimen plant in a lawn. The flowers are white and the heart-shaped, dark green leaves turn red-purple in autumn.

↕3m (10ft) ↔4m (12ft) ❋ ❋ ❋ H5 ☼ ☼ ○ ◐

SHRUBS FOR GROUND COVER

- *Calluna vulgaris* 'Gold Haze' p.310
- *Ceanothus thyrsiflorus* var. *repens* p.311
- *Cotoneaster dammeri* p.312
- *Cotoneaster horizontalis* p.312
- *Cotoneaster salicifolius* 'Gnom' p.312
- *Euonymus fortunei* 'Emerald Gaiety' p.312
- *Hebe pinguifolia* 'Pagei' p.313
- *Helianthemum* 'Wisley Primrose' p.313
- *Juniperus procumbens* p.313
- *Juniperus squamata* 'Blue Carpet' p.314
- *Lonicera pileata* p.314
- *Picea abies* 'Reflexa' p.315
- *Potentilla fruticosa* 'Dart's Golddigger' p.92
- *Prunus laurocerasus* 'Zabeliana' p.315
- *Santolina chamaecyparissus* p.92
- *Vinca major* p.93
- *Vinca minor* 'La Grave' p.317

Small shrubs

Artemisia arborescens
Grown for its silver-grey, feathery foliage, this evergreen shrub is tolerant of exposed sites and is useful in a coastal garden. It carries clusters of small yellow flowers in summer and autumn, but is most valued for its elegant leaves. It is also suitable for a herb or rock garden.

↕1m (3ft) ↔1.5m (5ft) ❄ ❄ H4 ☀ ◊

Ballota 'All Hallows Green'
Originally from the Mediterranean, ballota thrives in dry, free-draining, sunny sites and makes an attractive edging plant. This cultivar forms a bushy evergreen subshrub with heart-shaped, lime green leaves. Small, pale green flowers appear in midsummer. Trim in spring to keep the shrub compact.

↕60cm (24in) ↔75cm (30in) ❄ ❄ ❄ H5 ☀ ◊

Berberis x *stenophylla* 'Corallina Compacta'
This is a compact cultivar of the much larger evergreen shrub, *B.* x *stenophylla*, which can be grown as an informal hedge. Like its parent, it has arching stems and narrow, spine-tipped, dark green leaves. In late spring, small clusters of pale orange flowers open from red buds along the branches.

↕30cm (12in) ↔30cm (12in) ❄ ❄ ❄ H5 ☀ ◐ ◊

Berberis thunbergii 'Aurea'
Create a splash of colour in the garden with this compact, deciduous berberis, which has vivid yellow young foliage, maturing to yellow-green. Pale yellow flowers are produced along the branches in mid-spring, followed by glossy red fruit. Suitable for hedging, but the leaves may scorch in full sun.

↕1.5m (5ft) ↔2m (6ft) ❄ ❄ ❄ H7 ◐ ◊

Berberis thunbergii f. *atropurpurea* 'Atropurpurea Nana'
A dwarf, dome-shaped berberis with rounded, red-purple leaves, a dense, twiggy habit, and small bright red berries that are attractive to birds. It tolerates polluted air and is a very adaptable shrub, ideal for a border or a rock garden.

↕60cm (24in) ↔75cm (30in) ❄ ❄ ❄ H7 ☀ ◐ ◊

Berberis thunbergii f. *atropurpurea* 'Helmond Pillar'
This deciduous barberry has distinctive columnar stems and dark wine-red leaves, which turn bright red in autumn. Tiny yellow flowers appear in spring, followed by red berries. Its upright habit makes it useful for filling gaps in a border.

↕1.2m (4ft) ↔60cm (24in) ❄ ❄ ❄ H7 ☀ ◐ ◊

Buxus sempervirens 'Elegantissima'
Mainly grown for its foliage, box is easily clipped into shape, making it perfect for edging and topiary. 'Elegantissima' is a variegated cultivar and makes a dome-shaped bush with small, narrow, white-margined evergreen leaves. Tiny, star-shaped flowers appear in spring.

↕1.5m (5ft) ↔1.5m (5ft) ❄ ❄ ❄ H6 ◐ ◊

Buxus sempervirens 'Suffruticosa'
This compact, very slow-growing selection of box is good for hedging or screens, and is one of the best types for the structure of a knot garden or parterre. Its dense habit makes it easy to trim into different shapes. It prefers partial shade, but can tolerate full sun if it is not allowed to get too dry.

↕1m (3ft) ↔1.5m (5ft) ❄ ❄ ❄ H6 ◐ ◊

Calluna vulgaris 'Gold Haze'
Heathers are robust plants and make good low-maintenance ground cover. There are many cultivars to choose from, all derived from *C. vulgaris*, a hardy, bushy, evergreen shrub that grows on acid soils in the wild. 'Gold Haze' has pale yellow leaves and short spikes of white bell-shaped flowers.

↕60cm (24in) ↔45cm (18in) ❄ ❄ ❄ H7 ☀ ◊

Calluna vulgaris 'Spring Cream'
A compact heather with mid-green leaves, which are tipped with cream in spring, this cultivar produces short spikes of white bell-shaped flowers that remain from midsummer until late autumn. Along with other heathers, it is attractive to bees. Grow on a moist, but free-draining sunny bank.

↕35cm (14in) ↔45cm (18in) ❀ ❀ ❀ H7 ☀ ◊

Caryopteris x clandonensis 'Worcester Gold'
The small but vivid blue flowers are the main attraction of *Caryopteris*. The cultivar 'Worcester Gold' has lavender-blue flowers, which are produced from late summer to early autumn on the current year's shoots. They stand out against a dense mound of warm yellow, deciduous foliage.

↕1m (3ft) ↔1.5m (5ft) ❀ ❀ H4 ☀ ◊

Ceanothus x delilianus 'Gloire de Versailles'
Also known as California lilac, ceanothus are grown for their abundant blue, pink, or white flowers. 'Gloire de Versailles' is a fast-growing, bushy, deciduous shrub with finely-toothed, mid-green leaves. From midsummer to autumn, it produces loose bunches of scented, powder blue flowers.

↕1.5m (5ft) ↔1.5m (5ft) ❀ ❀ H4 ☀ ◊

Ceanothus thyrsiflorus var. repens
Also known as creeping blueblossom, this is a useful, low-growing, evergreen ceanothus. It forms a natural mound of glossy mid-green foliage and, in late spring, produces an abundance of fluffy, pale to dark blue flowers. A perfect shrub for the front of a border or to clothe a sunny bank.

↕1m (3ft) ↔2.5m (8ft) ❀ ❀ ❀ (borderline) H4 ☀ ◖

Ceratostigma willmottianum
This loosely-domed, deciduous shrub produces clusters of pale to mid-blue flowers from late summer through to autumn. The pointed, bristly leaves are initially mid- to dark green with purple margins and then turn red in autumn. It needs a warm, sunny sheltered site to thrive.

↕1m (3ft) ↔1.5m (5ft) ❀ ❀ ❀ (borderline) H4 ☀ ◊ ◖

Cistus x dansereaui 'Decumbens'
Rock roses prefer a sunny site and can be grown in beds or containers. The flowers, usually white or pink, only last a day but are carried in profusion. 'Decumbens' is a low-growing, spreading, evergreen shrub that bears large white flowers with a crimson blotch at the base of each petal.

↕60cm (24in) ↔1m (3ft) ❀ ❀ H4 ☀ ◊

Cistus x purpureus
The narrow, green leaves of this rounded, evergreen shrub make a good foil for the single, crinkled, dark pink flowers, which appear in succession throughout summer. Each petal has a crimson mark at the base. The stems are upright and red-flushed. It is drought-tolerant and needs a sunny site.

↕1m (3ft) ↔1m (3ft) ❀ ❀ H4 ☀ ◊

Convolvulus cneorum
With its silky, silvery leaves and stems, this convolvulus is an asset even when not in bloom. The delicate flowers emerge from pink buds from late spring to summer, and are white and funnel-shaped with yellow centres. In colder areas, grow in a pot and move into a conservatory or greenhouse over winter.

↕60cm (24in) ↔90cm (36in) ❀ ❀ H4 ☀ ◊

SHRUBS FOR SPRING INTEREST

Small shrubs

Coronilla valentina subsp. **glauca**
The leaves of this bushy, rounded evergreen shrub are an attractive blue-green and fleshy. From late winter to early spring, and again in late summer, fragrant, yellow, pea-like flowers appear, followed by slim pods. Either grow it in a shrub border or at the base of a warm, sunny wall.

‡80cm (32in) ↔80cm (32in) ❀ ❀ H4 ☼ ◊

Cotoneaster dammeri
Evergreen cotoneasters offer colour and texture all year round, and are at their best in autumn when the berries develop. *C. dammeri* is vigorous and spreading with long arching stems, and makes excellent ground cover. Small, white flowers are borne in early summer, followed in autumn by round red berries.

‡20cm (8in) ↔2m (6ft) ❀ ❀ ❀ H6 ☼ ☀ ◊

Cotoneaster horizontalis
Grown for the herringbone pattern formed by its branching stems, this decorative shrub is best grown as ground cover or trained against a wall. Small white flowers appear in spring, followed by bright red fruits, which make a brilliant display. The glossy, dark green deciduous leaves turn red in autumn.

‡1m (3ft) ↔1.5m (5ft) ❀ ❀ ❀ H7 ☼ ☀ ◊

Cotoneaster salicifolius 'Gnom'
This dwarf, evergreen shrub makes a prostrate, dense dome, with wide-spreading branches bearing small, slender, dark green leaves. In early summer, white flowers are produced and these are followed by clusters of bright red fruits in the autumn. It is a good choice for ground cover.

‡30cm (1ft) ↔2m (6ft) ❀ ❀ ❀ H6 ☼ ◊

Daphne cneorum
A low-growing, evergreen shrub with trailing branches and dense clusters of scented, pale to deep rose-pink flowers in late spring. The leaves are small, leathery, and dark green. Grow it in a border near a path or window, where its fragrance will be appreciated. It resents transplanting.

‡30cm (12in) ↔to 1m (3ft) ❀ ❀ ❀ H5 ☼ ☀ ◊

Daphne odora 'Aureomarginata'
This evergreen species of daphne is one of the most fragrant flowering shrubs for a winter garden. The variegated cultivar 'Aureomarginata' has leaves with narrow yellow margins. Clusters of pink trumpet-shaped flowers appear from midwinter to early spring, followed by red fruit.

‡1.5m (5ft) ↔1.5m (5ft) ❀ ❀ H4 ☼ ☀ ◊

Euonymus fortunei 'Emerald Gaiety'
Poor soil and full sun suit many *E. fortunei* cultivars, making them useful shrubs for difficult sites. They make good ground cover, and can be fan-trained against a wall if supported. The evergreen 'Emerald Gaiety' is compact and bushy, with bright green leaves with white margins, tinged pink in winter.

‡1m (3ft) ↔1.5m (5ft) ❀ ❀ ❀ H5 ☼ ☀ ◊

Euphorbia characias subsp. **wulfenii** 'John Tomlinson'
This striking evergreen shrub produces upright stems with grey-green leaves one year, followed the next spring by large showy heads of small, bright, yellow-green cup-shaped flowers, which last from early spring to early summer.

‡1.2m (4ft) ↔1.2m (4ft) ❀ ❀ H4 ☼ ◊

Hebe 'Great Orme'
Adaptable shrubs, hebes will grow in a wide range of garden situations, from a mixed border to a rock garden. 'Great Orme' is an open, rounded, evergreen shrub with deep purplish shoots and glossy, dark green leaves. Spikes of deep pink flowers, fading to white, appear from midsummer to mid-autumn.

‡1.2m (4ft) ↔1.2m (4ft) ❀ ❀ H4 ☼ ☀ ◊ ◊

Hebe macrantha

This evergreen is bushy, initially open-branched and then later spreading, with oval, fleshy, bright green leaves. In early summer, large white flowers are produced in clusters of three. It is suitable for a container or rock garden, and needs little or no pruning.

‡60cm (24in) ↔ 90cm (36in) ❅ H4 ☼ ☀ ○ ◐

Hebe ochracea 'James Stirling'

Hebes with small leaves lying flat against the stems are known as whipcords and make good rock garden plants. 'James Stirling' forms a dense, small bush, and has rich ochre-yellow leaves, which look especially attractive in winter. Small white flowers are produced in late spring and early summer.

‡45cm (18in) ↔ 60cm (24in) ❅ H4 ☼ ☀ ○ ◐

Hebe pinguifolia 'Pagei'

An evergreen, semi-prostrate shrub, 'Pagei' has small, slightly cupped blue-green leaves. Short spikes of delicate pure white flowers emerge in profusion in late spring or early summer. It is an excellent plant for a rock garden or for ground cover, and needs little or no pruning. It flowers best in full sun.

‡30cm (12in) ↔ 90cm (36in) ❅ ❅ ❅ H5 ☼ ☀ ○ ◐

Hebe 'Red Edge'

A decorative small shrub, 'Red Edge' has grey-green leaves that have narrow red margins and veins when the foliage is young. Lilac-blue flowers, which fade to white, are produced in spikes in summer. It is mound-forming and makes an attractive plant for edging, or for the front of a border.

‡45cm (18in) ↔ 60cm (24in) ❅ H4 ☼ ☀ ○ ◐

Helianthemum 'Wisley Primrose'

Also known as rock roses, helianthemums are sun-loving, carpeting plants that thrive in a rock garden or on a sunny bank. 'Wisley Primrose' forms low hummocks of evergreen, grey-green foliage, and bears plenty of saucer-shaped, pale yellow flowers with deep yellow centres, throughout summer.

‡to 30cm (12in) ↔ to 45cm (18in) ❅ ❅ H4 ☼ ☀ ○

Helichrysum italicum subsp. serotinum

The curry plant is a low-growing, evergreen subshrub with woolly stems and intensely aromatic, slim, silver-grey leaves. From summer to autumn, it produces dark yellow flowers, which many designers remove if using the plant for its foliage. One of the best silver shrubs for a dry, sunny site.

‡60cm (24in) ↔ 1m (3ft) ❅ ❅ H4 ☼ ○

Juniperus x pfitzeriana 'Pfitzeriana Aurea'

Junipers are hardy conifers, tolerant of a wide range of soils and growing conditions. J. x pfitzeriana is a spreading shrub, eventually forming a flat-topped bush with tiered foliage. 'Pfitzeriana Aurea' has golden yellow leaves, which turn yellowish-green over winter. Junipers need little pruning.

‡90cm (36in) ↔ 2m (6ft) ❅ ❅ ❅ H6 ☼ ○

Juniperus procumbens

Creeping juniper is a dwarf species with long, stiff branches that intertwine to form a mat, making it excellent as ground cover and in rock gardens. It has needle-like, bluish-green leaves, and small brown or black berry-like cones. It grows best in a sunny, open position.

‡to 50cm (20in) ↔ to 2m (6ft) ❅ ❅ ❅ H7 ☼ ○

SHRUBS FOR SUMMER COLOUR

Small shrubs

Juniperus squamata 'Blue Carpet'
The wide-spreading stems of this vigorous, prostrate juniper create a wide, undulating, low mat of prickly foliage, making it an excellent plant for ground cover. The cultivar 'Blue Carpet' is fast-growing, with needle-like, aromatic leaves that are a bright steely blue.

↕30cm (12in) ↔2–3m (6–10ft) ❋ ❋ ❋ H7 ☼ ◊

Lavandula angustifolia 'Munstead'
This evergreen, compact, bushy lavender has narrow, aromatic, grey-green leaves. From mid- to late summer, dense spikes of small, fragrant blue-purple flowers are produced on long stalks. Lavenders prefer warm conditions but suit a variety of situations, from a shrub border to a rock garden.

↕45cm (18in) ↔60cm (24in) ❋ ❋ ❋ H5 ☼ ◊

Lavandula stoechas
French lavender is a compact shrub that blooms from late spring to summer. Dense spikes of fragrant dark purple flowers, topped by distinctive rose-purple bracts, are carried on long stalks above the silvery-grey leaves. It grows best in a warm, sunny site, and also makes a good container plant.

↕60cm (24in) ↔60cm (24in) ❋ ❋ H4 ☼ ◊

Lonicera nitida 'Baggesen's Gold'
This decorative, dense evergreen shrub has long arching shoots and masses of tiny bright yellow leaves. Small yellow-green flowers are borne in mid-spring and are occasionally followed by purplish fruits. Its golden foliage will brighten up a border, or it can be planted as a hedge.

↕1.5m (5ft) ↔1.5m (5ft) ❋ ❋ ❋ H6 ☼ ◊

Lonicera pileata
With its wide-spreading habit, the shrubby honeysuckle is a good plant for ground cover. It is a low-growing evergreen with narrow dark green leaves, and in late spring it produces tiny, funnel-shaped, creamy-white flowers, which are occasionally followed by purple fruits.

↕60cm (24in) ↔2.5m (8ft) ❋ ❋ ❋ H6 ☼ ◊

Origanum 'Kent Beauty'
A pretty addition to a rock garden or container, this decorative subshrub (a cultivar of the herb oregano) has slender trailing stems and smooth aromatic leaves. In late summer, hop-like clusters of pale pink flowers appear above rose-tinted green bracts. It prefers a sunny position.

↕10cm (4in) ↔to 20cm (8in) ❋ ❋ H4 ☼ ◊

Perovskia 'Blue Spire'
Russian sage forms a clump of grey-green toothed leaves. In late summer, grey-white upright stems carry elegant spires of small, tubular purple-blue flowers. An eyecatching plant for a border, it looks particularly effective when planted in groups. The frosty-looking stems are attractive in winter.

↕1.2m (4ft) ↔1m (3ft) ❋ ❋ ❋ H5 ☼ ◊

Phlomis fruticosa
A mound-forming evergreen shrub, Jerusalem sage has aromatic, wrinkled, grey-green leaves, which are woolly underneath, and produces short spikes of hooded dark yellow flowers from early to midsummer. It looks effective when massed in a border, and also suits a sunny gravel garden.

↕1m (3ft) ↔1.5m (5ft) ❋ ❋ ❋ H5 ☼ ◊

Phygelius × rectus 'African Queen'
This upright evergreen shrub has dark green leaves and graceful upward-curving branches. The pendent tubular flowers produced by the cultivar 'African Queen' are brightly coloured: pale red with orange-red lobes and yellow mouths. Deadhead regularly to encourage further flowering.

↕1m (3ft) ↔1.2m (4ft) ❋ ❋ ❋ H5 ☼ ◊ ◖

Picea abies 'Reflexa'

This is an unusual creeping variety of Norway spruce with red-brown bark and blunt, dark green needle-like leaves. The long trailing branches form a dense spreading carpet, making this an excellent conifer for ground cover. It needs a sunny position to thrive.

↕to 15cm (6in) ↔ indefinite ❋ ❋ ❋ H7 ☀ ○ ◐

Pinus mugo 'Mops'

The evergreen dwarf mountain pine forms a spherical mound of thick branches bearing dark green needles and brown cones. It grows best in a sunny position and would suit a rock garden or large container; the shrub's rounded shape also creates a cloud-like effect when it is planted en masse.

↕1m (3ft) ↔ 1m (3ft) ❋ ❋ ❋ H7 ☀ ○

Potentilla fruticosa 'Abbotswood'

In summer and early autumn, this low, domed shrub is covered with small white flowers, set against a background of divided, dark blue-green leaves. Shrubby potentillas are compact, bushy, deciduous plants and their long flowering season makes them ideal for a mixed border or a low hedge.

↕75cm (30in) ↔ 1.2m (4ft) ❋ ❋ ❋ H7 ☀ ○

Potentilla fruticosa 'Goldfinger'

There are numerous cultivars of shrubby potentilla to choose from, with flower colours ranging from white, yellow, and orange to shades of pink and red. 'Goldfinger' is covered in large, saucer-shaped, rich yellow flowers, from late spring to autumn, and has small deep green leaves.

↕1m (3ft) ↔ 1.5m (5ft) ❋ ❋ ❋ H7 ☀ ○

Prunus × cistena

Valued for its foliage and flowers, this ornamental cherry is a slow-growing, upright deciduous shrub with glossy oval leaves that are red when young, maturing to a rich purple-red. Delicate white flowers appear in late spring, before the leaves, and may be followed by dark purple fruit.

↕1.5m (5ft) ↔ 1.5m (5ft) ❋ ❋ ❋ H6 ☀ ○ ◐

Prunus laurocerasus 'Zabeliana'

The cherry laurel is an evergreen bushy shrub, which looks its best in spring when long spikes of cup-shaped, fragrant white flowers appear. 'Zabeliana' has a low, wide-spreading habit, making it suitable for ground cover. The flowers are followed by red, cherry-like fruits, which later turn black.

↕1m (3ft) ↔ 2.5m (8ft) ❋ ❋ ❋ H5 ◐ ☀ ○ ◐

Rhododendron 'Golden Torch'

This small evergreen shrub has medium-sized leaves and is popularly grown for its trusses of flowers, which emerge as salmon-pink buds and open to funnel-shaped, pale creamy-yellow blooms in late spring and early summer. Rhododendrons need acid soil and some shade to thrive.

↕1.5cm (5ft) ↔ 1.5m (5ft) ❋ ❋ ❋ H6 ◐ ○ ◐

Rhododendron 'Kure-no-yuki'

A dwarf azalea with a compact habit, 'Kure-no-yuki' has small leaves and produces clusters of pure white flowers in mid-spring. Azaleas prefer sheltered conditions in deep, acid soil and do best in a woodland garden in dappled shade. This cultivar would make a pretty feature in a Japanese garden.

↕1m (3ft) ↔ 1m (3ft) ❋ ❋ ❋ H5 ◐ ○ ◐

AUTUMN- AND WINTER-FLOWERING SHRUBS

- Azara microphylla (late winter to early spring) p.300
- Chimonanthus praecox 'Grandiflorus' (winter) p.300
- Cornus mas (winter) p.300
- Coronilla valentina subsp. glauca (late winter to early spring) p.312
- Elaeagnus × ebbingei 'Gilt Edge' (mid- to late autumn) p.301
- Hamamelis × intermedia 'Pallida' (mid- and late winter) p.302
- Jasminum nudiflorum (winter to early spring) P.307
- Mahonia japonica (late autumn to early spring) P.307
- Mahonia × media 'Charity' (late autumn to late winter) p.302
- Sarcococca hookeriana var. digyna (winter) p.317
- Viburnum × bodnantense (late autumn to spring) p.309

Small shrubs

Rosa ANNA FORD
There are roses for virtually every situation, but whether they are grown in pots, against a wall, or in a border, most prefer a sunny site. This is a compact, dwarf floribunda rose with dark green leaves and semi-double, orange-red blooms that appear over a long season from summer to autumn.

↕45cm (18in) ↔40cm (16in) ✿ ✿ ✿ H6 ☼ ◊ ◗

Rosa 'Golden Wings'
This bushy, spreading shrub rose is suitable for hedging or a border. It has prickly stems and light green leaves, and bears cupped, fragrant, single pale yellow flowers from summer to autumn. A position in full sun will encourage repeat flowering. Apple green hips follow the flowers.

↕1.1m (3.5ft) ↔1.3m (4.5ft) ✿ ✿ ✿ H6 ☼ ◊ ◗

Rosa PEARL DRIFT
A vigorous shrub rose, spreading in habit, PEARL DRIFT produces clusters of lightly scented, semi-double, pale pink flowers from summer to autumn, against a background of glossy dark green leaves. It is ideal for a mixed cottage-style border, and is also sold under the official cultivar name of 'Leggab'.

↕1m (3ft) ↔1.2m (4ft) ✿ ✿ ✿ H6 ☼ ◊ ◗

Rosa 'The Fairy'
Suited to a border or a container, 'The Fairy' is a small shrub rose with a dense cushion-forming habit. The thorny stems are covered with small, glossy leaves, and from late summer to autumn it produces sprays of small, double, pink flowers.

↕60–90cm (24–36in) ↔60–90cm (24–36in)
✿ ✿ ✿ H6 ☼ ◊ ◗

Rosa WILDEVE
This robust rose has long, arching stems and forms a bushy shrub. The flower buds are pink, and open to fully-double, apricot-flushed pink fragrant blooms, which appear from late spring to early summer. Grow WILDEVE in a mixed border, or use for hedging. Its official cultivar name is 'Ausbonny'.

↕1.1m (3.5ft) ↔1.25m (2.5ft) ✿ ✿ ✿ H7 ☼ ◊ ◗

Rosmarinus officinalis
Rosemary is a tough evergreen Mediterranean shrub, grown for its aromatic leaves. It forms an attractive upright plant with slim, leathery leaves, and produces tubular, purple-blue to white flowers from mid-spring to early autumn. It needs a well-drained site and suits a rock or herb garden.

↕1.5m (5ft) ↔1.5m (5ft) ✿ ✿ ✿ H4 ☼ ◊

Ruta graveolens
This evergreen subshrub, also known as common rue, is grown for its aromatic, deeply divided blue-green leaves and is sometimes used as a medicinal herb. Cup-shaped yellow flowers appear in summer. 'Jackman's Blue' is a compact cultivar with intensely glaucous foliage.

↕1m (3ft) ↔75cm (30in) ✿ ✿ ✿ H5 ☼ ◊

Salvia microphylla
From late summer to autumn this salvia bears crimson flowers among its mid- to deep green leaves. It makes a colourful addition to a late season border or herb garden, but needs a sunny site to produce its best flower display.

↕90–120cm (36–48in) ↔60–100cm (24–39in)
✿ ✿ H4 ☼ ◊

Salvia officinalis 'Purpurascens'
The aromatic downy leaves of this shrubby evergreen or semi-evergreen perennial are purple when young, and later greyish-green. Purple sage is used as a culinary herb but is also decorative in a gravel garden or mixed border. Blue purple flowers are borne on spikes in early and midsummer.

↕to 80cm (32in) ↔1m (3ft) ✿ ✿ H4 ☼ ◊

Salvia officinalis 'Tricolor'

This cultivar of the common sage has grey-green leaves with creamy-white margins, flushed pink when young. It makes a compact plant and colours best in a sunny site. The leaves are aromatic and can be used for culinary purposes, while the flowers are attractive to bees and butterflies.

↕ to 80cm (32in) ↔ 1m (3ft) ❄ ❄ ❄ H5 ☀ ◊

Santolina pinnata subsp. neapolitana 'Sulphurea'

An evergreen shrub native to the Mediterranean, santolina forms a low, domed shape. The primrose-yellow, tubular flowers form button-like heads on long stems above narrow, feathery, grey-green leaves. It is useful as edging, and as part of a Mediterranean-style scheme.

↕ 75cm (30in) ↔ 1m (3ft) ❄ ❄ ❄ H5 ☀ ◊

Sarcococca hookeriana var. digyna

The robustness of this winter-flowering evergreen makes it a useful shrub for difficult sites in the garden, as it will tolerate dry shade and air pollution, and needs very little attention. It has slender, tapered dark green leaves and is prized for its highly fragrant white flowers, followed by black fruit.

↕ 1.5m (5ft) ↔ 2m (6ft) ❄ ❄ ❄ H5 ☀ ☀ ◊ ●

Viburnum × burkwoodii 'Anne Russell'

This compact, deciduous or semi-evergreen shrub produces clusters of intensely fragrant white flowers from mid- to late spring. 'Anne Russell' is suited to growing in a shrub border or woodland garden; plant it close to a seating area or pathway to make the most of its spring scent.

↕ 1.5m (5ft) ↔ 1.5m (5ft) ❄ ❄ ❄ H6 ☀ ☀ ◊ ●

Viburnum davidii

This evergreen shrub forms a dome of dark green gleaming foliage on branching stems. The flowers appear above the deeply veined, oval leaves in late spring, producing flattened heads of small white blooms. Where male and female plants are grown together, metallic-blue fruits form on the female.

↕ 1–1.5m (3–5ft) ↔ 1–1.5m (3–5ft) ❄ ❄ ❄ H5 ☀ ☀ ◊ ●

Vinca minor 'La Grave'

Woodland plants in the wild, periwinkles bear decorative, star-shaped flowers on slender stems. The evergreen foliage and pretty flowers make attractive ground cover, although they can be invasive and may need cutting back regularly. 'La Grave' (also seen as 'Bowles's Blue') has lavender-blue flowers.

↕ 10–20cm (4–8in) ↔ indefinite ❄ ❄ ❄ H6 ☀ ☀ ◊ ●

Weigela florida 'Foliis Purpureis'

This is a dark-leaved cultivar of the deciduous, arching shrub Weigela florida. Funnel-shaped flowers, deep pink on the outside and pale pink to white inside, are produced in late spring and early summer, and look striking against the tapered bronze-green foliage. Grow in a mixed or shrub border.

↕ 1m (3ft) ↔ 1.5m (5ft) ❄ ❄ ❄ H6 ☀ ☀ ◊

Yucca filamentosa 'Bright Edge'

A dramatic architectural plant, the yucca suits a hot, dry site, making it a good specimen plant for a warm courtyard. Yucca filamentosa produces stems of bell-shaped white flowers, tinged green, from mid- to late summer. The leaves of 'Bright Edge' have broad yellow margins.

↕ 75cm (30in) ↔ 1.5m (5ft) ❄ ❄ ❄ H5 ☀ ◊

EVERGREEN SHRUBS

- Aucuba japonica 'Crotonifolia' p.304
- Azara microphylla p.300
- Berberis darwinii p.304
- Berberis julianae p.304
- Camellia japonica 'Bob's Tinsie' p.304
- Camellia reticulata 'Leonard Messel' p.300
- Ceanothus 'Concha' p.304
- Choisya × dewitteana 'Aztec Pearl' p.305
- Cotoneaster lacteus p.301
- Daphne bholua 'Jacqueline Postill' p.305
- Elaeagnus × ebbingei 'Gilt Edge' p.301
- Escallonia 'Apple Blossom' p.305
- Fatsia japonica p.306
- Itea ilicifolia p.302
- Ligustrum ovalifolium 'Aureum' p.302
- Olearia macrodonta p.303
- Osmanthus × burkwoodii p.308
- Pieris japonica 'Blush' p.308
- Rhamnus alaternus 'Argenteovariegata' p.303
- Skimmia × confusa 'Kew Green' p.309

Climbers

Actinidia kolomikta
This deciduous climber's main attraction is the masses of purple-tinged young leaves, which later turn dark green with distinctive pink and silver splashes. Small, slightly scented white flowers appear in early summer. Although it is slow to establish, it is well worth the wait.

↕5m (15ft) ❄ ❄ ❄ H5 ☼ ◊

Akebia quinata
Also known as the chocolate vine, *A. quinata* is a vigorous semi-evergreen with attractive leaves and strong, twining stems. Clusters of cup-shaped, purplish female flowers in spring are followed by unusual sausage-shaped fruits. Grow against a wall or train into a tree or pergola.

↕10m (30ft) ❄ ❄ ❄ H6 ☼ ☼ ◊ ◊

Ampelopsis brevipedunculata
This vigorous, deciduous climber is valued for its attractive foliage and ornamental berries. The small summer flowers are green, and are followed by eye-catching, round, pinkish purple berries, which later turn a clear blue. Ideal for a warm, sheltered wall since fruiting is best in a sunny site.

↕5m (15ft) ❄ ❄ ❄ H6 ☼ ☼ ◊ ◊

Campsis x tagliabuana 'Mme Galen'
The trumpet creeper is a fast-growing, deciduous climber, which clings by aerial roots. In late summer or early autumn, 'Mme Galen' bears clusters of tubular, reddish-orange flowers that look striking against the rich green divided leaves. It may take a few seasons to establish.

↕3–5m (10–15ft) ❄ ❄ H4 ☼ ◊ ◊

Clematis armandii
This popular clematis is a vigorous climber and one of the hardiest of the evergreen species, bearing glossy, dark green leaves and producing masses of small, white scented flowers in early spring. It prefers a sunny, sheltered site and will clothe a wall or shed with ease.

↕3–5m (10–15ft) ❄ ❄ H4 ☼ ☼ ◊ ◊

Clematis 'Bill MacKenzie'
A vigorous, scrambling clematis, 'Bill MacKenzie' has small, single, yellow lantern-like nodding flowers in late summer and autumn, followed by large silky seedheads. The plant needs support from wires or netting, or leave it to scramble through shrubs and trees.

↕7m (22ft) ❄ ❄ ❄ H6 ☼ ☼ ◊ ◊

Clematis 'Étoile Violette'
From midsummer to late autumn, this deciduous viticella clematis produces masses of small, nodding, deep violet flowers with cream stamens. Flowers are produced on the current year's growth. 'Étoile Violette' can be grown through other shrubs or on a wall or fence.

↕3–5m (10–15ft) ❄ ❄ ❄ H6 ☼ ☼ ◊ ◊

Clematis florida var. florida 'Sieboldiana'
This deciduous or semi-evergreen clematis bears showy, single creamy white flowers with a distinctive domed cluster of purple stamens in late spring or summer. It does best in a warm, sunny sheltered location where its roots are shaded and moist. It is also suitable for growing in large containers.

↕2–2.5m (6–8ft) ❄ ❄ H3 ☼ ☼ ◊ ◊

Clematis 'Huldine'
A vigorous, deciduous, summer-flowering clematis, well suited to walls and fences. The small, cup-shaped, almost translucent white flowers with pale mauve margins and a mauve stripe beneath appear in summer. They are particularly attractive in sunshine when the stripes are more evident.

↕3–5m (10–15ft) ❄ ❄ ❄ H6 ☼ ☼ ◊ ◊

Clematis 'Markham's Pink'

This early-flowering macropetala clematis is vigorous and prolific, producing masses of bell-shaped, double, rich pink flowers from spring to early summer, followed by silky seedheads in autumn. Try growing through a shrub or small tree, or against a wall or fence.

↕2.5–3.5m (8–11ft) ❀ ❀ ❀ H6 ☼ ☼ ◊ ◖

Clematis montana var. rubens

White-flowered *Clematis montana* is a popular favourite: easy to grow, vigorous, and very adaptable to a wide variety of garden conditions. Many cultivars are available, including this pale pink flowering form, which bears a mass of four-petalled flowers with cream anthers in late spring and early summer.

↕10m (30ft) ❀ ❀ ❀ H5 ☼ ☼ ◊ ◖

Clematis 'The President'

A free-flowering early clematis, 'The President' produces large, single, rich blue-purple flowers in summer, followed by spiky seedheads. It suits pergolas and fences but its compact habit also makes it ideal for large containers. It makes a good partner for climbing roses that flower at the same time.

↕2–3m (6–10ft) ❀ ❀ ❀ H6 ☼ ☼ ◊ ◖

Eccremocarpus scaber

The Chilean glory flower is an evergreen, perennial, fast-growing climber with attractive ferny leaves. In warmer areas it will quickly clothe a trellis or pergola, or scramble through a large shrub or small tree. From late spring to autumn, spikes of orange-red tubular flowers appear.

↕3–5m (10–15ft) ❀ ❀ H3 ☼ ◊

Hardenbergia violacea

The purple coral pea is a strong-growing Australian native and does best in a sunny position outdoors, but is suitable for a greenhouse in cold regions. From late winter to early summer, clusters of violet pea-like flowers appear against the leathery rich green leaves.

↕2m (6ft) or more ❀ ❀ H3 ☼ ☼ ◊

Hedera colchica 'Sulphur Heart'

The Persian ivy cultivars 'Sulphur Heart' and 'Dentata Variegata' have similar large light green leaves with cream splashes. 'Sulphur Heart' (also known as 'Paddy's Pride') grows more rapidly, however, and the slightly more elongated leaves are splashed with creamy yellow.

↕5m (15ft) ❀ ❀ ❀ H5 ☼ ◊ ◖

Hedera helix 'Oro di Bogliasco'

This striking ivy, also known as 'Goldheart', has dark, glossy evergreen leaves with a gold central splash. A self-clinging climber, it makes an excellent wall ivy, slow to establish but then fast-growing. Unlike most variegated ivies, it will tolerate shade.

↕8m (25ft) ❀ ❀ ❀ H5 ☼ ☼ ◊ ◖

Hedera helix 'Parsley Crested'

As its name suggests, this ivy has dark green leaves with waved and crested margins. A vigorous, evergreen self-clinging climber with thick upright stems, it is hardy, easy to grow, and ideal for garden walls and fences, although its aerial roots may damage old brickwork.

↕2m (6ft) ❀ ❀ ❀ H5 ☼ ☼ ◊ ◖

CLIMBERS FOR SPRING AND SUMMER FLOWERS

Climbers

Humulus lupulus 'Aureus'
Hops make a good choice for shady walls and fences, although *H. lupulus* 'Aureus' produces its best leaf colour in sun. This strong-growing, herbaceous perennial climber has yellow-green, boldly lobed leaves and hairy, twining stems; spikes of female flowers (hops) appear in late summer.

‡6m (20ft) ❋ ❋ ❋ H6 ☼ ◐ ◊ ◖

Hydrangea anomala subsp. *petiolaris*
The climbing hydrangea is vigorous and produces large, open lacecap heads of creamy-white flowers in summer, on a background of broad, rounded leaves. The stems have rich brown peeling bark. Young plants need support until they are established; they then climb by self-clinging aerial roots.

‡15m (50ft) ❋ ❋ ❋ H5 ☼ ◐ ◊ ◖

Jasminum officinale 'Argenteovariegatum'
Strong-growing and semi-evergreen, climbing jasmine has pretty, ferny foliage and bears clusters of strongly scented, white star-shaped flowers in summer. The variegated cultivar 'Argenteovariegatum' has finely divided, grey-green leaves with cream margins.

‡12m (40ft) ❋ ❋ ❋ H5 ☼ ◐ ◊

Lonicera periclymenum Serotina Group
A twining, vigorous climber, the late Dutch honeysuckle can be grown alone or through a small tree or shrub. The spring foliage is lush and new shoots are purple when young. In summer, it produces long-tubed fragrant creamy white flowers streaked with dark red-purple.

‡7m (22ft) ❋ ❋ ❋ H6 ☼ ◐ ◊

Parthenocissus henryana
This deciduous ornamental vine, sometimes known as the Chinese Virginia creeper, clings to surfaces by the adhesive tips of its tendrils, making it a useful climber for growing on a wall. It produces the best colour in partial shade, its silver-veined leaves turning a rich red in autumn before they fall.

‡10m (30ft) ❋ ❋ ❋ (borderline) H4 ☼ ◐ ◊ ◖

Parthenocissus tricuspidata 'Veitchii'
Also known as Boston ivy, *P. tricuspidata* is vigorous and woody, and will clothe a wall or other support quite quickly, clinging without assistance. The cultivar 'Veitchii' is noted for its autumn colour, when the mid-green ivy-like leaves turn a deep red-purple before falling.

‡20m (70ft) ❋ ❋ ❋ H5 ☼ ◊

Passiflora caerulea
A good climber for a sunny, warm wall or fence, the blue passion flower is fast-growing, with rich green divided leaves. The striking flowers are usually white, with purple, blue and white coronas. The orange-yellow fruits that follow are decorative, but not edible.

‡10m (30ft) or more ❋ ❋ H4 ☼ ◊ ◖

Rosa 'Compassion'
A hybrid tea rose, 'Compassion' is an upright, freely branching climber with dark green leaves. The flowers are rounded and fully double, salmon pink tinged with apricot, and fragrant. They appear from summer to autumn; deadheading will prolong the flowering season. It is a good choice for a wall.

‡3m (10ft) ❋ ❋ ❋ H6 ☼ ◊ ◖

Rosa 'Félicité Perpétue'
This rambler is a semi-evergreen rose with long, slender stems and dark green leaves. The summer flowers are fully double, pale pink in bud and opening to faintly pink-tinged white. It is a beautiful rose for an arch or arbour, or it can be grown through a shrub or small tree.

‡to 5m (15ft) ❋ ❋ ❋ H6 ☼ ◊ ◖

Rosa 'Golden Showers'

Cupped, double to semi-double, lightly fragrant yellow flowers are borne on this upright climbing rose from summer to autumn, providing a long-lasting display against the glossy leaves. 'Golden Showers' will tolerate shady conditions, and works well with blue- and purple-flowered clematis.

↕ to 3m (10ft) ※ ※ ※ H6 ☼ ☼ ◊ ◗

Schizophragma integrifolium

Schizophragmas are slow-growing and mainly cultivated for their hydrangea-like blooms – flattened heads of creamy-white flowers with conspicuous, oval cream-coloured bracts, which appear in summer among the pointed green leaves. The plant will attach itself to a wall surface by aerial roots.

↕ 12m (40ft) ※ ※ ※ H5 ☼ ☼ ◊

Solanum crispum 'Glasnevin'

Vigorous and scrambling, S. crispum is a good choice for a warm, sunny wall or fence. The cultivar 'Glasnevin' produces sprays of long-lasting, deep purple-blue, star-shaped flowers from summer to autumn, and is evergreen in warmer areas. It is ideal for training through a shrub or small tree.

↕ 6m (20ft) ※ ※ H3 ☼ ☼ ◊ ◗

Solanum laxum 'Album'

Known as the potato vine, S. laxum is a scrambling semi-evergreen or evergreen climber which produces clusters of lightly fragrant flowers over a long season from summer to autumn. The cultivar 'Album' is a white-flowered form of the normally blue-flowered plant.

↕ 6m (20ft) ※ ※ H3 ☼ ◊ ◗

Tropaeolum speciosum

The flame nasturtium has fleshy, twining stems and long-stalked divided leaves, and is an excellent plant to train into trees, shrubs or hedges, where its brilliant colour will contrast with the green foliage. Long-spurred scarlet flowers appear from summer into autumn, followed by spherical blue fruits.

↕ to 3m (10ft) or more ※ ※ ※ H5 ☼ ☼ ◗

Vitis coignetiae

This ornamental vine is grown for its decorative foliage and vivid autumn colour. It is a vigorous, deciduous climber with large, heart-shaped leaves, brown-felted beneath, that turn bright red in autumn. Small, inedible, blue-black grapes appear at the same time. Train into a tree or shrub, or over a pergola.

↕ 15m (50ft) ※ ※ ※ H5 ☼ ☼ ◊

Vitis vinifera 'Purpurea'

An ideal climber for a warm, sunny wall or fence, the claret vine is a vigorous form of the grape vine, but is grown for its autumn foliage rather than the inedible grapes. It is a woody deciduous vine with toothed leaves which are grey at first, then mid-purple, turning a very deep purple in autumn.

↕ 7m (22ft) ※ ※ ※ H5 ☼ ☼ ◊ ◗

Wisteria floribunda 'Multijuga'

Showy, pendent spikes of pea-like early summer flowers make wisterias popular with garden designers. W. floribunda (Japanese wisteria) is a vigorous, twining climber with pretty leaves, available as a range of cultivars: 'Multijuga' bears fragrant, lilac-blue blooms; 'Alba' has white flowers.

↕ 9m (28ft) or more ※ ※ ※ H6 ☼ ☼ ◊ ◗

CLIMBERS FOR FOLIAGE INTEREST AND COLOUR

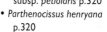

Tall perennials

Acanthus spinosus
From late spring through to midsummer, majestic spikes of white flowers sheltered by purple bracts rise from a bed of prickly, dark green leaves. This clump-forming perennial prefers rich soil and makes a striking architectural plant. Cut stems last well in flower arrangements.

↕1.5m (5ft) ↔60–90cm (24–36in) ❁ ❁ ❁ H5 ☼ ◐ ◌

Aconitum 'Spark's Variety'
Upright stems bearing deep violet, hooded flowers, well above the dark green, deeply divided leaves, identify this as one of the monkshoods. The flowers appear from mid- to late summer and perform best in moist, fertile soil, in a woodland garden or border. Taller plants may need staking. All parts are poisonous.

↕1.2–1.5m (4–5ft) ↔45cm (18in) ❁ ❁ ❁ H7 ◐ ◌

Agastache foeniculum
With its aniseed-scented leaves and spikes of violet-blue flowers from midsummer to early autumn, anise hyssop suits a mixed border. For paler-coloured flowers, try the cultivar 'Alabaster', which has delicate white blooms. Both prefer a sunny site, thriving in rich, fertile soil.

↕90–150cm (3–5ft) ↔30cm (12in) ❁ ❁ H3 ☼ ◌

Anemone × hybrida
The Japanese anemone bears semi-double, pink flowers on wiry stems from late summer to mid-autumn. The white-flowered 'Honorine Jobert' will shine in any border and like the other Japanese anemones, prefers rich soil. It dislikes cold, wet conditions during winter months.

↕1.2–1.5m (4–5ft) ↔indefinite ❁ ❁ ❁ H7 ☼ ◐ ◌

Asphodeline lutea
The yellow asphodel strikes a dominant pose in the border as its rocket-like spikes of star-shaped flowers stand above other late-spring perennials. Eye-catching blue-green leaves stud the length of each flower stem. Most well-drained soils will suit this clump-forming perennial.

↕1.5m (5ft) ↔30cm (12in) ❁ ❁ ❁ (boderline)H4 ☼ ◌

Cephalaria gigantea
The giant scabious needs a sizeable border for the best display of its tall flower stems bearing pale yellow, ruffled blooms in summer. Make the most of them by planting at the back of a border against a dark background, such as a conifer hedge or fence, for contrast.

↕to 2.5m (8ft) ↔60cm (24in) ❁ ❁ ❁ H7 ☼ ◐ ◌ ◌

Cirsium rivulare 'Atropurpureum'
The deep crimson flowers of this clump-forming perennial, coupled with its prickly green leaves, should make thistles more popular border plants than they are. Suited to damp conditions in a wild garden, they attract insects during the flowering season from early to midsummer.

↕1.2m (4ft) ↔60cm (24in) ❁ ❁ ❁ H7 ☼ ◌ ◌

Crambe cordifolia
Looking like a mass of confetti, the tiny white flowers of this perennial appear suspended in mid-air. The coarseness of the rich green leaves is softened by a cloud of blooms from late spring to midsummer. Crambes are suited to a wild garden and will tolerate coastal conditions. The flowers attract bees.

↕to 2.5m (8ft) ↔1.5m (5ft) ❁ ❁ ❁ H5 ☼ ◌

Cynara cardunculus
Few plants produce such large flowerheads as the cardoon. Fierce-looking bracts sit below brush-like flowerheads of blue-purple florets to create a dazzling summer and early autumn display. Protect plants from strong winds and in cold areas, mulch around the plant base.

↕1.5m (5ft) ↔1.2m (4ft) ❁ ❁ ❁ H5 ☼ ◌

Delphinium Blue Fountains Group
This hardy delphinium is the ideal choice for a windy garden as the plant is compact; the flowers grow to just 80cm (32in) high and do not need staking. Early summer blooms appear in short spikes in a variety of shades of blue, mauve-blue, and white. Suited to a cottage garden or mixed border.

‡80–100cm (32–39in) ↔ 60cm (24in) ❀ ❀ ❀ H5 ☼ ◗

Delphinium Pacific Hybrids
A cottage garden favourite, this tall perennial comes in a range of colours, including blue, pink, white, and violet. After the midsummer flowering, cut back the stems to encourage another flush of double flowers in late summer and early autumn. Protect from strong winds.

‡1.2–2m (4–6½ft) ↔ 90cm (36in) ❀ ❀ ❀ H6–H5 ☼ ○

Dierama pulcherrimum
The delightful name of angel's fishing rod perfectly suits this elegant perennial whose pendent, pink bells move gracefully in the slightest breeze against narrow, grass-like, green leaves. This combination looks good in the middle of a border or as edging alongside a pathway.

‡1–1.5m (3–5ft) ↔ 60cm (24in) ❀ ❀ H4 ☼ ○

Dryopteris wallichiana
Wallich's wood fern, named after the Danish plant collector, Nathaniel Wallich, is a deciduous fern that sports a shuttlecock-like array of young, green fronds with rusty-brown, furry mid-ribs, in spring. Provide shelter, shade and a generous depth of rich, moist soil.

‡90cm (36in) or more ↔ 75cm (30in) ❀ ❀ ❀ H5 ☼ ◗

Echinops bannaticus
The globe thistle is a good plant for a wild garden; it is very attractive to bees, with its spherical, blue flowerheads held above a spiny mass of grey-green leaves from mid- to late summer. The dense clumps can be divided from autumn to spring. The variety 'Taplow Blue' has powder-blue flowers.

‡0.5–1.2m (1½–4ft) ↔ 75cm (30in) ❀ ❀ ❀ H7 ☼ ○

Ensete ventricosum
This monster-sized plant brings a touch of the exotic to a garden. The Abyssinian banana has large, paddle-like, bright green leaves with bright red mid-ribs below. White flowers are borne in summer. Provide winter protection against frost.

‡6m (20ft) or more ↔ to 5m (15ft)

❀ H2 (min. 7°C/45°F) ☼ ☀ ◗

Eryngium agavifolium
An Argentinian plant, sea holly makes a dramatic silhouette in a border. Long, sword-shaped leaves, sharply toothed along their length, form rosettes from which the flowering stems emerge. The stalkless, greenish-white flowers form cone-like stuctures. Stems can be dried for flower arrangements.

‡1–1.5m (3–5ft) ↔ 60cm (24in) ❀ ❀ H4 ☼ ○ ◗

Foeniculum vulgare 'Purpureum'
This attractive perennial is well known for its aniseed-flavoured seeds and feathery mid-green leaves, which are used in cooking. Flat flowerheads of small yellow flowers appear from mid- to late summer. 'Purpureum' is hardier than the species and has striking bronze-purple foliage.

‡1.8m (6ft) ↔ 45cm (18in) ❀ ❀ ❀ H5 ☼ ○

PERENNIALS FOR ARCHITECTURAL INTEREST

Tall perennials

Helianthus 'Lemon Queen'
Sunflowers are always a good choice for the back of a border and this variety is no exception. Pale yellow flowers with a slightly darker eye mark this out as one of the more subtly coloured choices. Expect a long-lasting display from late summer to mid-autumn.

‡1.7m (5½ft) ↔ 1.2m (4ft) ❋ ❋ ❋ (borderline) H4 ☼ ◊ ◗

Helianthus 'Monarch'
The almost spidery blooms of this sunflower make it stand out from the usual crowd. Predominantly yellow with a pale brown eye, the semi-double flowers appear from early to mid-autumn. Long, hot summer months will promote a beautiful flowering display.

‡to 2m (6ft) ↔ 1.2m (4ft) ❋ ❋ ❋ H5 ☼ ◊ ◗

Inula magnifica
This fast-growing, clump-forming plant needs plenty of space in the garden. Large, frilly-petalled flowers are formed, up to 20 at a time, in late summer above a foil of dark green leaves with softly hairy undersides. Ideal for a wild garden, the plant likes sun but will tolerate damp soil.

‡to 1.8m (6ft) ↔ 1m (3ft) ❋ ❋ ❋ H6 ☼ ◊ ◗ ◆

Leucanthemella serotina
This large-flowered daisy makes excellent cutting material, lasting well in the vase. It is a vigorous plant, with stout stems that should not need staking, and prefers a moist situation with full sun or partial shade. It is useful for illuminating darker areas of the garden.

‡to 1.5m (5ft) ↔ 90cm (36in) ❋ ❋ ❋ H7 ☼ ☼ ◗ ◆

Leucanthemum × superbum 'Wirral Supreme'
One of the Shasta daisies, 'Wirral Supreme' boasts fully double flowerheads of white petals with pale yellow eyes. These are formed from early summer until early autumn, providing a long-lasting display. A strong growing plant, it needs moderately fertile soil and sunshine to perform well.

‡to 90cm (36in) ↔ 75cm (30in) ❋ ❋ ❋ H5 ☼ ☼ ◊ ◗

Macleaya microcarpa 'Kelway's Coral Plume'
This pink-flowered plume poppy is at its peak in early and midsummer, when large, open floral sprays sit above a sea of grey-green leaves. A tall, showy plant, it is best sited on its own, forming an eye-catching screen, or at the back of a large mixed border. Macleayas can be invasive.

‡to 2.2m (7ft) ↔ 1m (3ft) or more ❋ ❋ ❋ H6 ☼ ◊ ◗

Melianthus major
Grown more for its grey-green, tooth-edged leaves than its flowers, the honey bush is tolerant of sea air and is a good choice for coastal gardens. Use as an architectural focus or place it in strategic positions around the garden where its angular features can be admired. It is not frost hardy.

‡2–3m (6–10ft) ↔ 1–3m (3–10ft) ❋ ❋ H3 ☼ ◊ ◗

Musa basjoo
The Japanese banana can grow to 5m (15ft) and even flower and produce fruit (unpalatable, however) in cooler climates. It is ideal as a specimen plant, or can be used as the centrepiece of a tropical display. Strong winds can shred the leaves, so try to provide some protection.

‡to 5m (15ft) ↔ to 4m (12ft) ❋ H2 ☼ ◗

Phormium tenax Purpureum Group
Long, fibrous, sword-shaped leaves burst forth from the base of the New Zealand flax. The red-purple foliage contrasts well with paler phormiums or grasses. Alternatively, use it on its own to dominate a border. The plant likes fertile soil in full sun; mulch the base in winter in frost-prone areas.

‡2.5–2.8m (8–9ft) ↔ 1m (3ft) ❋ ❋ ❋ H5 ☼ ◊ ◗

Romneya coulteri 'White Cloud'
This plant will eventually become a woody perennial once it becomes established. Large white petals with a bobble of yellow stamens in the centre create a winning display. Protect plants from cold, strong winds, and in frost-prone areas, choose a site against a warm wall.

‡1–2.5m (3–8ft) ↔ indefinite ❀ ❀ H5 ☼ ◊

Salvia uliginosa
Native to South America, the bog sage comes into its own from late summer to mid-autumn, when square stems bearing clear blue flowers emerge above mid-green, toothed leaves. As the name suggests, bog sage is a moisture-loving plant. It is tall and suited to the back of a sunny border.

‡to 2m (6ft) ↔ 90cm (36in) ❀ ❀ H4 ☼ ◔

Symphyotrichum 'Ochtendgloren'
The long-lasting, purple-pink, daisy-like flowers of this aster are held on branching stems in late summer. It is a strong-growing plant, producing neat clumps that do not need to be regularly divided. It brightens up borders, can be grown in containers, and is also good for cutting.

‡1.2m (4ft) ↔ 80cm (32in) ❀ ❀ H4 ☼ ◊

Thalictrum flavum subsp. glaucum
The yellow meadow rue is a clump-forming perennial that spreads by means of underground stems or rhizomes. Its blue-green foliage is offset by the pale sulphur-yellow flowers formed in summer. The variety 'Illuminator' is taller than the subspecies and has bright green leaves.

‡to 1m (3ft) ↔ 60cm (24in) ❀ ❀ ❀ H7 ☼ ◔

Valeriana phu 'Aurea'
The leaves of this plant are soft yellow when young, turning green to lime green by summer. The leaves at the base of the stem are scented. Small white flowers appear in early summer to complete the display. A woodland plant in the wild, valerian suits a cottage garden or any informal scheme.

‡to 1.5m (5ft) ↔ 60cm (24in) ❀ ❀ ❀ H5 ☼ ☼ ◔

Verbascum 'Cotswold Queen'
Synonymous with cottage gardens, this semi-evergreen perennial will brighten any summer border with its prominent spikes of yellow, saucer-shaped flowers. In a garden exposed to the elements, this tall plant will probably need staking. Many Verbascum species are short-lived.

‡1.2m (4ft) ↔ 30cm (12in) ❀ ❀ ❀ H7 ☼ ◊

Verbena bonariensis
A popular plant, this verbena comes into its own when grown with grasses, allowing its branched flowerheads to punctuate a border display. It can be grown at the back of beds, but its slim stems also look striking at the front. It flowers from midsummer to early autumn.

‡to 2m (6ft) ↔ 45cm (18in) ❀ ❀ H4 ☼ ◊ ◔

Veronicastrum virginicum
From summer to autumn, the dainty flower spikes of this perennial bring white, pink, and purple shades to border plantings. For a pure white-flowered variety, look for V. virginicum 'Album' and grow it with dark foliage plants to bring out its best attributes.

‡to 2m (6ft) ↔ 45cm (18in) ❀ ❀ ❀ H7 ☼ ☼ ◊ ◔

PERENNIALS FOR ATTRACTING WILDLIFE

- Aquilegia vulgaris 'William Guiness' p.326
- Centaurea dealbata 'Steenbergii' p.327
- Cirsium rivulare 'Atropurpureum' p.322
- Crambe cordifolia p.322
- Digitalis x mertonensis p.328
- Doronicum 'Little Leo' p.335
- Echinacea 'Art's Pride' p.328
- Echinops bannaticus p.323
- Geranium 'Brookside' p.328
- Geranium macrorrhizum p.329
- Geranium 'Nimbus' p.329
- Geranium phaeum p.329
- Helenium 'Moerheim Beauty' p.329
- Knautia macedonica p.330
- Monarda 'Squaw' p.331
- Nepeta grandiflora 'Dawn to Dusk' p.331
- Nepeta 'Six Hills Giant' p.331
- Pulmonaria 'Diana Clare' p.337

Medium-sized perennials

Achillea 'Lachsschönheit'
Feathery foliage and large, flat heads of salmon-pink flowers (the plant is also seen labelled 'Salmon Beauty') make this clump-forming perennial a good choice to grow with wild flowers or in a mixed border. It is one of the Galaxy Hybrids series, which offers a wide range of colours.

↕75–90cm (30–36in) ↔ 60cm (24in) ❋ ❋ ❋ H7 ☼ ◊ ◗

Achillea 'Taygetea'
Large, creamy-yellow flowerheads appear in summer and autumn, providing perfect landing pads for summer-visiting insects looking for a source of nectar. Finely-cut, greyish-green leaves appear along the length of the stems, acting as a contrasting foil to the flowers.

↕60cm (24in) ↔ 45cm (18in) ❋ ❋ ❋ H7 ☼ ◊ ◗

Agapanthus Headbourne Hybrids
These plants were first raised in the 1940s by the Hon. Lewis Palmer in his Hampshire garden, using South African seed. The resulting hybrids have larger flowers than most African lilies, and are hardier. Grow in a mixed border or in pots.

↕60–90cm (24–36in) ↔ 90cm (36in)

❋ ❋ H4 ☼ ◊ ◗

Anaphalis triplinervis
These are easy garden plants to grow and are very effective in a border where the emphasis is on white and silver. The clusters of flowers, borne from mid- to late summer, have papery white bracts, and make good cut flowers.

↕80–90cm (32–36in) ↔ 45–60cm (18–24in)

❋ ❋ ❋ H7 ☼ ◊

Aquilegia vulgaris 'William Guiness'
There are many granny's bonnets to choose from, but the exquisite colours of 'William Guiness' (here shown against a background of hosta leaves) make it a popular choice. Tall flower stems are carried above divided leaves; the plants are suited to cottage gardens or mixed borders.

↕90cm (36in) ↔ 45cm (18in) ❋ ❋ ❋ H7 ☼ ☼ ◊ ◗

Artemisia ludoviciana 'Silver Queen'
Grown predominantly for its downy silver leaves, this artemisia is good for contrast in a mixed border or as an element in a white and silver planting scheme. Brownish-yellow flowerheads emerge from midsummer to autumn. The variety 'Valerie Finnis' has more deeply cut leaf margins.

↕75cm (30in) ↔ 60cm (24in) ❋ ❋ ❋ H6 ☼ ◊

Asplenium scolopendrium Crispum Group
The Hart's tongue fern is evergreen, with wavy-edged fronds, making it a year-round decorative asset in the garden. For the lushest plants, choose a position in dappled shade with moist, rich soil to prevent sun scorching. A mixed woodland border would be ideal.

↕30–60cm (12–24in) ↔ 60cm (24in) ❋ ❋ ❋ H6 ☼ ◊ ◗

Astelia chathamica
Dense clumps of arching, silver scaly leaves make this an attractive plant for a border or container. Pale yellowish-green flowers appear on long stalks from mid- to late spring, followed, on female plants, by orange berries. Do not allow roots to become over-wet during the winter months.

↕1.2m (4ft) ↔ to 2m (6ft) ❋ ❋ H3 ☼ ◗

Astrantia 'Hadspen Blood'
Astrantias are well suited to areas of dappled shade in the garden. The cultivar 'Hadspen Blood' is clump-forming, with deeply cut, mid-green leaves and clusters of dark red flowers surrounded by equally dark red bracts.

↕30–90cm (12–36in) ↔ 45cm (18in)

❋ ❋ ❋ H7 ☼ ☼ ◗

Astrantia major 'Sunningdale Variegated'
This astrantia is remarkable for its leaves, which are unevenly margined with soft yellow and cream. Although the pale pink flowers are attractive, the foliage is the main feature. Trim back the leaves to encourage new growth and choose a sunny site for best variegation.

↕30–90cm (12–36in) ↔45cm (18in) ❀ ❀ ❀ H7 ☼ ☀ ◐

Athyrium filix-femina
It is clear why the Victorians found ferns so charming when you see the lady fern at its best. Its large, very finely cut fronds, sometimes with red-brown stalks, suit dappled corners of the garden. Shady, sheltered areas or a woodland setting provide the perfect growing conditions.

↕to 1.2m (4ft) ↔ to 90cm (36in) ❀ ❀ ❀ H6 ☀ ◐

Campanula 'Burghaltii'
In midsummer, pendent, lavender-coloured bells, opening from blue-purple buds, dangle from the stems of this mound-forming perennial, against a background of heart-shaped leaves. The plant prefers neutral to alkaline conditions to thrive. Alternatively, grow it in a large container.

↕60cm (24in) ↔30cm (12in) ❀ ❀ ❀ H7 ☼ ☀ ◊ ◐

Campanula glomerata 'Superba'
The erect stems of this bellflower bear clusters of deep purple, bell-shaped flowers throughout the summer. Prolong the flowering season by cutting plants back to the top of the leaves after the first flush of blooms. This variety is vigorous and can even be invasive.

↕60cm (24in) ↔ indefinite ❀ ❀ ❀ H7 ☼ ☀ ◊ ◐

Centaurea dealbata 'Steenbergii'
Tolerant of dry conditions, knapweed is a magnet for bees and butterflies. The rich pink flowers with feathery petals can be cut for indoor displays when they appear in summer. The plant looks attractive in wild parts of the garden, or as part of a cottage garden scheme.

↕60cm (24in) ↔60cm (24in) ❀ ❀ ❀ H7 ☼ ◊

Clematis integrifolia
This herbaceous perennial carries flowers on the current year's shoots, from midsummer to late autumn. The mid-blue flowers have slightly twisted 'petals' and cream anthers, and are followed by silvery seedheads which provide an extended season of interest. The plant may need supporting.

↕60cm (24in) ↔60cm (24in) ❀ ❀ ❀ H6 ☼ ☀ ◊

Clematis tubulosa 'Wyevale'
This clematis grows as a free-standing shrub. The late summer flowers are pale blue, scented and are produced in clusters, resembling hyacinth blooms. Fluffy, silvery seedheads follow, which are also decorative. The plant prefers a chalky soil in full sun or part shade.

↕75–130cm (2½–4½ft) ↔ 1m (3ft) ❀ ❀ ❀ H5 ☼ ☀ ◊

Digitalis grandiflora
The yellow foxglove forms sturdy, imposing clumps of tall flower spikes with glossy leaves, and is best sited where it will make an impact. Large, tubular flowers with speckled throats radiate outwards. Choose dappled shade under trees or a sheltered part of the garden for best results.

↕to 1m (3ft) ↔45cm (18in) ❀ ❀ ❀ H6 ☀ ◐

EARLY-FLOWERING PERENNIALS

Medium-sized perennials

Digitalis x mertonensis
This cross between the yellow foxglove and common foxglove has resulted in a free-flowering perennial bearing large pink tubular flowers in late spring and early summer. An excellent plant for attracting bees. Self-sown seedlings will appear around the parent plant.

↕to 90cm (36in) ↔30cm (12in) ❋ ❋ ❋ H5 ☀ ◐ ◌

Dryopteris erythrosora
This slowly spreading fern from China and Japan emerges from the soil as coppery-red young fronds. These gradually turn pink and then silvery-green with age, forming a lacy network over the ground. Keep soil around the roots moist and site in a sheltered area. It makes a striking plant for a border.

↕60cm (24in) ↔40cm (16in) ❋ ❋ H4 ☀ ◐ ◌

Echinacea 'Art's Pride'
The narrow orange petals of this coneflower surround a prominent, rust-coloured, cone-shaped disc which appeals to all kinds of insects. Flowers are also slightly scented. Grow in a mixed border, or wildlife or cottage garden. With their sturdy stems, echinaceas also make good cut flowers.

↕60–90cm (24–36in) ↔45cm (18in) ❋ ❋ ❋ H5 ☀ ◌

Echinacea purpurea 'Alba'
This is the white version of the popular, purple-flowered *Echinacea purpurea*. Large, reflexed, white petals surround a central yellow cone on long flower stems, from midsummer to autumn. These plants work well with grasses and slim-stemmed perennials, such as *Verbena bonariensis*.

↕to 1.5m (5ft) ↔45cm (18in) ❋ ❋ ❋ H5 ☀ ◌

Eremurus stenophyllus
The lovely tapering flower spikes of foxtail lilies emerge and bloom in summer. Staking may be required to prevent the tall stems blowing over. Provide a site with free-draining soil, and mulch around the crowns with garden compost in autumn. Suited to the back of a garden border.

↕1m (3ft) ↔60cm (24in) ❋ ❋ ❋ H6 ☀ ◌

Euphorbia griffithii 'Dixter'
This is a striking herbaceous perennial that contrasts well with other green-leaved euphorbias. Its copper-tinted, dark green leaves make an effective background to the orange bracts that surround the inconspicuous true flowers. The best colour comes from plants grown in dappled shade.

↕75cm (30in) ↔1m (3ft) ❋ ❋ ❋ H7 ☀ ◐ ◌

Euphorbia x martinii
With unusual flowers in a mixture of greens and reds, produced on the previous year's shoots, this euphorbia would be a welcome addition to any garden. It flowers over a long season from spring to midsummer and is a very adaptable plant, tolerating sun and shade.

↕1m (3ft) ↔1m (3ft) ❋ ❋ ❋ H5 ☀ ◐ ◌

Euphorbia schillingii
Pale yellow flowerheads perch above a mass of wiry, leafy stems on this strong-growing herbaceous perennial. Plant it with other border perennials, choosing colours carefully to bring out the subtleties of this late summer- to autumn-flowering plant. Provide rich soil in dappled shade.

↕1m (3ft) ↔30cm (12in) ❋ ❋ ❋ H5 ☀ ◐ ◌

Geranium 'Brookside'
This densely growing perennial is ideal for border edges; it is a vigorous, spreading plant and makes attractive ground cover, for sun or part-shade. Abundant violet-blue flowers with pale centres appear in summer, held above a mass of finely divided green leaves.

↕60cm (24in) ↔45cm (18in) ❋ ❋ ❋ H7 ☀ ◐ ◌

Geranium macrorrhizum
This plant has strongly aromatic, toothed, sticky leaves that turn an attractive red in autumn. Clusters of flat pink flowers with protruding stamens are borne in early summer from a mass of sprawling stems. This is a good plant for ground cover or underplanting in a shady site.

‡50cm (20in) ↔ 60cm (24in) ✿ ✿ ✿ H7 ☼ ◊

Geranium 'Nimbus'
A very vigorous and floriferous geranium that becomes a sea of blue when the lavender blue flowers appear in summer. This plant is very tolerant of shade and is a good choice for darker borders or corners that receive little direct sunlight. Clip to encourage repeat flowering.

‡to 1m (3ft) ↔ 45cm (18in) ✿ ✿ ✿ H7 ☼ ☀ ◊

Geranium phaeum
The dusky cranesbill is undemanding in its garden requirements. It will tolerate sun but is also a useful plant for deep shade. Dark maroon flowers with white eyes are produced in early summer. For a brighter-flowered geranium, try G. psilostemon, with its black-centred magenta flowers.

‡80cm (32in) ↔ 45cm (18in) ✿ ✿ ✿ H7 ☼ ☼ ☀ ◊

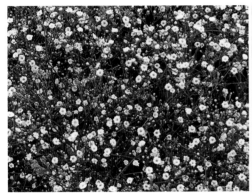

Gypsophila paniculata 'Bristol Fairy'
Also known as baby's breath, branching gypsophila creates a cloud of blossom as a profusion of tiny double-white flowers emerge in midsummer. It looks marvellous in a mixed border and also provides good cut flowers. 'Bristol Fairy' has double white flowers but may be shorter-lived than G. paniculata.

‡1.2m (4ft) ↔ 1.2m (4ft) ✿ ✿ ✿ H7 ☼ ◊

Helenium 'Moerheim Beauty'
Copper-red daisy flowers, each with a distinctive round central disc, are borne in early to late summer, filling the garden with warm colour. Deadhead through the season to encourage repeat flowering. The plant's striking colour and form mix well with either hot colours or pastel shades.

‡90cm (36in) ↔ 60cm (24in) ✿ ✿ ✿ H7 ☼ ◊ ◑

Helleborus argutifolius
The Corsican hellebore is a valuable plant for the designer in search of winter interest. A glossy-leaved perennial, it comes into flower in late winter and early spring, and the blooms are an unexpected pale green. It grows well in most conditions but will not thrive in acid soils.

‡to 1.2m (4ft) ↔ 90cm (36in) ✿ ✿ ✿ H5 ☼ ☼ ◑

Helleborus foetidus
The stinking hellebore is named for the unpleasant smell its leaves give off when crushed. However, the greenish-white flowers the plant bears in mid-winter and early spring make up for this downside. Other good varieties to choose from include the Wester Flisk Group, with red-tinted main stems.

‡to 80cm (32in) ↔ 45cm (18in) ✿ ✿ ✿ H7 ☼ ☀ ◑

Hemerocallis 'Buzz Bomb'
Originally bred in 1961, this brightly coloured daylily is a strong grower, flowering in midsummer. The large, orange-red blooms with yellow throats are carried above strap-like green leaves over a long season. Grow in a mixed or herbaceous border in full sun for maximum effect.

‡60cm (24in) ↔ 60cm (24in) ✿ ✿ ✿ H6 ☼ ◊ ◑

LATE-FLOWERING PERENNIALS

Medium-sized perennials

Hemerocallis 'Marion Vaughn'
A late afternoon-flowering daylily, 'Marion Vaughn' is a dependable evergreen with clear lemon-yellow flowers and bright green strap-like foliage, making a crisp addition to a mixed border. It looks good growing in a drift with other daylilies. Full sun will promote best flowering.

↕85cm (34in) ↔75cm (30in) ❄ ❄ ❄ H6 ☼ ◊ ◐

Hosta 'Francee'
Hostas are shade-loving foliage plants and versatile in the garden, suited to ground cover, containers, a woodland garden or mixed border. Olive-green heart-shaped leaves with a variable white margin make 'Francee' a popular choice. Lavender-blue flowers emerge in summer.

↕55–70cm (22–28in) ↔1m (3ft) ❄ ❄ ❄ H7 ☼ ◊ ◐

Hosta 'Royal Standard'
This clump-forming perennial does well in shade but will also tolerate some sun. Its unmarked pale green leaves have prominent ribs and provide an excellent foil for the funnel-shaped, fragrant white flowers that appear in late summer. It is fast-growing and vigorous.

↕60cm (24in) ↔1.2m (4ft) ❄ ❄ ❄ H7 ☼ ◊ ◐

Hosta sieboldiana var. elegans
With its heavily puckered, blue-green leaves, this large hosta makes a dramatic border plant. It tolerates shade although a very dark position will subdue the production of lilac-coloured flowers in early summer. Place a group of hostas together for a stunning foliage effect.

↕1m (3ft) ↔1.2m (4ft) ❄ ❄ ❄ H7 ☼ ◊ ◐

Knautia macedonica
Similar to a scabious, this knautia carries purple-red pincushion flowerheads, held above the foliage on branching stems, from mid- to late summer. It is attractive to bees and butterflies and ideally suited to a wildflower or cottage garden. It is fairly drought-tolerant.

↕60–80cm (24–32in) ↔45cm (18in) ❄ ❄ ❄ H7 ☼ ◊

Kniphofia 'Bees' Sunset'
This is a yellow-orange variety of the deciduous plant familiarly known as the red hot poker. Upright, fleshy stems support a bottlebrush-like array of the downward-pointing, tubular flowers from early to late summer. Grow in the herbaceous border in groups for a dramatic display.

↕90cm (36in) ↔60cm (24in) ❄ ❄ ❄ H5 ☼ ☼ ◊ ◐

Kniphofia 'Percy's Pride'
This cultivar of the red hot poker produces long spikes of greenish-yellow flowers, maturing to cream, which emerge in late summer and early autumn on long, fleshy stems. The unusual flower colour makes it suitable for a colour-themed border using white, green and pale yellow.

↕to 1.2m (4ft) ↔60cm (24in) ❄ ❄ ❄ H6 ☼ ☼ ◊ ◐

Lamprocapnos spectabilis f. alba
When in flower, the graceful, arching stems of the bleeding heart (or Dutchman's breeches) look like a miniature washing line. New shoots appear in spring with rose pink or white flowers. 'Alba' is a less vigorous selection with pure white blooms. It will tolerate some sun if the roots are kept moist.

↕to 1.2m (4ft) ↔45cm (18in) ❄ ❄ ❄ H6 ☼ ◐

Liatris spicata 'Kobold'
The spikes of deep purple flowerheads on this plant are unusual in that the flowers open from the top downwards. 'Kobold' flowers from late summer to early autumn and suits a mixed border, but needs regular moisture to thrive. Stems can be cut for a cheerful indoor display.

↕70cm (30in) ↔45cm (18in) ❄ ❄ ❄ H7 ☼ ◊ ◐

Lupinus Band of Nobles Series 'Chandelier'

If space allows, grow lupins in drifts, allowing complementary colours to sit close to one another. The pale yellow, pea-like blooms of clump-forming 'Chandelier' appear in early and midsummer and are ideal for a mixed or herbaceous border in a cottage-style or informal design.

↕90cm (36in) ↔75cm (30in) ❋ ❋ ❋ H5 ☼ ☀ ◊

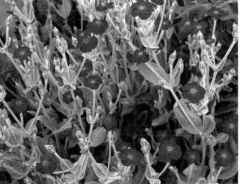

Lychnis coronaria

Known by the common names of dusty miller and rose campion, this short-lived perennial has soft silvery-grey stems and leaves. Late summer sees a long succession of rounded vermilion blooms. It self-seeds freely. For a pure white form, choose 'Alba'.

↕80cm (32in) ↔45cm (18in) ❋ ❋ ❋ H7 ☼ ☀ ◊

Lysimachia ephemerum

Woodland or streamside plants in the wild, these pretty herbaceous perennials are best suited to a damp border, bog garden or pond margin. In early and midsummer, erect spikes of saucer-shaped white flowers rise above mid-green tapered leaves. Plants may need protection in winter.

↕1m (3ft) ↔30cm (12in) ❋ ❋ ❋ H6 ☼ ☀ ◖ ◆

Lythrum salicaria 'Feuerkerze'

Masses of tiny star-shaped, intensely red-purple flowers cover the slender upright spikes of this purple loosestrife cultivar, making a beautiful display from midsummer to early autumn. The leaves are downy. The plant needs moisture and suits a damp border or bog garden.

↕to 90cm (36in) ↔45cm (18in) ❋ ❋ ❋ H7 ☼ ◖

Monarda 'Squaw'

Bergamots are grown for their long-lasting, colourful flowers which appear from midsummer to early autumn. 'Squaw' is particularly striking, with its bright scarlet flowers held above dark bracts. Bergamots attract bees and butterflies and are ideal for a wildflower garden.

↕to 1.2m (4ft) ↔45cm (18in) ❋ ❋ ❋ H4 ☼ ☀ ◊ ◖

Nepeta grandiflora 'Dawn to Dusk'

As the name grandiflora suggests, the pale mauve-pink flowers on this catmint are larger than usual. Grow this cultivar near a path or garden seat to make the most of the distinctive fragrance released as the leaves are crushed. The plants are attractive to bees – and cats.

↕65cm (26in) ↔30cm (12in) ❋ ❋ ❋ H6 ☼ ☀ ◊

Nepeta 'Six Hills Giant'

This is a vigorous perennial bearing masses of lavender-blue flowers throughout the summer months. Be prepared for it to take up some space in the border. The leaves are light grey, and noticeably aromatic when touched. Clumps can be divided in spring or autumn to rejuvenate plants.

↕90cm (36in) ↔60cm (24in) ❋ ❋ ❋ H7 ☼ ☀ ◊

Origanum laevigatum 'Herrenhausen'

Strongly aromatic leaves and bright clusters of pink flowers characterize this woody perennial, an ornamental cultivar of the culinary herb marjoram. The leaves are purple-flushed when young and in winter, and the flowers appear from late spring to autumn. Suited to a herb garden or border margin.

↕60cm (24in) ↔45cm (18in) ❋ ❋ ❋ H6 ☼ ◊

PERENNIALS FOR FOLIAGE INTEREST

Medium-sized perennials

Paeonia 'Sarah Bernhardt'
The powder-pink, showy double flowers of 'Sarah Bernhardt' are large by peony standards and come into their own in summer. The individual blooms are heavy and will need supporting. They are also good for cutting.

↕95cm (38in) ↔50–70cm (20–28in)
❀ ❀ ❀ H6 ☼ ◑ ◌ ◗

Papaver Oriental Group 'Black and White'
The bold, beautiful flowers of the Oriental poppy make an immediate impact. There are many cultivars; the large, ruffled petals of 'Black and White', each with a black blotch at the base, are papery white and surround a boss of dark stamens.

↕45–90cm (18–36in) ↔60–90cm (24–36in)
❀ ❀ ❀ H7 ☼ ◌

Penstemon 'Alice Hindley'
A favourite with many gardeners, foxglove-like penstemons are reliable and rewarding to grow. Large, tubular bell-like flowers open in succession along upright stems from midsummer to autumn. There are many cultivars; the flowers of 'Alice Hindley' are pale lilac-blue. Feed well.

↕90cm (36in) ↔45cm (18in) ❀ ❀ ❀ H4 ☼ ◑ ◌

Penstemon 'Andenken an Friedrich Hahn'
This hardy, vigorous, bushy penstemon carries elegant spikes of bright garnet-red flowers in profusion from midsummer through to mid-autumn, above masses of narrow green leaves. Deadheading will significantly prolong the flowering display.

↕75cm (30in) ↔60cm (24in) ❀ ❀ ❀ H5 ☼ ◑ ◌

Persicaria amplexicaulis 'Firetail'
This semi-evergreen perennial is a robust, undemanding garden plant. From midsummer to early autumn, the lush green foliage is joined by tall, rigid stems bearing small, bright red bottlebrush flowers. Grow as border plants, as ground cover, or naturalize in a woodland garden.

↕to 1.2m (4ft) ↔to 1.2m (4ft) ❀ ❀ ❀ H7 ☼ ◑ ◗

Persicaria bistorta 'Superba'
A long-flowering, semi-evergreen plant with rounded spikes of soft pink, miniature blooms, which present a good show all summer and well into autumn. Grow behind 'Firetail' (*left*) for interesting contrast. Divide particularly vigorous clumps in spring or summer to control their size and spread.

↕to 90cm (36in) ↔90cm (36in) ❀ ❀ ❀ H7 ☼ ◑ ◌ ◗

Phlomis russeliana
This sage-like plant looks very effective grown in a large group in a border. The pale yellow, hooded flowers begin to appear in late spring and continue until autumn, with the best show of colour in early summer. The cut stems are good for dried arrangements.

↕to 90cm (36in) ↔75cm (30in) ❀ ❀ ❀ H6 ☼ ◑ ◌

Phlox paniculata 'Balmoral'
Sweetly scented perennial phlox is a staple of the traditional cottage garden, grown for its late summer and early autumn flowers. This vigorous, herbaceous cultivar has large trusses of pale pink, flattened flowers, and is a good choice for a mixed border. It does best in rich soil.

↕90cm (36in) ↔60–100cm (24–39in) ❀ ❀ ❀ H7 ☼ ◑ ◗

Phlox paniculata 'Norah Leigh'
Variegated forms of phlox are a relatively new phenomenon. The tapering leaves of 'Norah Leigh' have green mid-ribs but are mainly creamy-white with splashes of green. Clusters of pale lilac flowers with deeper pink centres are borne from summer to autumn over a long season.

↕to 90cm (36in) ↔60–100cm (24–39in) ❀ ❀ ❀ H7 ☼ ◗

Phormium cookianum subsp. hookeri 'Tricolor'
The mountain flax from New Zealand comes in a number of forms. Here, the narrow, arching, strap-like leaves are green with cream and red margins. Yellow-green flowers emerge in summer on long, stiff stems, although it is for the foliage that the plant is grown. Ideal for a coastal garden.

↕1.2m (4ft) ↔3m (10ft) ❀ ❀ H4 ☼ ☼ ◐

Potentilla atrosanguinea
These pretty plants have attractive leaves with silver-haired undersides, but are mainly grown for their bright, saucer-shaped flowers, which vary from yellow through to rich red and bloom over a long season from spring to autumn. The plants thrive in a cool but bright position.

↕45–90cm (18–36in) ↔60cm (24in) ❀ ❀ ❀ H7 ☼ ◊

Rudbeckia fulgida var. sullivantii 'Goldsturm'
Coneflowers are popular late-season plants, producing quantities of yellow flowerheads with dark eyes, held on bristly stems, from late summer to mid-autumn. The rich green leaves are tapering and toothed. Pair 'Goldsturm' with *Verbena bonariensis* and grasses for a dramatic display.

↕to 60cm (24in) ↔45cm (18in) ❀ ❀ ❀ H6 ☼ ☼ ◊

Rudbeckia laciniata 'Goldquelle'
The deeply cut green leaves make an effective background for this double-flowered, lemon-yellow coneflower. 'Goldquelle' makes a fine addition to the late summer border and will continue flowering until the middle of autumn. Rudbeckias and grasses make a happy combination in a large border.

↕to 90cm (36in) ↔45cm (18in) ❀ ❀ ❀ H6 ☼ ☼ ◊

Sedum 'Matrona'
Fleshy leaves, initially green and later flushed purple, and dark red stems form the backdrop to the flattened heads of tiny pink star-like flowers in late summer. The dried flower heads add structure and interest to the winter garden.

↕60–75cm (24–30in) ↔to 30–45cm (12–18in)
❀ ❀ ❀ H6 ☼ ◊

Sisyrinchium striatum 'Aunt May'
Excellent front-of-the-border plants, sisyrinchiums also suit a gravel garden. Less vigorous than the green-leaved species, the cultivar 'Aunt May' has cream-edged, grey-green, narrow leaves. In summer, the stiff flower stems are studded with small pale yellow flowers.

↕50cm (20in) ↔25cm (10in) ❀ ❀ ❀ (borderline) H4 ☼ ◊

Symphyotrichum ericoides 'White Heather'
A reliable and easy-to-grow perennial, 'White Heather' produces sprays of small daisy blooms at the end of summer, prolonging the season of interest in the garden. A sunny site will ensure an extended spell of flowering. To increase the stock, divide larger plants in spring.

↕1m (3ft) ↔30cm (12in) ❀ ❀ ❀ H7 ☼ ◊

Symphyotrichum novae-angliae 'Andenken an Alma Pötschke'
These Michaelmas daisies bear rich cerise-pink blooms in profusion from late summer to mid-autumn. Mix varieties together to create your own aster display or plant among other perennials for late summer colour.

↕1.2m (4ft) ↔60cm (24in) ❀ ❀ ❀ H7 ☼ ☼ ◐

PERENNIALS FOR DAMP SOIL CONDITIONS

- *Aconitum* 'Spark's Variety' p.322
- *Adiantum venustum* p.334
- *Alchemilla mollis* p.334
- *Astelia chathamica* p.326
- *Astrantia major* 'Hadspen Blood' p.326
- *Athyrium filix-femina* p.327
- *Dicentra* 'Bacchanal' p.335
- *Dryopteris wallichiana* p.323
- *Ensete ventricosum* p.323
- *Helleborus argutifolius* p.329
- *Helleborus x hybridus* 'Pluto' p.336
- *Inula magnifica* p.324
- *Pachysandra terminalis* p.336
- *Persicaria bistorta* 'Superba' p.332
- *Pulmonaria* 'Diana Clare' p.337
- *Salvia uliginosa* p.325
- *Symphyotrichum novae-angliae* 'Andenken an Alma Pötschke' p.333
- *Valeriana phu* 'Aurea' p.325
- *Veronicastrum virginicum* p.325

Small perennials

Adiantum venustum
The evergreen Himalayan maidenhair fern is a decorative plant for a shady wall crevice or a damp, shady corner. It looks delicate but is in fact surprisingly robust. Old growth should be removed in late winter before new pink croziers unfurl in spring, developing into fresh green fronds.

↕15cm (6in) ↔ indefinite ❄ ❄ ❄ H7 ☼ ◊

Ajuga reptans
Spikes of deep blue flowers emerge from the low-growing, dark green leaves of this evergreen perennial from late spring to early summer. The plant spreads rapidly and makes excellent ground cover. For a less invasive form try 'Catlin's Giant', which has large bronze-purple leaves.

↕15cm (6in) ↔ 60–90cm (24–36in) ❄ ❄ ❄ H7 ☼ ◊

Alchemilla mollis
Dependable and drought-tolerant, lady's mantle is grown for its pretty foliage and frothy sprays of tiny greenish-yellow flowers, which appear from early summer to autumn and are good for cutting. Deadhead after flowering to prevent self-seeding. Plant it at the front of a border or in a gravel garden.

↕to 60cm (24in) ↔ 75cm (30in) ❄ ❄ ❄ H7 ☼ ☼ ◊

Anemone nemorosa 'Bracteata Pleniflora'
A striking variant of the wood anemone, this creeping perennial bears semi-double white flowers, held over a ruff of closely packed green leaves, from spring to early summer. The plant thrives in rich soil. Dappled shade brings out its best features.

↕8–15cm (3–6in) ↔ 30cm (12in) or more ❄ ❄ ❄ H5 ☼ ◊ ◊

Anthemis punctata subsp. cupaniana
This Sicilian daisy naturally prefers a sunny site, such as an open, well-drained rock garden. Flowers are long-lasting and bloom over a long season from late spring to late summer. The plant forms a tight mat at ground level and in winter the silvery-grey leaves turn grey-green.

↕30cm (12in) ↔ 90cm (36in) ❄ ❄ ❄ H4 ☼ ◊

Arum italicum subsp. italicum 'Marmoratum'
A truly exotic-looking plant whether in leaf, flower, or fruit, lords and ladies is excellent for filling in gaps in border displays. The glossy green leaves are veined with white, while the pale cream spathes give way to stalks of bright orange berries. It's at its best in a sheltered site.

↕30cm (12in) ↔ 15cm (6in) ❄ ❄ ❄ H6 ☼ ☼ ◊

Aster amellus 'Veilchenkönigin'
A clump-forming perennial, this aster produces a mass of tiny, violet-purple, daisy-like flowers in late summer, which are attractive to butterflies. The mid-green leaves are narrow and slightly hairy. Divide plants in spring and replant the strongest sections for most vigorous regrowth.

↕30–60cm (12–24in) ↔ 45cm (18in) ❄ ❄ ❄ H7 ☼ ◊

Athyrium niponicum var. pictum
These graceful, deciduous ferns (also known as lady ferns) are easy to grow and will thrive in a shady, sheltered border or woodland garden, as long as there is sufficient moisture. The arching fronds are light green or greyish, sometimes flushed purple, with a purple midrib.

↕to 30cm (12in) ↔ indefinite ❄ ❄ ❄ H6 ☼ ◊

Bergenia 'Morgenröte'
Known by many as elephant's ears because of the large, rounded, leathery green leaves, bergenias are robust, clump-forming evergreen plants. 'Morgenröte' produces clusters of bright reddish-pink flowers at the top of strong purple stems from mid- to late spring.

↕40cm (18in) ↔ 60cm (24in) ❄ ❄ ❄ H6 ☼ ☼ ◊ ◊

Brunnera macrophylla 'Dawson's White'

This relative of borage is a good choice for a woodland garden. The heart-shaped leaves, carried on stiff stalks, have irregular, creamy-white edges. In mid- and late spring, sprays of small blue flowers are borne above the foliage. Choose a cool, partly shaded site in rich soil.

↕ 45cm (18in) ↔ 60cm (24in) ❊ ❊ ❊ H6 ☼ ◊ ◑

Calamintha grandiflora 'Variegata'

A plant for the woodland garden or a cool, sheltered position, this calamint has toothed, pale green leaves, speckled creamy-white, which are aromatic when crushed. From summer to autumn, pink-mauve, two-lipped flowers emerge above and level with the topmost leaves.

↕ 30cm (12in) ↔ 45cm (18in) ❊ ❊ ❊ H5 ☼ ◊ ◑

Coreopsis verticillata 'Moonbeam'

A row of this brightly coloured plant will make a fine edging for a border. Finely cut leaves mingle together with a profusion of yellow, star-like flowers in early summer. A sunny position will promote the best show. Deadhead to encourage flowering.

↕ to 50cm (20in) ↔ 45cm (18in) ❊ ❊ H4 ☼ ☼ ◊

Dianthus 'Bovey Belle'

This hardy pink bears clove-scented, bright pink double blooms on long stems above silver-grey strappy foliage in summer, making an impact in a mixed border or raised bed. Deadhead regularly to promote further flowering. Pinks make long-lasting cut flowers.

↕ 25–45cm (10–18in) ↔ 40cm (16in) ❊ ❊ ❊ H6 ☼ ◊

Dicentra 'Bacchanal'

Layer upon layer of deeply divided, grey-green leaves make an effective foil for the delicate, crimson, heart-shaped flowers, dangling from arching stems, which appear from mid- to late spring. 'Bacchanal' is one of the darkest cultivars. These are shade-loving plants and suit a moist, shady border.

↕ 45cm (18in) ↔ 60cm (24in) ❊ ❊ ❊ H5 ☼ ◑

Doronicum 'Little Leo'

A spring-flowering perennial, this bright daisy is as attractive to wildlife as it is to gardeners. Bees, butterflies, and hoverflies are regular visitors to the large yellow blooms. Plant in small groups at the front of borders, or in containers. The flowers are also good for cutting.

↕ 25cm (10in) ↔ 30–60cm (12–24in) ❊ ❊ ❊ H5 ☼ ◑ ◊

Epimedium x perralchicum

Strong-growing woodland plants, epimediums make excellent ground cover under trees or shrubs. This hybrid has both interesting foliage – bronze when young, deep green when mature – and pretty, bright yellow flowers, borne on leafless stems in spring. It is also drought-tolerant.

↕ 40cm (16in) ↔ 60cm (24in) ❊ ❊ ❊ H6 ☼ ◊ ◑

Eryngium bourgatii 'Oxford Blue'

This is one of the smaller sea hollies. It is a herbaceous, clump-forming plant with dark green lower leaves and, in summer, spiny, silvery flower stems bearing silver-blue, thistle-like flowers, surrounded by blue-tinged bracts. The flower stems can be dried for indoor arrangements.

↕ 15–45cm (6–18in) ↔ 30cm (12in) ❊ ❊ ❊ H5 ☼ ◊

SHADE-TOLERANT PERENNIALS

- *Aconitum* 'Spark's Variety' p.322 (damp shade)
- *Adiantum venustum* p.334 (damp shade)
- *Anemone nemorosa* 'Bracteata Pleniflora' p.334 (damp shade)
- *Asplenium scolopendrium* Crispum Group p.326 (damp shade)
- *Athyrium niponicum* var. *pictum* p.334 (damp shade)
- *Dicentra* 'Bacchanal' p.335 (damp shade)
- *Dryopteris erythrosora* p.328 (damp shade)
- *Epimedium* x *perralchicum* p.335 (dry shade)
- *Euphorbia griffithii* 'Dixter' p.328 (damp shade)
- *Geranium macrorrhizum* p.329 (dry shade)
- *Geranium* 'Nimbus' p.329 (dry shade)
- *Helleborus* x *hybridus* 'Pluto' p.336 (dry shade)
- *Hosta* cultivars p.330 (damp shade)
- *Lamprocapnos spectabilis* f. *alba* p.330 (damp shade)
- *Pachysandra terminalis* p.336 (dry shade)

Small perennials

Euphorbia epithymoides

This euphorbia forms a loose, dome-shaped mound of lime-green stems that carry canary-yellow flowerheads from mid-spring to midsummer. The blooms are at their brightest when young. The plant dies down in winter, re-emerging the next year with a batch of fresh young shoots.

‡40cm (16in) ↔ 60cm (24in) ❀ ❀ ❀ H6 ☼ ◐ ◊ ◖

Geranium clarkei 'Kashmir White'

Cranesbills make versatile, undemanding garden plants. The 'Kashmir' cultivars are spreading, herbaceous perennials with dissected green foliage; they come in blue, pink, purple, and white. This cultivar produces large, whitish summer flowers with pale lilac-pink veining. Divide vigorous plants in spring.

‡to 45cm (18in) ↔ indefinite ❀ ❀ ❀ H6 ☼ ◐ ◊

Geum 'Lady Stratheden'

Also sold as Goldball, this cultivar yields large, semi-double, rich yellow flowers over a long period throughout summer, brightening up any border. A clump-forming perennial, it has rounded and kidney-shaped leaves, both of which are hairy to the touch. The plant comes true from seed.

‡40–60cm (16–24in) ↔ 60cm (24in) ❀ ❀ ❀ H7 ☼ ◊

Helleborus x hybridus 'Pluto'

A named cultivar among a group of highly variable hellebores, 'Pluto' is a clump-forming perennial noted for its striking flowers. They are purple on the outside but green-tinged purple within and appear during winter when little else is offering garden interest.

‡45cm (18in) ↔ 45cm (18in) ❀ ❀ ❀ H7 ◐ ◖

Heuchera 'Plum Pudding'

Compact, evergreen perennials, heucheras are useful for all-year-round interest. This cultivar has purple ruffled leaves with deeper purple veins. Small white flowers are held aloft on thin wiry stems in late spring. Grow alongside silvery-leaved 'Pewter Moon' to show both off to good effect.

‡65cm (26in) ↔ 50cm (20in) ❀ ❀ ❀ H6 ☼ ◐ ◊ ◖

Heuchera 'Red Spangles'

Throughout the summer, the stems of 'Red Spangles' rise from among green foliage, supporting small, tubular, rich scarlet-crimson flowers, adding a splash of vibrant colour to the garden. Grow where it can spill over onto a path, or include it in a mixed or shrub border.

‡50cm (20in) ↔ 25cm (10in) ❀ ❀ ❀ H5 ☼ ◐ ◊ ◖

Lamium maculatum 'White Nancy'

Spreading, low-growing plant, excellent for ground cover. The toothed leaves are silver with a green edge, while the summer flowers are pure white. Grow to cover bare soil and to suppress weeds. 'Red Nancy' has silver leaves with purplish-red flowers.

‡to 15cm (6in) ↔ to 1m (3ft) or more
❀ ❀ ❀ H7 ☼ ◐ ◊ ◖

Oenothera fruticosa 'Fyrverkeri'

From late spring to late summer, the large, bright yellow flowers of this evening primrose appear on upright stems above the purple-brown-flushed leaves below. The flowers bloom during the day and are short-lived but are borne over a long season. The plant will perform best in a sunny site.

‡30–90cm (12–36in) ↔ 30cm (12in) ❀ ❀ ❀ H5 ☼ ◊

Pachysandra terminalis

Good for ground cover, this tough evergreen perennial is grown for its foliage and will spread freely given enough moisture. It has coarsely toothed, glossy, dark green leaves and tiny white flowers, which are carried in spikes in early summer. A useful plant for shady sites.

‡20cm (8in) ↔ indefinite ❀ ❀ ❀ H5 ◐ ◖

Polypodium x mantoniae 'Cornubiense'
The finely dissected fronds of this ground cover fern easily cover the soil and break up the hard lines of path edges. New growth starts in spring with the fronds taking several weeks to unfurl. This is a handsome, resilient plant for a damp and shady spot in the garden.

‡30cm (12in) ↔ indefinite ❆ ❆ ❆ H7 ☼ ☀ ◌ ◍

Pulmonaria 'Diana Clare'
An early spring-flowering perennial, 'Diana Clare' is easy to grow and needs very little attention once established. In late winter and spring, clusters of violet-blue flowers, striped red, open above green leaves marked with silver. It makes good ground cover where the soil is not too dry.

‡30cm (12in) ↔ 45cm (18in) ❆ ❆ ❆ H7 ☼ ☀ ◍

Rhodanthemum hosmariense
Plants that flower from spring until autumn are much prized in the garden and this daisy-flowered, shrubby perennial amply fulfils this role. The leaves are silver and deeply lobed while the flowers are white-petalled with a yellow eye. A plant for a sunny border or rock garden with very free-draining soil.

‡10–30cm (4–12in) ↔ 30cm (12in) ❆ ❆ H4 ☼ ◌

Salvia nemorosa
Wrinkly green leaves form a neutral backdrop to the main attraction of purple, white, or pink flowers during the summer and autumn months. The flower stems stand stiff and upright and, when seen from a low viewpoint, create a sea of colour. Grow in sun or dappled shade in well-drained soil.

‡to 1m (3ft) ↔ 60cm (24in) ❆ ❆ ❆ H7 ☼ ☼ ◌

Sedum 'Vera Jameson'
A striking stonecrop to grow for colour impact. Purplish, fleshy leaves and stems sprawl sideways while rounded heads of rose-pink flowers are held aloft in late summer and early autumn. Mix with silvers and greys to accentuate the bold colouring; grow in a rock garden or at a border edge.

‡20–30cm (8–12in) ↔ 45cm (18in) ❆ ❆ ❆ H7 ☼ ◌

Sempervivum tectorum
The common houseleek creates starry patterns over the ground as its tight red rosettes hug the soil. Grow in old sinks, troughs, or terracotta pots to show off the architectural shapes. Reddish-purple flowers are borne in summer. A gritty, well-drained compost and full sun are desirable.

‡15cm (6in) ↔ 50cm (20in) ❆ ❆ ❆ H7 ☼ ◌

Veronica gentianoides
This pretty veronica is grown for its spikes of pale blue, early-summer flowers held on erect stems above a mound of glossy, bright green foliage. In hot-hued borders it makes a contrast with reds and oranges, and is also effective when planted in drifts on its own. It performs best in moist soil.

‡45cm (18in) ↔ 45cm (18in) ❆ ❆ ❆ H7 ☼ ☼ ◌ ◍

Veronica spicata subsp. incana
Also known as the silver speedwell, this perennial marries silver hairy leaves with spikes of purple-blue flowers, making it a good choice for a border comprising cool colours. Summer flowering, it is mat-forming and will spread, so clip back if necessary after the blooms have faded.

‡30cm (12in) ↔ 30cm (12in) ❆ ❆ H4 ☼ ☼ ◌ ◍

PERENNIALS FOR CONTAINERS

Bulbs, corms, and tubers

Allium caeruleum
An early summer-flowering ornamental onion, this allium has alluring ice-blue rounded flowerheads. The mid-green leaves disappear before the flowers open, leaving solitary "lollipops" punctuating the border display on stiff stems. Well-drained soil helps to prevent bulbs rotting in winter.

↕60cm (24in) ↔2.5cm (1in) ❄ ❄ ❄ H5 ☼ ◊

Allium cristophii
Huge, rounded flowerheads made up of many star-like, pinkish-purple blooms ensure this plant's place as a designers' favourite. A scattering of these bulbs among low-growing plants adds unexpected interest in early summer. The dried seedheads are spectacular in indoor arrangements.

↕30–60cm (12–24in) ↔15cm (6in) ❄ ❄ ❄ H5 ☼ ◊

Allium hollandicum 'Purple Sensation'
The deep purple, spherical flowerheads of 'Purple Sensation' look stunning when planted with silver-leaved, shorter plants. This is a summer-flowering bulb that will self-sow around the garden, although the resulting seedlings may not be so richly coloured. The blooms make decorative dried flowers.

↕1m (3ft) ↔7cm (3in) ❄ ❄ ❄ H6 ☼ ◊

Anemone blanda 'White Splendour'
Quick to establish and form a carpet, this white anemone brings a gleam of light to gardens in spring. For a different colour, try 'Radar', which has magenta flowers with a white eye, or 'Pink Star', with bright pink blooms. All look delightful in large drifts below spring-flowering trees.

↕15cm (6in) ↔15cm (6in) ❄ ❄ ❄ H6 ☼ ☼ ◊

Canna 'Durban'
Vividly coloured foliage and bright, "hot" flowers, which appear from late summer to autumn, make cannas an exotic addition to mixed borders. The deep purple, paddle-shaped leaves sometimes have contrasting midribs. Cannas look very attractive in containers, adding a tropical element to a patio.

↕1.2m (4ft) ↔60cm (24in) ❄ ❄ H3 ☼ ◊ ◗

Canna 'Striata'
A statement plant for a bed or border, 'Striata' has broad, rich green leaves striped with yellow, and showy, bright orange flowers, carried on dark red-purple stems, from midsummer to early autumn. As with most cannas, in cold areas rhizomes should be lifted to overwinter in a frost-free place.

↕1.5m (5ft) ↔50cm (20in) ❄ ❄ H3 ☼ ◊

Convallaria majalis
Lily-of-the-valley is a creeping perennial loved for its sweetly fragrant, white, bell-shaped flowers. Dark green leaves are upward-pointing, with leafless flowerstalks rising among them in late spring. The plant relishes moist, fertile soil in either full or partial shade. All parts are toxic.

↕23cm (9in) ↔30cm (12in) ❄ ❄ ❄ H7 ☼ ☼ ◗

Crinum × powellii
A very decorative plant, this lily produces flared trumpet blooms, up to ten at a time, at the top of rigid stems from late summer until mid-autumn. It suits a position at the base of a sheltered, sunny wall. In cooler areas, provide a deep winter mulch. For a pure white form, choose the cultivar 'Album'.

↕1.5m (5ft) ↔30cm (12in) ❄ ❄ ❄ H5 ☼ ◊ ◗

Crocosmia × crocosmiiflora 'Gerbe d'Or'
In a sunny border, the lemon-yellow trumpets of this South African plant will shine brightly against a background of bronze-tinted, mid-green foliage. Split the clumps every few years for a good supply of flowers. Crocosmias make excellent cut flowers and can be grown solely for this purpose.

↕75–90cm (30–36in) ↔45cm (18in) ❄ ❄ ❄ H5 ☼ ◊ ◗

Crocosmia x crocosmiiflora 'Venus'

The dense green, strappy foliage of this crocosmia is attractive even before the red blooms appear in summer. As each flower opens, a distinctive deep yellow throat is revealed. Overgrown clumps can be split and divided in spring and used to expand your border display.

↕70cm (28in) ↔45cm (18in) ❋ ❋ H4 ☼ ◊ ◖

Crocosmia masoniorum x C. x crocosmiiflora 'Firebird'

A strong-growing crocosmia, 'Firebird' has tapering, strap-like foliage, joined in summer by arching stems of bright orange-red flowers with speckled throats. It tolerates drier conditions than many crocosmias, and flowers freely.

↕80cm (32in) ↔30–45cm (12–18in) ❋ ❋ H4 ☼ ◊ ◖

Crocus goulimyi

This is one of the autumn-flowering crocuses, producing scented, long-tubed, lilac flowers at the same time as the leaves. It can be naturalized in a lawn in drifts, grown around the edges of mixed borders, or planted in containers on a patio (use a gritty potting mix to ensure free drainage).

↕10cm (4in) ↔5cm (2in) ❋ ❋ ❋ H6 ☼ ◊

Crocus tommasinianus

Silvery-lilac to purple petals are the distinguishing features of this late winter- to early spring-flowering crocus. Grow in naturalized drifts in grassy areas or in small clumps in terracotta pots on a windowsill. For a white-flowered selection, try *Crocus tommasinianus* f. *albus*.

↕8–10cm (3–4in) ↔2.5cm (1in) ❋ ❋ ❋ H6 ☼ ◊

Cyclamen hederifolium

These fluted pink flowers are carried above the soil surface in mid- to late autumn before the appearance of any foliage. The triangular or heart-shaped leaves are dark green with intricate silver patterning. The plant self-seeds freely and suits a site under trees or shrubs in partial shade. Mulch annually.

↕10–13cm (4–5in) ↔15cm (6in) ❋ ❋ ❋ H5 ☼ ◊

Dahlia 'Bishop of Llandaff'

The vivid red, semi-double flowers of this dahlia look dramatic against the black-red foliage, making it a striking addition to a mixed border from summer to autumn. 'Bishop of Llandaff' also suits containers. In frost-prone areas, tubers should be lifted after the first frost and stored in a cool, dry place.

↕1.1m (3½ft) ↔45cm (18in) ❋ ❋ H3 ☼ ◊

Dahlia 'David Howard'

The dark green-purple leaves and stems make an excellent foil for the large, double, burnt orange flowers of this dahlia. Stems can be used for indoor arrangements, and regular cutting will encourage further flowering. Site in a sunny border. See *D.* 'Bishop of Llandaff' for overwintering advice.

↕75cm (30in) ↔60cm (24in) ❋ ❋ H3 ☼ ◊

Dahlia 'Gay Princess'

Waterlily dahlias are so-called because of the flowerhead form, which is double and resembles a waterlily. This cultivar has lilac-pink blooms in summer and autumn, above rich green foliage. At 1.5m (5ft) tall it can be planted behind shorter perennials in a border or grown for cut flowers.

↕1.5m (5ft) ↔75cm (30in) ❋ ❋ H3 ☼ ◊

BULBS, CORMS, AND TUBERS FOR SPRING COLOUR

A range of bulbous plants will provide spring colour, including tulips, daffodils (*Narcissus*), crocuses, snowdrops (*Galanthus*), winter aconites (*Eranthis*) and hellebores.

Bulbs, corms, and tubers

Eranthis hyemalis
Buttercup-yellow cup-shaped flowers, surrounded by a collar of deeply-cut green leaves, are a welcome sight in the depths of winter. Relatives of buttercups, winter aconites rapidly spread by way of their underground tubers. Plant where the soil does not dry out in summer.

↕5–8cm (2–3in) ↔ 8cm (3in) ❄ ❄ ❄ H6 ☀ ◌ ◑

Erythronium dens-canis
The European dog's-tooth violet produces heavily marked green leaves and dainty nodding flowers from winter to early spring, in colours ranging from white through to pink. The plant likes well-drained soil in dappled shade, and looks attractive grown underneath deciduous trees or shrubs.

↕10–15cm (4–6in) ↔ 10cm (4in) ❄ ❄ ❄ H5 ☀ ◌

Eucomis bicolor
The pineapple lily from South Africa needs full sun and rich soil in order to flourish. Maroon-spotted stems appear among the leaves in late summer, bearing pale green flowers with purple markings. It will grow best in a sheltered bed against a warm wall. Mulch dormant bulbs in very hard winters.

↕30–60cm (12–24in) ↔ 20cm (8in) ❄ ❄ H4 ☀ ◌

Fritillaria imperialis
Tall, stately, and strong-growing, the crown imperial stands regally in the centre of an island bed or within a mixed border or rock garden. Clusters of orange flowers, yellow if you choose the cultivar 'Maxima Lutea', radiate from the top of tall stems in early summer.

↕to 1.5m (5ft) ↔ 25–30cm (10–12in) ❄ ❄ ❄ H7 ☀ ◌

Fritillaria meleagris
A native of English grasslands, the snake's head fritillary looks stunning when planted en masse in grassy areas, each petal featuring a distinctive chequered pattern. These spring-flowering bulbs in pinkish-purple or white can be mixed to create a patchwork effect.

↕to 30cm (12in) ↔ 5–8cm (2–3in) ❄ ❄ ❄ H5 ☀ ☀ ◑

Galanthus 'Atkinsii'
The cold season would not be the same without snowdrops, and there are plenty of cultivars to choose from. They flower from late winter and can be planted in grass or in small pots on their own. Lift and divide clumps when the leaves die back. 'Atkinsii' is vigorous, with slender green-marked flowers.

↕20cm (8in) ↔ 8cm (3in) ❄ ❄ ❄ H5 ☀ ◌ ◑

Galtonia viridiflora
A hyacinth relative from South Africa, galtonia has funnel-shaped, pale green flowers which add glistening highlights to a border. The flowers appear in late summer, suspended from tall arching stems. In very cold areas, lift the bulbs over winter and store in a cool spot indoors.

↕to 1m (3ft) ↔ 10cm (4in) ❄ ❄ H3 ☀ ◌ ◑

Hyacinthoides non-scripta
This is the English bluebell rather than the more upright-growing Spanish species. Plant the bulbs in broad drifts under trees in dappled shade for maximum impact in spring. Flowers are traditionally blue, although pink or white forms can be found. It can become invasive if planted in the border.

↕20–40cm (8–16in) ↔ 8cm (3in) ❄ ❄ ❄ H6 ☀ ◌ ◑

Hyacinthus orientalis 'Blue Jacket'
Famed for their exquisitely perfumed flowers, hyacinths are very easy to grow. They are available in a range of colours and the bulbs can be planted as spring bedding, singly in pots, or even rooted in water on a windowsill indoors. 'Blue Jacket' has navy-blue, waxy flowers with purple veins.

↕20–30cm (8–12in) ↔ 8cm (3in) ❄ ❄ H4 ☀ ◌

Iris 'Golden Alps'
This cream and yellow, tall bearded iris should be planted with its lower stem and rhizome just above soil level. Sword-shaped green leaves form a fan, while summer flowers are held high on sturdy stems. Bearded irises come in a range of colours, and all are ideal for a sunny, mixed border.

↕90cm (36in) ↔60cm (24in) ✤ ✤ H4 ☼ ◌

Iris pallida 'Variegata'
The long, tapering, yellow-striped leaves of this iris surround a succession of showy, scented blue flowers in late spring and early summer. This is a perfect plant for a hot border or exposed site where the sun can bake the soil surface. Lift clumps, divide, and replant in early autumn.

↕to 1.2m (4ft) ↔45–60cm (18–24in) ✤ ✤ ✤ H6 ☼ ◌

Iris 'Superstition'
Purple-brown and blue-black combine here with dramatic effect in this deeply coloured, tall bearded iris. Plant with pale-coloured selections such as 'White Knight' to create a contrasting combination. The dark flowers are also fragrant, and appear almost black in fading light.

↕90cm (36in) ↔60cm (24in) ✤ ✤ ✤ H7 ☼ ◌

Leucojum aestivum 'Gravetye Giant'
Similar to a large snowdrop, the summer snowflake is an attractive plant for damp areas of the garden. Nodding white flowers with green petal tips emerge in spring; the narrow green leaves providing a subtle backdrop. 'Gravetye Giant' is robust and will grow quite tall next to water.

↕90cm (36in) ↔8cm (3in) ✤ ✤ ✤ H7 ☼ ◌ ◖

Lilium 'African Queen'
Place some pots of these by your back door and you will be greeted by deliciously fragrant, bright orange trumpet flowers every time you step outside from mid- to late summer. This lily can also be grown in a border, if the flowers are in the sun while the roots are kept shaded.

↕1.5–2m (5–6ft) ↔25cm (10in) ✤ ✤ ✤ H6 ☼ ◌

Lilium 'Black Beauty'
Lilies with this flower form are known as turk's caps because of the way the petals curve back on themselves, revealing pollen-laden anthers. 'Black Beauty' is a vigorous type and can be positioned among herbaceous plants in the border, or grown in containers for a movable midsummer display.

↕1.4–2m (4½–6ft) ↔25cm (10in) ✤ ✤ ✤ H6 ☼ ◌

Lilium Citronella Group
An Asiatic lily, this group contains yellow-flowered plants with speckled, recurved petals. In midsummer, tall flower spikes are held erect and the buds unfold to reveal the showy blooms. It is vigorous, but needs feeding if it is to flourish. It also makes a superb cut flower.

↕1.2–1.5m (4–5ft) ↔25cm (10in) ✤ ✤ ✤ H6 ☼ ◌

Lilium martagon
Scatter bulbs of the common turk's-cap lily around a mixed border and plant them where they land. The pretty flowers, which have recurved purple petals with dark markings, appear from early to midsummer. The flowers of Lilium martagon var. album are pure white.

↕0.9–2m (3–6ft) ↔20cm (8in) ✤ ✤ ✤ H6 ☼ ☀ ◌

BULBS, CORMS, AND TUBERS FOR SUMMER COLOUR

Bulbs, corms, and tubers

Lilium Pink Perfection Group
First introduced in 1950, the large, pinkish-red trumpets of this lily hybrid soon caught the attention of keen gardeners. In midsummer, short flower stems are laden with lightly-scented blooms with protruding orange anthers. Choose a sunny site with some shade for the roots for best results.

‡1.5–2m (5–6ft) ↔25cm (10in) ❀ ❀ ❀ H6 ☼ ◊

Lilium regale
The large, white, trumpet-shaped flowers of the regal lily are purple on the outside and held in clusters on tall stems, creating an eye-catching display in midsummer. The lilies are very fragrant and are ideal for use in mixed borders or as cut flowers. The stems may need staking.

‡0.6–2m (2–6ft) ↔25cm (10in) ❀ ❀ ❀ H6 ☼ ◊

Lilium 'Star Gazer'
Both the colour and the perfume of 'Star Gazer' attract attention and make this Oriental lily one of the most popular cut flowers ever developed. The pink and white flowers with speckled petals are upward-facing and robust, and appear in midsummer. Plant in a border or in a stylish container.

‡1–1.5m (3–5ft) ↔25cm (10in) ❀ ❀ ❀ H6 ☼ ◊

Muscari armeniacum 'Blue Spike'
This is a double-flowered form of the common grape hyacinth. Fleshy green narrow leaves form a carpet as small fat spikes of blue flowers push their way through in spring. The plant can become invasive, so restrict its spread by growing it in a container. Choose a site in full sun.

‡20cm (8in) ↔5cm (2in) ❀ ❀ ❀ H6 ☼ ◊ ◊

Muscari latifolium
The flowers of this grape hyacinth seem to be wearing little hats. Blue flowerspikes are topped by small, paler-coloured flowers, while the leaves are mid-green and more flattened than those of *Muscari armeniacum* (*left*). Attractive in drifts at the front of a border, it is also good for a rock garden.

‡20cm (8in) ↔5cm (2in) ❀ ❀ ❀ H6 ☼ ◊ ◊

Narcissus 'Bridal Crown'
'Bridal Crown' has sweetly-scented double white blooms with pale orange centres. The flowers cluster together at the top of the stems and appear in early spring. Plant bulbs during autumn in well-drained soil in a sunny border, or in a container. 'Bridal Crown' makes a pretty cut flower.

‡40cm (16in) ↔15cm (6in) ❀ ❀ ❀ H6 ☼ ◊

Narcissus poeticus var. recurvus
Known as the old pheasant's eye, this late spring-flowering daffodil differs from *Narcissus poeticus* in having backward-curving petals. Pure white petals surround a yellow eye, which has a dainty orange frilled edge. It can be naturalized in a lawn, and is also good for cut flowers for the house.

‡35cm (14in) ↔5–8cm (2–3in) ❀ ❀ ❀ H6 ☼ ◊

Narcissus 'Tête-à-Tête'
Tiny flowers on short stems make this a favourite spring bulb for planting at the front of borders, in rock gardens, and in containers of all shapes and sizes. Plant en masse for the best effect, as small clumps can look insignificant. Container-grown plants can be grown on a windowsill indoors.

‡15cm (6in) ↔5cm (2in) ❀ ❀ ❀ H6 ☼ ◊

Narcissus 'Thalia'
This delicately beautiful daffodil carries two milky-white flowers per stem. Mid-spring sees these emerge from papery buds to lighten border plantings or provide early interest in a "white" border. Grow in a tall container and place against a painted wall to make a bold statement.

‡35cm (14in) ↔8cm (3in) ❀ ❀ ❀ H6 ☼ ◊

Nectaroscordum siculum subsp. bulgaricum
The flowers on this onion relative are green, white, and burgundy. Grouped in sprays of 10–30 on top of tall stems, they make an attractive display in early summer. Grow in a wild garden or herbaceous border where the flowers will catch the eye. Deadhead to prevent it spreading.

‡to 1.2m (4ft) ↔ 30–45cm (12–18in) ❀ ❀ ❀ H5 ☀ ◖

Nerine bowdenii
South Africa has given gardeners worldwide many wonderful plants and this spectacular bulb is no exception. Stems of vivid pink, spidery flowers appear from bare soil in autumn. Nerines look good in groups at the foot of a sunny, light-coloured wall. Provide a deep mulch in winter in very cold areas.

‡45cm (18in) ↔ 8–12cm (3–5in) ❀ ❀ ❀ H5 ☼ ◊

Scilla siberica
The Siberian squill produces bright blue, pendent flowers in spring, giving the garden a dash of colour. The bulbs can be grown in groups in a rock garden, between paving stones or at the front of herbaceous and mixed borders. Plant in full sun or part shade, and water well when in growth.

‡10–20cm (4–8in) ↔ 5cm (2in) ❀ ❀ ❀ H6 ☼ ☀ ◊

Trillium grandiflorum
A vigorous plant for a shady woodland area, wake robin forms clumps of dark green, rounded leaves with distinctive, three-petalled white flowers in spring and summer. The cultivar 'Flore Pleno' is slower-growing and has double flowers.

‡to 40cm (16in) ↔ 30cm (12in) or more
❀ ❀ ❀ H5 ☀ ☀ ◊

Tulipa 'Flaming Parrot'
This late spring-flowering tulip has fringed yellow petals, each with a distinctive red blaze. Inside is a cluster of black anthers. Grow as a single variety in formal beds or in drifts, merging with other colours. Alternatively, plant a number of the bulbs in a tall pot or container in a sunny position.

‡55cm (22in) ↔ 15cm (6in) ❀ ❀ ❀ H6 ☼ ◊

Tulipa 'Prinses Irene'
The orange petals of this striking tulip look like they have been painted with delicate brush strokes of purple. Flowering in mid-spring, 'Prinses Irene' is effective when grouped in swathes in a border or planted as part of a container display with decorative grasses. It can also be cut for indoor arrangements.

‡35cm (14in) ↔ 15cm (6in) ❀ ❀ ❀ H6 ☼ ◊

Tulipa 'Queen of Night'
Popular because it is so deeply coloured and satiny, this late spring-flowering tulip looks striking if planted among purple and black-leaved perennials and low shrubs, or with grey or silver-leaved plants. Alternatively, use it in front of a pale-painted fence or wall for contrast.

‡60cm (24in) ↔ 15cm (6in) ❀ ❀ ❀ H6 ☼ ◊

Tulipa 'Spring Green'
This Viridiflora Group tulip sports a green feathery flash on each of its ivory-white petals and adds an elegant touch to a mixed or colour-themed border. Plant where it can be appreciated at close quarters, as it is only 40cm (16in) high when flowering in late spring.

‡40cm (16in) ↔ 10cm (4in) ❀ ❀ ❀ H6 ☼ ◊

BULBS, CORMS, AND TUBERS TO USE FOR SCENT

Plant groups with a range of scented cultivars include many daffodils (Narcissus), crocuses, lilies, some snowdrops (Galanthus), Leucojum (snowflake), hyacinths, cyclamen, and freesias.

- Convallaria majalis p.338
- Crocus goulimyi p.339
- Hyacinthus orientalis 'Blue Jacket' p.340
- Leucojum aestivum 'Gravetye Giant' p.341 (light scent)
- Lilium 'African Queen' p.341
- Lilium 'Black Beauty' p.341
- Lilium Citronella Group p.341
- Lilium martagon p.341
- Lilium Pink Perfection Group p.342
- Lilium 'Star Gazer' p.342
- Narcissus 'Bridal Crown' p.342
- Narcissus poeticus var. recurvus p.342

Grow bulbs in pots by the house or in drifts for maximum appreciation.

Grasses, sedges, and bamboos

Acorus calamus 'Argenteostriatus'

An undemanding evergreen, the sweet rush, or sweet flag, thrives in damp or boggy soils, making it the perfect plant for the shallows of a pond edge. Like all acorus, it is non-invasive, and its strong cream variegation will remain vivid, even in deep shade.

↕45cm (18in) ↔ 45cm (18in) ❄ ❄ ❄ H7 ☼ ◐ ◊ ◆

Anemanthele lessoniana

Fine-leaved pheasant's-tail grass has a pleasing arching habit. In summer, it produces purplish flower spikes; in winter, the evergreen leaves turn an eye-catching orange-brown. Leave the seedheads – hungry birds will quickly tidy them up during winter. The plant may need protection in cold areas.

↕1m (3ft) ↔ 1.2m (4ft) ❄ ❄ H4 ☼ ◐ ◊ ◊

Arundo donax var. versicolor

The striking variegation of the evergreen giant reed (the white stripes turn a creamy yellow in summer) makes it a popular choice, although it is less vigorous than the green form and not as hardy. In cold areas, enjoy it outdoors in summer, then bring it under cover for the winter; grow it in a pot for flexibility.

↕2.2m (7ft) ↔ 2m (6ft) ❄ ❄ H4 ☼ ◐ ◊ ◆

Briza maxima

One of the most attractive of the annual grasses, quaking grass is easy to grow from seed (sow into individual modules for the best results). The nodding flowerheads rattle in the lightest breeze, making it clear how the common name arose. The stems dry well for flower arranging.

↕30cm (1ft) ↔ 23cm (9in) ❄ ❄ ❄ H6 ☼ ◊ ◊ ◆

Calamagrostis x acutiflora 'Overdam'

Use the striped feather reed to make a strong vertical accent in prairie-style planting. As the leaves emerge in spring, there is a pink tinge to the green and white variegation; cutting the foliage back in late summer will encourage a second flush of new growth. Unfussy, the plant tolerates most soils.

↕1m (3ft) ↔ 1.2m (4ft) ❄ ❄ ❄ H6 ☼ ◊

Carex buchananii

This striking evergreen sedge from New Zealand has slender, coppery-brown leaves with a hint of a curl. It is stiffly upright when young, becoming more arching with age, and it contrasts well with golden sedges and blue grasses. In early spring, comb out any dead leaves with a fork, or cut them back.

↕60cm (2ft) ↔ 60cm (2ft) ❄ ❄ ❄ H5 ☼ ◐ ◊ ◊

Carex elata 'Aurea'

Deservedly one of the most widely grown sedges, Bowles' golden sedge produces a broad spray of vibrant yellow leaves, edged in green. In summer there is the added bonus of feathery brown flower spikes. A compact, deciduous plant, it produces its best colour in partial shade.

↕75cm (30in) ↔ 1m (3ft) ❄ ❄ ❄ H6 ☼ ◐ ◊ ◆

Carex oshimensis 'Evergold'

The low-arching habit of this neat evergreen sedge makes it a useful plant for containers or as ground cover in shade, where its long golden yellow and thinly striped green leaves add a touch of light colour. Like many sedges, it is happy in boggy soil and makes a decorative addition to poolside plantings.

↕50cm (20in) ↔ 45cm (18in) ❄ ❄ ❄ H7 ☼ ◐ ◊ ◆

Carex testacea

In full sun, the hair-thin, olive-yellow leaves of this sedge develop orange tints. In midsummer, small brown flower spikes appear. A New Zealand plant, it forms dense, evergreen mounds, but it is not as hardy as its relatives and it may need winter protection in cold areas.

↕45cm (18in) ↔ 1m (3ft) ❄ ❄ ❄ H5 ☼ ◐ ◊ ◊

Cortaderia selloana 'Aureolineata'
Ideal for small gardens, this dwarf pampas is half the size of the parent species, and has broad leaves with golden edges that become more richly coloured as the season progresses. The colourful leaves and silky plume-like flowerheads add a dramatic highlight to late summer borders and gravel gardens.

↕1.5m (5ft)↔1.5m (5ft) ❀ ❀ ❀ H6 ☀ ◌ ◖

Cortaderia selloana 'Pumila'
Hardier and more free-flowering than the taller species, this dwarf pampas grass mixes surprisingly well in a border. Long-lasting golden-brown plumes are produced in summer on stout stems. Combing through the leaves with a hand fork in winter will keep the clump looking tidy.

↕2m (6ft)↔2m (6ft) ❀ ❀ ❀ H6 ☀ ◌ ◖

Deschampsia flexuosa 'Tatra Gold'
Wavy hair grass forms slowly spreading tufts of fine evergreen leaves. 'Tatra Gold' grows well in moist shade, where its acid-green leaves look almost luminous. In summer, it produces a shimmering haze of red-brown flowers. Plant it in large drifts among bright leaved sedges for a dramatic effect.

↕15cm (6in) ↔ 15cm (6in) ❀ ❀ ❀ H6 ☀ ☀ ◖

Elymus magellanicus
Blue wheatgrass is so-named because of its wonderful blue colour – it looks stunning against a gravel mulch – and the herringbone flowerheads that look like ears of wheat. It forms slow-spreading, rather sprawling clumps of evergreen leaves that need winter protection in cold areas.

↕45cm (18in)↔45cm (18in) ❀ ❀ ❀ H6 ☀ ◌ ◖

Fargesia murielae
A tough plant for tough situations, this evergreen bamboo copes well with dry soils and exposed sites, and makes an effective windbreak or screen. The closely spaced, arching canes are slow-spreading, and it won't engulf its neighbours. Use it at the back of a border or in a container.

↕4m (12ft) ↔ 4m (12ft) ❀ ❀ ❀ H5 ☀ ☀ ◌ ◖

Festuca glauca 'Elijah Blue'
One of those useful plants that look good all year round, the silvery-blue, needle-like leaves of this fescue form neat, round mounds. In summer, the plant produces spikes of small blue flowers that age to brown. It is particularly effective grown as a container plant, contrasting well with terracotta and metal.

↕30cm (12in) ↔ 60cm (24in) ❀ ❀ ❀ H5 ☀ ☀ ◌ ◖

Hakonechloa macra 'Aureola'
A beautiful slow-growing, deciduous grass from Japan that deserves to be the centrepiece in a container or a dry gravel border. The low-arching, golden yellow leaves, which are thinly striped with lime green, develop a warm reddish tinge in autumn. Cut back in early spring to encourage new growth.

↕25cm (10in) ↔ 1m (3ft) ❀ ❀ ❀ H7 ☀ ☀ ◖

Imperata cylindrica 'Rubra'
Japanese blood grass is undisputedly one of the finest foliage plants – fluffy white flowerspikes are a bonus in summer. Position it carefully, so the crimson-tipped, upright leaves are backlit by the sun. In cold areas, grow it in a container and bring under cover during winter.

↕45cm (18in) ↔ 1.8m (6ft) ❀ ❀ ❀ H4 ☀ ☀ ◖

GRASSES, SEDGES, AND BAMBOOS FOR CONTAINERS

Grasses, sedges, and bamboos

Lagurus ovatus

A popular garden plant because of its fluffy flowerheads, the hare's-tail grass is a tufted annual that can be grown easily from seed sown in situ in spring. The soft, hairy spikelets, pale green at first, maturing to pale cream, form in summer and can be cut for indoor displays.

↕to 50cm (20in) ↔30cm (12in) ❋ H4 ☀ ◊

Miscanthus sinensis 'Gracillimus'

A dainty-looking subject for a grass garden or mixed border, maiden grass produces a shock of narrow green leaves with white midribs. After the late summer flush, the curved leaves take on a bronzy hue as temperatures cool. Leave in place as a structural element through the winter.

↕1.3m (4½ft) ↔1.2m (4ft) ❋ ❋ ❋ H6 ☀ ◊ ◊

Miscanthus sinensis 'Kleine Silberspinne'

An attractive ornamental grass with colourful, curving plumes, this miscanthus does not grow as tall as the species. In late summer and early autumn, silky white and red flower spikes appear, turning to silver as they age and lasting all winter. Cut down to ground level in spring before new growth emerges.

↕1.2m (4ft) ↔1.2m (4ft) ❋ ❋ ❋ H6 ☀ ◊ ◊

Miscanthus sinensis 'Malepartus'

One of the easiest of the miscanthus to establish, 'Malepartus' looks good spilling onto a lawn or path edge where it can be seen at close quarters. Feathery reddish-brown flowerheads, maturing to cream, appear from late summer to autumn among the cascading green foliage.

↕2m (6ft) ↔2m (6ft) ❋ ❋ ❋ H6 ☀ ◊ ◊

Miscanthus sinensis 'Silberfeder'

This cultivar is grown mainly for its autumn show of red-tinged, creamy flowers that last well and are held above narrow, green foliage. 'Silberfeder' needs space to be seen at its best and a site that doesn't get waterlogged. Plant in front of a dark-leaved hedge for a perfect backdrop.

↕2.5m (8ft) ↔1.2m (4ft) ❋ ❋ ❋ H6 ☀ ◊ ◊

Miscanthus sinensis 'Zebrinus'

Easily confused with the more upright-growing *M. sinensis* 'Strictus', 'Zebrinus' has a more lax habit and spreads more readily. The unusual horizontal bands of pale cream variegation make it an interesting subject for a grass garden or large zinc planter. The brown deciduous foliage offers winter interest.

↕to 1.2m (4ft) ↔1.2m (4ft) ❋ ❋ ❋ H6 ☀ ◊ ◊

Molinia caerulea subsp. *caerulea* 'Variegata'

This is a densely tufted perennial with boldly variegated green and cream leaves. From spring through to autumn, purple-tinged flowers are borne on yellow flower stems. The whole plant matures to a pale bronzy-brown in autumn, an effect that looks striking in a gravel garden.

↕45–60cm (18–24in) ↔40cm (16in) ❋ ❋ ❋ H7 ☀ ☀ ◊

Ophiopogon planiscapus 'Nigrescens'

Few plants are as deeply coloured as this clump-forming, tufted perennial. Although not strictly a grass, its appearance and habit make it a useful plant in schemes where grasses predominate. It also looks dramatic in pale-coloured containers. Small, pale purplish-white flowers appear in summer.

↕20cm (8in) ↔30cm (12in) ❋ ❋ ❋ H5 ☀ ☀ ◊ ◊

Panicum virgatum 'Heavy Metal'

A deciduous perennial grass with stiff, upright, steely grey-green leaves. In favourable conditions, the foliage will turn yellow in autumn, gradually fading to pale brown in winter. Wispy flowerheads bearing purple-green flowers emerge during summer. Plant in clumps of threes or fives for impact.

↕1m (3ft) ↔75cm (30in) ❋ ❋ ❋ H5 ☀ ◊

Pennisetum alopecuroides

Also known appropriately as the fountain grass, this evergreen perennial has narrow, mid-green leaves that tumble from the centre of the plant, joined in summer and autumn by flowing, bristly, decorative flowerheads. It needs a warm, sheltered site since it is not fully hardy.

↕0.6–1.5m (2–5ft) ↔0.6–1.2m (2–4ft) ❀ ❀ H3 ☼ ◊

Phalaris arundinacea var. picta

Gardeners' garters is a vigorous, spreading plant, useful for lightening a shady corner or in a cottage garden. Trim untidy leaves in late summer to maintain a neat look. New plantlets will spread if the clump is not kept in check, so grow in a container sunk into the ground if this is a concern.

↕to 1m (3ft) ↔indefinite ❀ ❀ ❀ H7 ☼ ☼ ●

Phyllostachys aureosulcata f. aureocaulis

A delightful mix of green-streaked yellow stems and green, tapering leaves make this evergreen bamboo a popular garden choice. The yellow-groove bamboo, as it is known, is a vigorous plant and is recommended for larger gardens, where it can be used as a screen. It can also be grown in containers.

↕3–6m (10–20ft) ↔indefinite ❀ ❀ ❀ H5 ☼ ☼ ◊ ◖

Phyllostachys nigra

The black bamboo is grown for its distinctive stems, which are initially green and then turn glossy black, contrasting well with the fresh green leaves. It has a tall, upright habit, so grow for impact in a border, or in blocks in a Modernist scheme.

↕3–5m (10–15ft) ↔2–3m (6–10ft)
❀ ❀ ❀ H5 ☼ ☼ ◊ ◖

Phyllostachys vivax f. aureocaulis

Like many bamboos, this is a vigorous, fast-growing plant. The bright yellow canes are flecked with green and it has slim, arching foliage. Plant it in a large container, or surround the plant's roots below soil level with an impenetrable barrier to control its spread.

↕to 8m (25ft) ↔4m (12ft) ❀ ❀ ❀ H5 ☼ ☼ ◊ ◖

Stipa gigantea

Giant feather grass is a fabulous plant for the garden, commanding a prime position in an island bed or mixed border in full sun. Tall, fluttering plumes of flowers emerge above the evergreen foliage in summer; the stems create a transparent screen, allowing shorter plants to be seen behind them.

↕to 2.5m (8ft) ↔1.2m (4ft) ❀ ❀ ❀ (borderline)H4 ☼ ◊

Stipa tenuissima

In summer, this neat, compact, deciduous perennial produces soft feathery stems with green flowerheads that fade to buff. The fine leaves gently wave in the slightest breeze, and contrast well with dark green foliage plants. The autumn seedheads are very attractive to birds.

↕60cm (24in) ↔30cm (12in) ❀ ❀ ❀ (borderline)H4 ☼ ◊

Uncinia rubra

The tough ochre-red leaves of this evergreen perennial are three-angled and upright, joined in mid- and late summer by dark brown flowers. It makes an unusual specimen for a gravel or scree garden where the soil is free-draining but not too dry. Protect from the elements in very cold winters.

↕30cm (12in) ↔35cm (14in) ❀ ❀ H3 ☼ ☼ ◊ ◖

EVERGREEN GRASSES, SEDGES, AND BAMBOOS

Water and bog plants

Actaea simplex Atropurpurea Group 'Brunette'
A herbaceous perennial for a damp, shady area in the garden, 'Brunette' has bronze, deeply-cut foliage and slender spires of fluffy, fragrant white flowers in late summer, which show up well against a dark background. Plant in moisture-retentive soil in a woodland or shady bog garden.

‡1.2m (4ft) ↔60cm (24in) ✳ ✳ ✳ H7 ☼ ～

Aruncus dioicus 'Kneiffii'
Fern-like foliage and tumbling flowerheads resembling small white caterpillars combine to create this striking plant. The flowers appear in summer and make a bright focal point in a bog garden or at a pond edge. It looks delicate, but is in fact robust and will tolerate full sun or part shade.

‡75cm (30in) ↔45cm (18in) ✳ ✳ ✳ H6 ☼ ☼ ～

Astilbe 'Fanal'
Producing feathery plumes of long-lasting, crimson flowers in early summer, 'Fanal' adds fiery interest to a garden with boggy soil. Finely cut, dark green leaves provide a suitable backdrop for the intense flower colour. Plant in groups of threes or fives to make a bold statement.

‡60–100cm (2–3ft) ↔60cm (24in) ✳ ✳ ✳ H7 ☼ ～

Astilbe 'Professor van der Wielen'
A plant that needs space to show off its full potential, this astilbe produces large, arching sprays of delicate creamy-white flowers in midsummer above fern-like foliage. Place at the back of a wet border or pond-edge planting scheme, and divide clumps every three to four years.

‡1.2m (4ft) ↔to 1m (3ft) ✳ ✳ ✳ H7 ☼ ～

Astilbe 'Willie Buchanan'
This astilbe cultivar produces a haze of pink when its tiny white flowers with red stamens, borne on fine, branching flower stems, open from mid- to late summer. Ideal for a pond or path edge, plant en masse for a wonderful floral display. The flowers attract beneficial insects.

‡23–30cm (9–12in) ↔20cm (8in) ✳ ✳ ✳ H7 ☼ ～

Butomus umbellatus
The flowering rush is a deservedly popular plant for pond margins, where it can immerse its feet in wet soil. The leaves are narrow and angled, bronze-purple when young, turning to mid-green. In late summer, delicate, pale pink, fragrant flowers are borne on slender stems.

‡1m (3ft) ↔unlimited ✳ ✳ ✳ H5 ☼ ≈⊥ 5–15cm (2–6in)

Caltha palustris
Marsh marigolds bring colour to pond margins as their intense yellow, cup-shaped blooms appear in late spring. Grow in planting baskets to control their spread. Try *C. palustris* var. *alba* for white flowers.

‡60cm (24in) ↔45cm (18in) ✳ ✳ ✳ H7 ☼ ☼
≈⊥ at water level

Darmera peltata
The umbrella plant is a slow-spreading perennial that looks good alongside streams and pond margins. Heads of white to pink flowers appear in late spring on long stems before the large, rounded green leaves appear. The foliage gradually turns red in autumn before dying down.

‡1.2m (4ft) ↔unlimited ✳ ✳ ✳ H6 ☼ ☼ ～

Eupatorium maculatum 'Atropurpureum Group'
A great plant for late summer and early autumn colour, this stately perennial bears clusters of small pink flowers on tall, purple stems. Toothed, purple-green leaves circle the stems right up to the flowerheads. It attracts bees and butterflies, and makes a superb addition to a wildlife bog garden.

‡2m (6ft) ↔1m (3ft) ✳ ✳ ✳ H7 ☼ ☼ ～

❀ ❀ ❀ H7–H5 fully hardy ❀ ❀ H4–H3 hardy in mild regions/sheltered sites ❀ H2 protect from frost over winter ❀ H1c–H1a no tolerance to frost ☼ full sun ☀ partial sun ☀ full shade ∿ bog plant ≈ marginal plant ≈ aquatic plant ⊥ planting depth

Filipendula rubra 'Venusta'

The queen of the prairies needs space to spread, so choose a planting position for this perennial carefully. Green jagged leaves sit below wiry stems bearing a frothy display of deep rose-pink flowers in early and midsummer. Use its height to form a screen at the back of a bog garden display.

↕2m (6ft) ↔unlimited ❀ ❀ ❀ H5 ☼ ☀ ∿

Gunnera manicata

A real giant of the bog garden with huge, rhubarb-like leaves, gunnera demands plenty of room, even for just one plant. A herbaceous perennial, it makes a dramatic statement at the waterside. Plant in permanently moist soil and cover the crowns with a dry mulch in hard winters.

↕4.5m (15ft) ↔3m (10ft) ❀ ❀ ❀ H5 ☼ ☀ ∿

Iris laevigata

This iris flourishes reliably in the wet soil in the shallows of ponds and streams. Blue-purple flowers crown green stems in early and midsummer, and sit among broad, sword-shaped, mid-green leaves. Clumps will spread steadily.

↕75cm (30in) ↔1m (3ft) ❀ ❀ ❀ H6 ☼
≈ ⊥ 10–15cm (4–6in)

Iris pseudacorus 'Variegata'

This is the variegated-leaved version of the well-known yellow flag iris. Pale yellow stripes decorate the green, upright leaves when young; the yellow blooms appear in summer. A spreading iris, it needs restricting if it is not to become invasive. Plant in a basket at the margins of a pond.

↕1m (3ft) ↔75cm (30in) ❀ ❀ ❀ H7 ☼ ≈ ⊥ 15cm (6in)

Iris 'Butter and Sugar'

Bred from the Siberian iris, 'Butter and Sugar' bears shapely flowers with white upper petals and butter-yellow lower petals from mid- to late spring. Each stem is surrounded by green strappy foliage and can hold up to five blooms. Divide the tight clumps in spring or once flowers have faded.

↕50cm (20in) ↔25cm (10in) ❀ ❀ ❀ H7 ☼ ∿

Iris sibirica 'Perry's Blue'

This is a traditional cultivar producing closely spaced flower stems that carry mid-blue flowers with rusty-coloured veins. It flowers in early summer and will bring colour to the edges of small ponds and borders with boggy soil. Plant with lighter-flowered irises for duo-tone effect.

↕1m (3ft) ↔60cm (24in) ❀ ❀ ❀ H7 ☼ ∿

Iris versicolor 'Kermesina'

From eastern North America, the blue flag is a small iris for small ponds. In summer, the species has lavender-blue flowers with white markings, while 'Kermesina' bears red-purple blooms. The long, strappy leaves add architectural interest to a pond margin from spring until autumn when they die down.

↕75cm (30in) ↔60cm (24in) ❀ ❀ ❀ H7 ☼ ≈ ⊥ 5cm (2in)

Kirengeshoma palmata

An unusual plant for the bog garden, this clump-forming perennial has jagged green leaves with reddish-purple stems. Pale yellow, bell-shaped flowers hang from the slim stems above the foliage in late summer and early autumn. Plant in moist acid soil in a part-shaded sheltered site.

↕1.2m (4ft) ↔75cm (30in) ❀ ❀ ❀ H7 ☼ ∿

PLANTS FOR YOUR POND

Water and bog plants

Ligularia stenocephala 'The Rocket'
A plant of contrasts with jet black flower stems and bright yellow flowers, this bog lover is a must for larger gardens. The leaves form a carpet through which the flower spikes emerge from early to late summer. Choose a bright site but one that is shaded from the midday sun.

‡2m (6ft) ↔1.1m (3½ft) ❋ ❋ ❋ H6 ☼ ◐ ∼

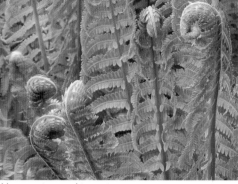

Matteuccia struthiopteris
The common names of shuttlecock fern and ostrich fern can be easily understood when the enormous finely dissected fronds emerge from the ground in spring. During late summer, fertile, narrow brown fronds cluster at the centre of the plant and last through winter. Grow in moist shade.

‡1.7m (5½ft) ↔ to 1m (3ft) ❋ ❋ ❋ H5 ◐ ∼

Myosotis scorpioides
Plant the water forget-me-not close to a pond edge, where its flowers can be seen clearly. The tiny blue blooms have white, pink, or yellow eyes and appear in early summer. The cultivar 'Mermaid' has a more compact habit.

‡45cm (18in) ↔unlimited ❋ ❋ ❋ H6 ☼ ◐
≈⊥ at water level

Nymphaea 'Darwin'
The almost peony-like, fragrant flowers of this waterlily are pale pink in the centre while the outermost petals are white with a tinge of pink. With its large, flat, dark green leaves and vigorous growth, 'Darwin' (also sold as Hollandia) is best suited to medium-sized to large ponds.

↔1.5m (5ft) ❋ ❋ H3 ☼ ≈⊥ 60–100cm (2–3ft)

Nymphaea 'Froebelii'
Tiny burgundy-red flowers with golden stamens open between the dark green leaves (bronze when young) Of 'Froebelii' to make a perfect miniature water lily. Ideal for small ponds, tubs, or half-barrels, it will put on a beautiful flower display from midsummer to autumn.

↔75cm (30in) ❋ ❋ ❋ H5 ☼ ≈⊥ 30–45cm (12–18in)

Nymphaea 'Gonnère'
A stunning water lily for medium-sized ponds, 'Gonnère' sends up pure white fragrant flowers with yellow stamens from mid- to late summer. The circular lily pads are bronze when young but soon turn a light pea-green. Grow in full sun for the best results.

↔1.5m (5ft) ❋ ❋ ❋ H5 ☼ ≈⊥ 60–75cm (24–30in)

Nymphaea 'Marliacea Chromatella'
This is a very old cultivar that has stood the test of time. Lemon-yellow flowers, with broad incurved petals and deep yellow centres, are produced from mid- to late summer and appear between floating olive-green leaves with bronze markings. Plant in a medium-sized pond or pool in full sun.

↔1.5m (5ft) ❋ ❋ ❋ H5 ☼ ≈⊥ 60–100cm (2–3ft)

Osmunda regalis
The royal fern makes an arresting sight at the edge of a pond with its toes just in the water. It is deciduous, producing a crop of fresh, mid-green sterile fronds that gracefully unfurl each spring. In summer, upright, fertile, tassel-like fronds form in the centre of the plant. This fern needs space to spread.

‡2m (6ft) ↔4m (12ft) ❋ ❋ ❋ H6 ☼ ◐ ∼

Pontederia cordata
Pretty from a distance, this plant is exquisite close-up. The pickerel weed is a marginal plant with bright green, lance-shaped leaves with spikes of starry blue flowers in late summer. There is also a white-flowered cultivar, 'Alba'.

‡0.9–1.3m (3–4½ft) ↔60–75cm (24–30in)
❋ ❋ ❋ H5 ☼ ≈⊥ 30cm (12in)

Primula alpicola
Originally from Tibet, this moisture-loving primula flowers in midsummer with fragrant, white, yellow or violet, tubular blooms on whitish stems. The deciduous leaves are mid-green and have toothed or scalloped margins. Plant in a bog garden or in soil that stays reliably damp.

↕50cm (20in) ↔30cm (12in) ✿ ✿ ✿ H7 ☼ ∼

Primula beesiana
A semi-evergreen candelabra primula, *P. beesiana* has vivid magenta flowers in summer. The spherical flowerheads appear at intervals up greenish-white stems, giving rise to the plant's common name. Plant in a boggy border, or at a pond edge, in large groups with ferns to create a colourful, textured display.

↕60cm (24in) ↔60cm (24in) ✿ ✿ ✿ H6 ☀ ∼

Primula 'Inverewe'
In summer, up to 15 bright red flowers appear on each white stem on this semi-evergreen candelabra primula. The mid-green leaves are oval with toothed margins. The plant is a vigorous grower that prefers partial shade, but will tolerate full sun as long as the roots are kept moist.

↕75cm (30in) ↔60cm (24in) ✿ ✿ ✿ H5 ☼ ☀ ∼

Rheum palmatum 'Atrosanguineum'
This ornamental rhubarb needs a large garden to accommodate its metre-long, toothed leaves and huge plumes of cerise-pink summer flowers. The young leaves are purple, but fade to green as they age. The soil has to be deep, moist, and very fertile to sustain healthy growth.

↕to 2.5m (8ft) ↔to 1.8m (6ft) ✿ ✿ ✿ H6 ☼ ☀ ∼

Rodgersia pinnata 'Superba'
Grown for its foliage, the young, purplish-bronze leaves of this plant mature to dark green with distinctive veins, giving a puckered appearance. From mid- to late summer, clusters of tiny bright pink flowers reach above the leaves, followed by brown seedheads. Protect from cold winds.

↕to 1.2m (4ft) ↔75cm (30in) ✿ ✿ ✿ H6 ☼ ☀ ∼

Sanguisorba canadensis
This is a tall plant that needs to be placed at the back of a bog garden or moist border. It produces lush green foliage on branching stems, and long, bottlebrush-like spikes of small white flowers, which open from the bottom upwards, in late summer and early autumn. Divide clumps in spring or autumn.

↕to 2m (6ft) ↔1m (3ft) ✿ ✿ ✿ H7 ☼ ☀ ∼

Typha minima
An ideal plant for small ponds or tubs, this perennial has clusters of narrow vertical leaves, which are joined in late summer by cylindrical flower spikes. The flower stalks can be cut and used in indoor arrangements.

↕to 75cm (30in) ↔30–45cm (12–18in) ✿ ✿ ✿ H7 ☼
≈⏁ 30cm (12in)

Zantedeschia aethiopica
One of the most exotic-looking marginal plants, the arum lily brings grace and style to ponds and bog gardens. Large pure white flowers, which gleam against the bright green foliage, open from late spring through to midsummer. Grow in shallow water, dividing the rootstock if necessary in spring.

↕90cm (36in) ↔90cm (36in) ✿ ✿ H4 ☼ ≈⏁ 15cm (6in)

PLANTS FOR BOGGY SOIL

Materials guide

Hard landscaping materials provide the essential structure that every garden needs to create a usable space. As well as their practical functions, walls, paving, fences, and structures also help to shape the overall design, forming a permanent framework for the more ephemeral planting. Factors to consider when choosing materials include their cost, colour range, ease of installation, durability, and their environmental impact – look online for options and check readers' reviews of those you select. This at-a-glance directory shows you what materials are available and their essential properties.

Surfaces

Bricks
Clay bricks are timeless and can be laid in a variety of patterns. The colour range is determined by the clay and the firing; also the higher the temperature (and the cost), the more durable the brick. For paths and patios, bricks must be frostproof and hardwearing; house bricks are not suitable.

£–££ ♦♦ ((reds, buffs, browns, blue/greys

Concrete blocks
In place of bricks you can use less costly concrete blocks, which come in a wide range of sizes, shapes, colours, and textures. You can also buy blocks set on a fabric backing ("carpet stones") or moulded into a slab for easy laying. Concrete blocks can easily take the weight of a car and are ideal for driveways.

£ ♦♦ ((concrete can be dyed almost any colour

Granite setts
Fast disappearing from our city streets, granite setts have great charm and are increasingly available from reclamation yards for use in the garden – where they make a hardwearing surface for paths and drives. Individual setts vary in size and depth, which can make levelling and fitting them together a challenge.

££ ♦♦ ((blue/greys, pink, black

Terracotta tiles
These offer the warmth and colour of the Mediterranean, but most are not frostproof. Their porosity creates a safe, nonslip surface, but makes them vulnerable to staining, so apply a sealant. Available in a huge range of sizes and shapes, the colours are determined by the kiln firing of natural clays.

£–£££ ♦ ((orange, red, mellow yellow

Stone and tiles
You can have some fun with mixed coloured materials – here, granite setts, terracotta and glazed tiles. If you have a handful of expensive tiles, this is a great way to eke them out. Laying the blocks and tiles on a dry mortar mix will help you to adjust the different levels and avoid an uneven surface.

£–£££ ♦♦ ((various

Crazy paving
A 1970s favourite, crazy paving is brought up-to-date by using just one type of stone – here, reclaimed Yorkstone. It makes a hardwearing surface for paths, patios, and drives, although laying a random pattern isn't as easy as it appears and you may need professional help to achieve a decorative mosaic effect.

£–££ ♦♦ ((large range

££££ high cost   ££ medium cost   £ low cost   ◆◆ high durability   ◆ low durability   ⁄⁄⁄ colour options

Granite
A popular stainproof surface for kitchens, polished granite is diamond-hard and tough enough for use in the garden. It comes in a huge range of colours; some also include speckled and streaked detailing. Affordable composite and terrazzo (granite chips bonded with cement and polished) are available.

££–£££ ◆◆ ⁄⁄⁄ black and greens to pinks, reds, cream

Limestone
A sedimentary rock, limestone often has shells and fossils embedded in it. Riven stone (*shown here*) is popular in gardens because it is split in a way that leaves a roughened, nonslip surface. Limestone darkens when it is wet and it can stain, so consider sealing it. Available as composite.

££–£££ ◆◆ ⁄⁄⁄ grey, white, pale red, yellow or black

Marble
More familiar in sunnier climes, marble is increasing in popularity as a sophisticated landscaping material. When polished, it has a lustrous quality that will smarten up any patio. The characteristic veining is caused by mineral impurities. Consider sealing. Available as composite.

££–£££ ◆◆ ⁄⁄⁄ white, black, grey, green, pink, red, brown

Sandstone
Made up of small mineral grains, sandstone is easy to cut and lay. The import market has made available a wide range of colours and patterns, including streaking and stripes. The colour darkens when wet. Reclaimed sandstone paving is a less expensive option. Sealing is advisable. Available as composite.

££–£££ ◆◆ ⁄⁄⁄ gold, jade, rose, brown, grey, white, black

Slate
Stylish and modern, slate is a hardwearing fine-grained stone. Unless polished, it's nonslip, even when wet, making it ideal for pathways. Note the colour darkens when wet. Various surface textures are available, including rough cut (visible saw marks), sandblasted, and polished (called "honed"). Consider sealing.

££–£££ ◆◆ ⁄⁄⁄ black, blue-grey, green, purple

Travertine
Popular as a building material since Roman times, travertine is a dense form of calcium carbonate. Pure travertine is white, but impurities add colour. The characteristic pitting is caused by gases trapped in the molten rock. The best quality travertine has smaller holes that are infilled and polished.

££–£££ ◆◆ ⁄⁄⁄ white, pink, yellow, brown

Yorkstone
Most of Britain's cities are paved with this hardwearing fine-grained sandstone. The colour, which darkens when wet, depends on where it was quarried in Yorkshire. Reclaimed and composite paving slabs with a nonslip, riven surface (*as shown*) are available. Consider sealing.

££–£££ ◆◆ ⁄⁄⁄ grey, black, brown, green or red tinged

"Green" cement
The chance to employ greener, cleaner landscaping materials is an exciting prospect. This type of cement decomposes air pollutants by means of a photocatalytic reaction, and is used to make composite stone. When mixed with recycled granite, it produces a hardwearing surface that helps improve air quality.

££ ◆◆ ⁄⁄⁄ various

ENVIRONMENTAL ISSUES

Our purchasing power as consumers can have a huge impact on the environment, especially when choosing materials for the garden.

- Wood and stone that's been transported halfway around the world has a large carbon footprint, so first check what's available from local quarries. If you do decide to use imported stone, check that it isn't produced by child labourers.

- Soft- and hardwoods should be from a sustainably managed source. Look for accreditation from a recognized authority, such as the Forest Stewardship Council (FSC), or try to use recycled wood. The Greenpeace *Good Wood Guide* will also help you make an informed decision.

- Low-solvent or water-based paints and wood preservatives are a responsible choice.

Surfaces

Patio kit
Used as a centrepiece for a patio or path, this stylized sun comes in kit form ready to fit together like a jigsaw. Other popular designs include fish, butterflies, and geometrical patterns. Usually made from hardwearing moulded composite stone, it can add a decorative note to a patio.

££ ◆◆ /// various stone colours

Flooring kit
Composite stone flooring kits allow you to experiment with different textures, while maintaining uniformity of colour and material. What looks like a complex pattern of blocks, cobbles, and slivers of stone is, in fact, a much simpler collection of moulded slabs, which are quick and easy to lay.

£–££ ◆◆ /// various stone colours

Metal grille
Parallel steel tracks (one shown here) follow the route of car tyres on a driveway, creating a modern, strong, safe surface for parking; when the car is not there, the ground cover beneath is revealed. Commission a specialist blacksmith or metalworker to make a similar grille to suit your needs.

£££ ◆◆ /// shiny metallic

Wooden decking tiles
Choose decking tiles with battens attached on the underside and lay them straight on to a level concrete or asphalt surface. Made from softwood, they are lightweight and ideal for roof terraces, balconies, and patios. When they start to wear, just lift the damaged squares and replace like carpet tiles.

£ ◆ /// oil or stain tiles

Wooden decking
Hardwoods, such as balau (shown) and oak, are a popular choice for decks. They are warp- and weather-resistant and more durable than softwoods. Most decks, however, are made from pressure-treated softwoods, which are less costly and also available as kits. If well maintained, they should last 20 years.

£–£££ ◆◆ /// oil or stain

Plastic decking
Made from recycled waste, plastic decking is weatherproof, UV stable, rot-proof, and low maintenance. Construction is the same as when using wood, the difference is in the aftercare. It needs no oiling or re-treating, just an occasional hose down. There is a good range of colour and texture.

£–££ ◆◆ /// "natural" wood, green, black, blue

Wooden sleepers
Old railway sleepers are no longer available; saturated in creosote and bitumen, they are now considered a health risk. You can buy untreated timber lookalikes (often oak) that are just as heavy to lift and as hard to cut – you will need a chain saw. Good for stepping stones, but slippery when wet.

£–££ ◆◆ /// natural wood, could be stained

Concrete sleepers
Made from cast concrete, these composite sleepers are amazingly realistic and very hardwearing. They come in varying lengths (minimizing cutting) but, like paving slabs, the depth is consistent, making them easy to lay on a bed of mortar. The wood-grain pattern provides a sure grip in the wet.

££ ◆◆ /// "natural" wood

Bark
Bark provides a springy surface for paths and play areas. Fine shredded is kinder on children's knees, but will break down and need replacing more frequently than coarse chipped bark. You can lay it directly on soil (it acts as a soil improver), but for best results, spread it over a weed-suppressing membrane.

£ ◆ /// usually brown; dyed chips are also available

£££ high cost   ££ medium cost   £ low cost   ◆◆ high durability   ◆ low durability   /// colour options

Gravel
Gravel comes in a wide range of colours and sizes and is a tough, quick-to-lay surface for paths and drives. Spread in a thick layer over a weed-suppressing membrane, or, to stop it spilling everywhere, use a honeycomb gravel containment mat. Guests – welcome or not – are announced by loud crunching.

£ ◆◆ /// wide range of stone colours

Cobblestones
Laying a cobblestone path – whether patterned or plain – is a painstaking exercise, but, if you have the patience, the result is worth the effort. Set the cobbles on a bed of mortar, then brush a dry mortar mix into the joints for a hardwearing surface. Use only smooth rounded stones; others are hard to walk on.

£ ◆◆ /// white, creams, greys, blacks, browns

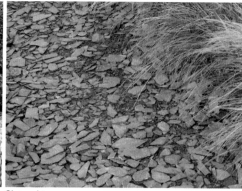

Slate chips
If you use slate chips on a well-trodden path, they will crack and slowly break down. Renewing them every few years, however, is a small price to pay for the beautiful colour that provides a foil for edging plants. Lay over a weed-suppressing membrane. Sharp pieces of slate are not child- or pet-friendly.

£ ◆ /// grey with green, blue, purple, or plum tones

Paddlestones
Usually large pieces of slate, paddlestones are tumbled to round off the edges. In Japanese-style gardens they are used as decorative paths designed to resemble a winding river bed. Smooth and flat, they are fairly easy to walk on, but they are best reserved for areas of light traffic.

££ ◆◆ /// grey with green, blue, purple, or plum tones

Self-binding gravel
Soil and small stone particles are usually washed off gravel, but in this form they are retained and help bind the gravel together to form a more solid surface. Tamp down a thick layer over a solid bed of hardcore to form a hardwearing surface that is easy to walk on.

£–££ ◆◆ /// grey, gold, plum, red, green

Decorative shell
Shells are much too fragile to walk on, and should only be used as decorative surfaces. They are a waste product from the shellfish industry, and have a lovely light-reflective quality. Lay them over a weed-suppressing membrane and use them in Mediterranean-style or seaside gardens as a foil for plants.

£ ◆ /// cream, grey, pink, soft brown

Shredded rubber
As a decorative mulch, shredded rubber can look quite chic. Its spongy quality also makes it ideal for play surfaces, but it does have quite a distinctive odour (that deters cats) and is therefore unsuitable for areas close to seating and dining tables. It does not rot, so won't need replacing.

£ ◆◆ /// grey-black

Glass pebbles
These glass pebbles form a colourful, light-reflective surface, but, be warned, they are easy to slip on when wet and should only be used as a decorative detail on paths or patios. Lay them on a bed of mortar, brushing a dry mortar mix into the joints. Hose them down occasionally to retain their lustre.

£–£££ ◆◆ /// various

Coloured aggregates
Usually made from glass fragments that have been tumbled to remove the razor-sharp edges, aggregates can be used between plants, or for secondary paths – they are not suitable for play areas. Lay the aggregate over a weed-suppressing membrane and hose down occasionally to refresh the colours.

£ ◆◆ /// various

Walls and railings

Brick
Acting like a storage heater, brick walls absorb the sun's heat during the day and release it at night to create a mild microclimate. While walls make a garden feel protected, permeable screens are actually better at filtering winds (*see p.57*). Brick is cheaper than stone and just as durable.

£ ♦♦ *℮* yellow, red, blue-grey, mottled

Weathered stone
Structures made from aged and weathered natural stone look particularly effective in the gardens of period homes, especially when they match the house walls. Stone that has to be worked or shaped for a wall will add to the cost. Reconstituted (or composite) stone made from concrete is a more affordable option.

£–££ ♦♦ *℮* various natural stone colours

Mortared stone
Rough-hewn stone forms a structure that is as much a work of art as it is a wall. "Gluing" it together with mortar makes it easier to build than a dry stone wall, where each stone has to fit neatly within a specific space. Top with coping stones and point between the joints to prevent water and frost damage.

££ ♦♦ *℮* various natural stone colours

Dry stone wall
The materials (a tonne of stone per cubic metre), skill and time required to build a dry stone wall make it an expensive, though beautiful, option. Two parallel walls, built on foundation stones, are bound together with an infill of rubble; the meticulous placement of the stones negates the need for mortar.

££–£££ ♦♦ *℮* various stone colours

Gabion
Rocks, cobbles, bricks or tiles crammed into metal gabions, which are then wired together, create an instant, fairly inexpensive "dry stone" wall. The weight and strength of the filled cages makes them ideal for retaining as well as decorative walls. Gabions come in various sizes.

£ ♦♦ *℮* grey metal; depends on the filling

Knapped flint
Popular as a building material, flint is a tough silica that forms as "nodules" in chalk beds. Here, the flints have been "knapped", ie split in half, and set in lime putty (which retains a degree of flexibility and is resistant to cracking) to form a decorative facing on a brick or block wall.

££–£££ ♦♦ *℮* black and white

Mosaic wall
A mixture of terracotta and glazed tiles, cobbles, setts, and bricks, this wall is both colourful and tactile. In practical terms, the materials are set into a layer of rendering (a mix of cement and sand) covering a brick or block wall. For a neat finish, smooth out the pointing in between each piece.

£–££ ♦♦ *℮* as colourful as you wish to make it

Screen wall
Concrete blocks offer the strength of brick without cutting out the light. Prices are similar, too, but walls made from blocks are quicker to build. Use them for low patio walls, or to top an existing wall, adding extra height and privacy. Their open structure makes them effective windbreaks.

£ ♦♦ *℮* cement grey unless you paint them

Shell mosaic
Mosaics are a weatherproof decoration for the garden. Here, a low retaining wall has been brightened up with a collection of shells, fossils, and stones. The pieces are set into a thin skim of still-damp render (cement and sand). Once dried, a coat of water-based varnish helps protect the mosaic.

£ ♦ *℮* various, depending on the materials used

Shuttered concrete

For a textured finish, concrete is poured into moulds made from timber shuttering. Walls taller than knee height need foundations and steel reinforcement rods for strength. Red sand in the concrete mix gives a buff colour; yellow sand the usual grey; for stronger colours, use concrete dyes or paint.

£ ◆◆ // buff or grey; various if using dyes or paint

Rendered walls

Applying a skim of render (a mix of cement and sand) is a relatively quick – and inexpensive – way to tidy up rough block walls or crumbling brick. Once dry, you have a smooth blank canvas for applying exterior masonry paints. These come in a range of colours, from subtle to shocking – like this pink.

£ ◆◆ // various

Glass panels

Surrounding a patio, balcony or raised deck with glass panels provides a degree of shelter without blocking the view. For safety and strength, use toughened glass fixed to sturdy posts. Treat the glass with a silicon-based rain-repellent coating to make it easier to clean and to prevent smears.

££ ◆ // clear

Aluminium panels

Hide an ugly fence or view and provide an unfussy backdrop for planting with powder-coated aluminium panels. The coating is fade- and flake-resistant. At night, treat them like a projectionist's screen, creating shadow play with spotlights. For a cheaper option, paint sheets of marine ply.

£–££ ◆–◆◆ // various

Wooden block wall

Building a wall using random materials is a skilful job; like a 3-D jigsaw puzzle, each piece must fit neatly with its neighbour. Here, cedarwood offcuts and squares of rusted steel have been glued and screwed together and mounted on a sheet of marine ply, which, in turn, is fixed to a solid wall.

£–££ ◆–◆◆ // various

Wooden pallets

Use pallets to make a "wildlife wall", wiring them together and packing the gaps with moss, wool, and grass (nesting material for birds), and crocks, rotting wood, and hollow canes (homes for insects and amphibians). Usually made from pine, better quality pallets are available from specialist suppliers.

£ ◆ // natural wood shades

Corrugated iron

A maintenance-free fencing option, corrugated iron has one drawback – sharp edges. To cover these, use protective metal edging strips, and fix panels to sturdy posts to hold them steady in gusting winds. Galvanized metal (*shown here*) has a matt finish, while metal paints can add a splash of colour.

£–££ ◆◆ // metallic grey or, if painted, various

Iron railings

Off-the-peg cast-iron railings make an attractive divider in a garden. After a few years, however, they will need repainting. While "no-paint", plastic-coated metal seems a good idea, the coating eventually becomes brittle and chips off, allowing rust to get a hold.

£–££ ◆◆ // usually black or dark green

Bespoke ironwork

Many blacksmiths specialize in decorative metal work – this whimsical fence made from steel horseshoes is a bespoke commission. The shoes, which are mounted on horizontal metal bars, are painted to protect against rust and make an eye-catching feature, as well as a functional boundary.

££–£££ ◆◆ // usually black, especially if wrought iron

Screens and gates

Shiplap
This is one of the cheapest and most popular ready-made fencing options, though not the most durable. Even though the panels are pre-treated, it is best to apply a preservative every few years. The larch strips often warp, leaving small gaps. Available in standard fence panel sizes.

£ ♦ *often pre-stained orange, but will tone down*

Featheredge
Ready-made panels come in various sizes, but the design (vertical softwood timbers nailed on to horizontal rails at the top and bottom) makes it easy to construct. If fixed to strong post supports, the sturdy panels are good for boundaries. Best given a coat of preservative every few years, even if pre-treated.

£ ♦♦ *often pre-stained orange, but will tone down*

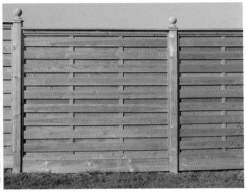

Hit and miss
While offering privacy, the alternating panels of hit and miss fencing are wind permeable, making it ideal for exposed sites. Attached to sturdy posts, it is unlikely to blow down, and the wood strips (fixed vertically or horizontally) are easy to replace. Buy ready-made or construct panels yourself.

£ ♦♦ *often pre-stained orange, but will tone down*

Chevron panel
Decorative panels are not usually strong enough for use as a boundary fence, but this chevron design, a variation on the sturdy hit and miss (*see above right*), is suitable. It is also ideal for dividing up a garden into rooms, or screening an ugly view, perhaps where the compost bins are stored.

££ ♦♦ *usually stained a subtle tan*

Trellis panel
Another hit and miss variant, but this time with an inset of trellis down the centre. It would make a good windbreak, but the lack of privacy could be a problem for a boundary. One way to mask the gaps would be to train a climber through the trellis, thereby creating a colourful display of flowers.

££ ♦♦ *usually stained a subtle tan*

Slatted wood
This fence creates a contemporary, durable screen that allows both light and wind to pass through. Use it to divide up the garden or to mask bins or a shed; it also doubles as a plant support for climbers. Paint or a wood stain will help protect the timber, and introduce colour into your design.

£ ♦ *natural wood or painted*

Picket fence
This simple wood fence has rustic charm, yet it also works well with a modern property. Leave it natural, or paint it to match your house or planting scheme. Its open structure and low profile makes it more of a visual boundary than a barrier to keep out unwanted visitors. Available ready-made.

££ ♦ *natural wood or painted*

Oak panel
This made-to-order fence is perfect for a country-style front garden, where you want the world to admire your planting design. The hardwood has a beautiful appearance and is best left unpainted, but a clear oil will preserve its colour; over time, if left untreated, oak develops lovely silver hues.

£–£££ ♦♦ *natural wood*

Chestnut paling
Often seen on farms, this fencing is naturally rot-resistant and perfect for a subtle, rustic barrier between a country garden and the natural landscape beyond. The wood pieces come on a roll and are linked, at the top and bottom, by a double row of twisted wires. This fence is fixed to wood rails for extra strength.

£ ♦♦ *natural wood*

£££ high cost   ££ medium cost   £ low cost   ◆◆ high durability   ◆ low durability   ⁄⁄ colour options

Willow hurdle

Surprisingly robust, willow hurdles make effective windbreaks. They can be woven to order, or are available in standard panel sizes. Willow makes a beautiful backdrop for naturalistic or cottage-style plantings, or fix it to the top of a wall to increase privacy. Protect with linseed oil.

££ ◆ ⁄⁄ golden brown

Willow screen

If you love the look of willow but want a more contemporary look, choose a framed willow screen – it provides a neat yet natural backdrop for planting. Good for privacy around the patio, the screen is clamped into a timber frame for extra strength, but the size range is limited. Treat with linseed oil.

£ ◆ ⁄⁄ golden brown

Bamboo/reed screen

Ideal for when you want an instant screen to block out an ugly view. You could also use it to make a roof for a pergola. For extra strength, attach it to an existing fence – it works especially well on chain-link. It's not suitable for exposed sites, and it will start to deteriorate after a few seasons.

£ ◆ ⁄⁄ soft browns

Formal hedge

While slow growers, such as yew and beech, may take a few years to thicken up, quick-fix conifers require endless cutting. It's tempting to buy established plants for instant results; but young "whips" are cheaper and quickly catch up. Plant thorny *Berberis*, *Pyracantha*, and *Rosa rugosa* to keep out intruders.

£ ◆◆ ⁄⁄ various

Fedge

The backbone of this lovely hedge is a chain-link fence with climbing plants grown through it. Results are not instant, but the low price makes this a good choice for a long boundary in an informal or wildlife garden. Plant a mixture of prickly plants for security, and flowering climbers for colour.

£ ◆◆ ⁄⁄ various

Living willow

Is it an art installation or is it a screen? Both really, and that is the fun of woven willow structures. Plant the young willow "whips" in winter or early spring in a sunny spot, then come summer, you can start weaving. To stop your screen maturing into a forest, prune back to the framework in late winter.

£ ◆ ⁄⁄ golden stems and lush green foliage

Gate in a fence

Choosing a gate that closely matches the fence panels gives a visually unbroken line for a crisp, clean design. If you can, position the gate in a gap between two whole fence panels – reducing the size of some panels, such as featheredge, involves complicated carpentry.

£ ◆ ⁄⁄ natural wood or painted

Bespoke gate

This spiral metal gate was made to order, but there are plenty of lovely designs available off-the-peg. Set between two sturdy steel posts, it makes a beautiful focal point in a country-style hedge. Regularly trim the foliage away from the hinges and the catch.

£–£££ ◆ ⁄⁄ wrought iron, rusting steel, painted metal

Wooden door

An arched wooden door set in a stone or brick wall is a design classic. Peeling paint and rusting fittings will only add to its charm. This door was custom-made to fit the space, but, if you are building a wall from scratch, it is worth checking out the sizes of off-the-peg doors before you start.

££ ◆ ⁄⁄ natural wood or painted

Structures and storage

Contemporary garden room
These garden rooms range from compact, relatively inexpensive structures to luxury state-of-the-art buildings that include the latest technology and equipment. Most are constructed from timber and glass, with heating and cooling systems, and an electricity supply connected to the house mains.

£–£££ ♦–♦♦ ⁄⁄⁄ natural wood, steel, glass

Garden office/studio
Usually made from wood, you can work in peace away from the hubbub of family life in these buildings. Ideal as an art studio, workshop, or home office. For comfort and to protect books, etc, opt for insulation and a heater fitted with a thermostat. Fit blinds, and a lock for security.

££–£££ ♦♦ ⁄⁄⁄ natural wood, painted, or stained

Traditional garden room
Built straight on to the house but surrounded by greenery, garden rooms allow you to enjoy the outdoors whatever the weather. A timber and brick construction with an insulated sheet metal roof makes the room more usable year round than the average glass conservatory, though not so light and airy.

£££ ♦♦ ⁄⁄⁄ brick, stone, rendered walls; stained wood

Colonial-style gazebo
Relatively small, this type of gazebo can be slotted in almost anywhere, eg, next to a pool or surrounded by pots of subtropical plants on a deck. Offers shaded seating for drinks or afternoon tea. Usually wooden with a thatched roof; some designs are more weatherproof with removable slatted sides.

££–£££ ♦ ⁄⁄⁄ wood shades and muted period colours

Garden shed
DIY or off-the-peg, a shed is a must for anyone needing extra storage or space for a hobby. Can be painted or stained a wide range of colours. Sheds made from shiplap (overlapping wood) panels may warp; tongue-and-groove models are more expensive but superior in quality. Usually with a felted roof.

£–££ ♦ ⁄⁄⁄ natural wood, painted, or stained

Green roof
A shed roof may need shoring up with extra timbers for it to take the weight of a planted roof. Before laying the sedum matting and moisture-retentive growing medium you will need to protect the roof with butyl or polythene sheeting. Green roofs provide good insulation and increase biodiversity.

££ ♦♦ ⁄⁄⁄ sedums and other succulents provide colour

Lean-to greenhouse
Space-saving design. Best for south- or west-facing walls, which act like storage heaters releasing warmth at night. Off-the-peg and bespoke models available in wood or aluminium, with glass or polycarbonate (the latter offers good insulation and safety). Cheap tubular steel frame models with plastic covers available.

£–£££ ♦–♦♦ ⁄⁄⁄ white/dark green, cedar, or painted

Obelisk
A sturdy wooden obelisk (this traditional design is topped with a finial) is a feature in its own right, adding extra height to a border as well as providing support for climbers. DIY or buy off-the-peg; they are made from wood or metal (the latter available in more decorative designs).

£–££ ♦–♦♦ ⁄⁄⁄ natural wood, painted, or stained

Willow arch
Easy to construct and adaptable for the smallest garden, use long "rods" of living willow (plant in winter) or buy dried and pre-soak to make them flexible and workable. Push into the ground, weave together, then tie the tops to form an arch. If the willow starts to sprout, prune it back in late winter.

£ ♦ ⁄⁄⁄ natural willow

£££ high cost   *££* medium cost   *£* low cost   ♦♦ high durability   ♦ low durability   ⟋⟋ colour options

Arbour seat
Self-assembly kits range in price and quality. Custom-built and corner models are available. In a sunny spot, the roof provides shade. Trellis sides and/or roof are ideal for scented climbers. Usually constructed in wood, but also available in wrought iron or a wood/metal mix.

£–£££ ♦–♦♦ ⟋⟋ natural wood, painted, or stained

Modern arbour with brazier
This designer piece with a Moorish flavour incorporates bench seating and a metal brazier – perfect for entertaining on summer evenings. Made from wood, the structure is a sculptural focus for a modern or period garden. A canvas awning would provide extra weather proofing.

£££ ♦♦ ⟋⟋ natural wood, painted, or stained

Traditional pergola
Easy to construct for a competent woodworker. Substantial uprights and horizontal supports can carry heavy climbers, such as grapevines, roses and wisteria. Creates dappled shade for a pathway or seating area. Flat-pack timber kits, wrought iron and bespoke models also available.

££–£££ ♦♦ ⟋⟋ natural wood, painted, or stained

Folly
A focal point, especially for period gardens. May be any design, but often hinting at a specific point in history. Examples include mock Gothic ruins, "ancient" stone circles, classical temples, rustic buildings, and grottoes. DIY construction possible, eg, with reclaimed masonry.

£–£££ ♦♦ ⟋⟋ depends on construction materials

Wendy house
From the simplest wooden box to a two-storey chalet with windowboxes, owning a Wendy house is every child's dream. Bespoke; mid-price, off-the-peg self-assembly; and cheaper click-together plastic are available. Ensure the base is stable. Paints and fixings must be child-safe.

£–£££ ♦–♦♦ ⟋⟋ natural wood; child-safe paints/stains

Children's play area
The best play structures are made to order and erected on site. When buying – especially self-assembly – look for the CE (European compliance) symbol and the European Safety Standard BSEN-71 for toys. Needs safe flooring material, ie, at least 15cm (6in) depth of play bark or a bonded-rubber surface.

££–£££ ♦♦ ⟋⟋ natural wood; child-safe paints/stains

Storage/tool box
A spacious mini shed for tools and lawn mowers, garden furniture or bicycles, can be made from panels of larch lap fencing bolted together, or bought ready-made (usually with a felted roof). It only needs to be as high as your tallest tool. Tuck away in a corner and paint green to blend in.

£–££ ♦♦ ⟋⟋ natural wood, painted, or stained

Recycling cupboard
A great way to disguise unsightly wheelie bins and plastic recycling boxes in a front garden. Wide-opening doors give good access. Make yourself or buy ready-made in wood, plastic, trellis screening, or even woven willow. This one has a green roof, further increasing its eco credentials.

£–££ ♦–♦♦ ⟋⟋ paint/stain to blend in or match house

Garden furniture storage
This bench seat opens to reveal a weatherproof box for storing loose cushions, throws and covers from garden furniture. Wood and plastic ready-made models available. Site next to the patio for convenience. Also useful as a toy box or compact tool storage for courtyard gardens.

£–££ ♦♦ ⟋⟋ natural wood, painted, or stained

Containers

Terracotta clay pots
Today's clay pots are mostly machine-moulded rather than hand-thrown, but you can still buy handmade pots from specialist potteries or antique shops. The higher the temperature of the firing, the greater the frost resistance – and cost. Clay is porous, and pots dry out quickly in hot sun.

£–££ ♦ 🍃 soft orange and sandy yellow clay

Terracotta-style trough
Versatile clay can be moulded to almost any shape; but take a good look, could this be plastic? These days it is hard to tell the two apart. While replicating the look of clay, plastic is lighter, frostproof, and usually cheaper. It's also better at keeping compost and plant roots moist during hot dry spells.

£–££ ♦ 🍃 clay colours or, if plastic, a huge colour range

Glazed ceramic
Glazing a clay pot transforms it. During the kiln firing, the glaze melts to coat the pot in a thin layer of glassy material. As a result, the pot becomes stronger, frost- and waterproof, if it is glazed inside and out, and, depending on the glaze, more colourful. Match your pots with planting for a unified display.

£–££ ♦♦ 🍃 huge colour range

Water feature
For water features, eg, bubble fountains and patio ponds, choose pots that are glazed (or at least glazed inside) to minimize water loss. This urn is set on a cobble-covered metal grille over a reservoir; water is pumped up through the drainage hole in the base to overflow back into the tank.

£–££ ♦ 🍃 huge colour range if glazed

Strawberry pot
Hand-thrown or moulded (the cheaper option) clay strawberry pots, with their "balcony" planting shelves, are also ideal for herbs. With this type of pot, big is best as the increased volume of compost prevents the plants drying out too quickly. May not be frostproof. Also available in plastic.

£–££ ♦ 🍃 usually terracotta

Stone urn
Whether empty or planted up, stone urns have a classic, timeless quality. You can find originals in reclamation yards at a price; but composite stone (ie, cast concrete) is a more affordable and widely available option. Stand an urn on a plinth and it instantly becomes a focal point.

££–£££ ♦♦ 🍃 natural stone colours

Cast concrete
Strong and cheap, concrete is a versatile material for making planters, like this rough-cast bowl. Containers made from concrete are available in both contemporary and classic designs, and, because they are very heavy, they make a good choice for top-heavy plants, such as trees and shrubs.

£ ♦♦ 🍃 concrete can be dyed almost any colour

Terrazzo
Hardwearing, easy to clean, and very tactile, terrazzo is the ideal material for contemporary containers. Granite or marble chips are bonded with cement, then polished to create a smooth surface – a technique that has been around since Roman times. Lightweight polyester terrazzo planters are available.

£–££ ♦♦ 🍃 marble and granite greys, white, and black

Weathering steel
Never has rust looked so good. Weathering steel, of which Cor-Ten is the best known brand, is a high-strength steel alloy. It is designed to develop a layer of rust that, ironically, helps to protect the metal underneath. Strong and durable, it is perfect for long-term plantings, and, as here, water features.

£££ ♦♦ 🍃 rusty orange

£££ high cost   ££ medium cost   £ low cost   ◆◆ high durability   ◆ low durability   ⋰⋰ colour options

Powder-coated metal
A much tougher, non-flaking finish than paint, powder coating (a mix of pigments and resin) is baked on to the surface of metal. Available in a huge range of colours and finishes, the coating inhibits rust. To protect the surface, clean with soapy water and a soft dry cloth, and avoid abrasive solvents.

£–££ ◆◆ ⋰⋰ huge colour range

Galvanized metal
The mottled patina of galvanized metal is created by "hot dipping" – a chemical process that coats steel and iron with rust-resistant zinc. Planters come in a range of styles and sizes; most are lightweight and single skinned. In winter, protect plant roots by wrapping the container with bubble plastic.

£ ◆◆ ⋰⋰ mottled matt grey

Lead planter
Lead is a soft, malleable metal that is easy to work. This planter is made from a sheet of lead hammered into shape; the raised pattern is formed by pressing it into a mould. Lead is toxic and shouldn't come into contact with food plants. Glass fibre lead-style planters are a "food-safe" option.

££–£££ ◆ ⋰⋰ grey

Wooden barrel
Traditionally made from oak, the wooden pieces (called staves) are shaped to fit tightly together and held in place with metal hoops. You may be lucky enough to find half wine or whisky barrels; cheaper replicas are also available. Best lined with plastic or butyl, especially if using as a patio pond.

£–££ ◆ ⋰⋰ wood with black metal bands

Wooden trough
Lightweight and insulating in winter, this rustic planter is made from woven hazel twigs set in a timber frame. For longevity, choose pressure-treated timber, and check that the planter is lined with plastic (with drainage holes at the bottom) to prevent compost and water leaking through the sides.

£ ◆ ⋰⋰ natural wood

Versailles planter
Relatively light for the volume of compost they contain, these planters were originally designed for the orange trees at Versailles so they could be brought indoors over winter. Lining them with plastic extends the life of both hard- and softwood planters. Good quality plastic imitations are also available.

£–££ ◆ ⋰⋰ natural wood or, if painted or plastic, various

Old boots
The more holes in the soles, the better the drainage! Fill the boots with compost, packing it firmly into the toe, and plant up. Be warned that even if you have enormous feet, boots still hold relatively little compost and plants are at risk of dehydrating in hot sun, so consider using water-retaining gel.

£ ◆ ⋰⋰ various fashion colours

Recycled kitchenware
Old colanders, chipped teapots, saucepans that have lost their handles – almost any old household vessel has planting potential for a sustainable garden design. Kitchen cupboards are an especially rich hunting ground. You may need to drill holes for drainage or go easy on the watering.

£ ◆ ⋰⋰ depends on your crockery and cookware

Car and lorry tyres
Get extra mileage out of old tyres by giving them a splash of paint and a new lease of life as a raised bed. Place the tyres straight on to the soil and fill with compost (line them first with plastic if you're growing food). The rubber absorbs the sun's heat and warms up the compost for early plantings.

£ ◆◆ ⋰⋰ black (brightened up with a splash of colour)

Understanding hardiness ratings

All plants in the Plant Guide (pp.292–351) have been assigned RHS hardiness ratings, using one of nine categories – H1a to H7 – determined by the lowest temperature range the plant is likely to withstand, along with various other factors, such as the relative exposure of the planting location. These ratings serve as a general guide to growing conditions and should be interpreted according to the table below. Bear in mind, however, that they are guidelines only, and many other factors will affect a plant's overall hardiness.

RATING	TEMPERATURE RANGE	CATEGORY	DEFINITION
H1a	warmer than 15°C (59°F)	Heated greenhouse – tropical	Grow as a house plant or under glass all year round.
H1b	10–15°C (50–59°F)	Heated greenhouse – subtropical	Can be grown outside in summer in hotter, sunny, and sheltered locations, but generally performs better as a house plant or under glass all year round.
H1c	5–10°C (41–50°F)	Heated greenhouse – warm temperate	Can be grown outside in summer throughout most of the UK while daytime temperatures are high enough to promote growth.
H2	1–5°C (34–41°F)	Cool or frost-free greenhouse	Tolerant of low temperatures, but will not survive being frozen. Except in frost-free, inner-city areas or coastal extremities, requires greenhouse conditions in winter. Can be grown outside once risk of frost is over.
H3	-5–1°C (23–34°F)	Unheated greenhouse/ mild winter	Hardy in coastal and relatively mild parts of the UK, except in hard winters and at risk from sudden, early frosts. May be hardy elsewhere with wall shelter or a good microclimate. Can often survive with some artificial protection in winter.
H4	-10–-5°C (14–23°F)	Average winter	Hardy throughout most of the UK apart from inland valleys, at altitude, and central/northerly locations. May suffer foliage damage and stem dieback in harsh winters in cold gardens. Plants in pots are more vulnerable.
H5	-15–-10°C (5–14°F)	Cold winter	Hardy in most places throughout the UK, even in severe winters. May not withstand open or exposed sites or central/northerly locations. Many evergreens will suffer foliage damage and plants in pots will be at increased risk.
H6	-20–-15°C (-4–5°F)	Very cold winter	Hardy in all of the UK and northern Europe. Many plants grown in containers will be damaged unless given some artificial protection in winter.
H7	colder than -20°C (-4°F)	Very hardy	Hardy in the severest European continental climates, including exposed upland locations in the UK.

Suppliers and useful contacts

When ordering decoration or a structure for your garden, it is important to research suppliers and styles carefully, and ensure that you choose someone whose work fits in with your design. Take accurate measurements of your space and plan placement carefully before commissioning a bespoke piece. Request quotes from several suppliers – the list below will provide a starting point for your enquiries – and, before you place your order, check that your chosen designer will deliver directly to you.

BUILDINGS

Breeze House
01538 398488
breezehouse.co.uk

The Caulfield Company
0113 387 3118
caulfieldcompany.co.uk

Contemporary Garden Rooms
contemporarygardenrooms.co.uk
01952 825 630

Crown Pavilions
01491 817 849
crownpavilions.com

Dunster House
01234 272 445
dunsterhouse.co.uk

Garden Affairs
01225 774 566
gardenaffairs.co.uk

The Garden Escape
0800 917 7726
thegardenescape.co.uk

The Garden Office
01296 328 555
thegardenoffice.co.uk

Garden Lodges
0800 043 4821
gardenlodges.co.uk

Green Retreats
01296 325 777
greenretreats.co.uk

Green Studios
01923 205 090
green-studios.com

The Qube
01604 785 786
theqube.co.uk

Room in the Garden
01730 816 881
roominthegarden.co.uk

Riverside Shepherd Huts
01527 821 848
riversideshepherdhuts.co.uk

Scotts of Thrapston
01832 732 366
scottsofthrapston.co.uk

FENCING AND WALLS

Bamboo Supplies Limited
01825 890 041
ukbamboosupplies.com

Elegant Gardens
020 7228 2443
elegantgardens.net

Forest
0333 003 0026
forestgarden.co.uk

Grange
01952 588 088
grange-fencing.com

Jacksons
0800 408 2234
jacksons-fencing.co.uk

FURNITURE DESIGNERS AND SUPPLIERS

Alexander Rose
01444 258 931
alexander-rose.co.uk

Barbed Limited
020 8878 1994
barbed.co.uk

Barlow Tyrie
01376 557600
teak.com

Bramblecrest
bramblecrest.com

Charlie Davidson Studio
00 46 705 494 721
charlie-davidson.com

Cox & Cox
0330 333 2123
coxandcox.co.uk

Design and Landscape
designandlandscape.co.uk

The Garden Furniture Centre Ltd
01564 793 652
gardenfurniturecentre.co.uk

Garpa
01273 486 400
garpa.co.uk

Gloster
00 49 413 128 7530
gloster.com

Go Modern
020 7731 9540
gomodern.co.uk

Green Meadow Furniture Ltd
01386 584918
greenmeadows-s.co.uk

Green Oak Furniture
01635 281786
greenoakfurniture.co.uk

Ingarden
01732 463 409
ingarden.co.uk

Mosaic & Stone
01342 892792
mosaicandstone.co.uk

Myburgh Designs
01428 741 768
myburghdesigns.com

New Dawn Furniture
01243 375535
newdawnfurniture.co.uk

Outer Eden
07961 443 407
outer-eden.co.uk

Panik
01908 307 020
panik-design.com

PJH Designs
01440 788 949
pjhgardenfurniture.com

Riverco Trading
01538 361 393
riverco.co.uk

Sitting Spiritually
01297 443 084
sittingspiritually.co.uk

Tristan Cockerill
07917 320 572
tristancockerill.com

Twentytwentyone
020 7837 1900
twentytwentyone.com

LANDSCAPE MATERIALS

Ashfield Group
01502 528 877
ashfieldgroup.com

Brett
01227 829 000
brett.co.uk

CED Limited
01708 867 237
ced.ltd.uk

Jewson
02476 608 235
jewson.co.uk

Marshalls
0370 120 7474
marshalls.co.uk

Natural Stone
01904 488 605
naturalstone.co.uk

Organicstone
01452 411 991
organicstone.com

Ormiston Wire
020 8569 7287
ormiston-wire.co.uk

Silverland Stone
01932 569 277
silverlandstone.co.uk

Stoneage
020 8362 1666
stoneagearchitectural.com

Stonemarket
0345 302 0603
stonemarket.co.uk

LIGHTING DESIGNERS AND SUPPLIERS

Garden Lighting By Design
0845 601 5763
gardenlightingbydesign.co.uk

Lighting for Gardens
01462 486 777
lightingforgardens.com

Lighting Styles
01780 767 617
lightingstyles.co.uk

Moonlight Design
020 8925 8639
moonlightdesign.co.uk

PLANT SUPPLIERS

Architectural Plants
(specialists in large hardy
and exotic plants)
01798 879 213
architecturalplants.com

Barcham (container tree
specialist)
01353 720 950
barcham.co.uk

Big Plant Nursery
01903 891 466
bigplantnursery.co.uk

Bloms Bulbs
01234 709 099
blomsbulbs.com

Burncoose Nurseries
01209 860 316
burncoose.co.uk

Claire Austin Hardy Plants
(herbaceous perennials
specialist)
01686 670 342
claireaustin-hardyplants.co.uk

Coblands
01452 742 445
coblands.co.uk

Crocus
01344 578 000
crocus.co.uk

David Austin Roses
01902 376 300
davidaustinroses.com

Fibrex Nurseries
01789 720 788
fibrex.co.uk

Hardy's Cottage Garden
Plants
01256 896 533
hardys-plants.co.uk

Hilliers Garden Centres
01794 368 944
hillier.co.uk

Hopleys Plants
01279 842 509
hopleys.co.uk

Kelways
01458 250 521
kelways.co.uk

Knoll Gardens (specialists in
grasses and perennials)
01202 873 931
knollgardens.co.uk

Majestic Trees
01582 843 881
majestictrees.co.uk

Mickfield Hostas
01449 711 576
mickfieldhostas.co.uk

Notcutts
0344 879 4166
notcutts.co.uk

Peter Beales Roses
01953 454 707
classicroses.co.uk

Plantagogo
01270 820 335
plantagogo.com

Raymond Evison Clematis
01481 245 942
raymondevisonclematis.com

Taylors Clematis
01302 700 716
taylorsclematis.co.uk

Tendercare
01895 835 544
tendercare.co.uk

Whitewater Nursery and
Plant Centre
0118 932 6487
whitewaterplantcentre.co.uk

Wisley Plant Centre
01483 211 113
rhs.org.uk/wisleyplantcentre

POTS AND CONTAINERS

Cadix UK
01440 713 704
cadix.co.uk

Iota
01934 522 617
iotagarden.com

Italian Terrace
01284 789 666
italianterrace.co.uk

Original Stone Troughs
0113 2841 184
stonetroughs.co.uk

Urbis Design
01759 373 839
urbisdesign.co.uk

Whichford Pottery
01608 684 416
whichfordpottery.com

The Worm That Turned
0345 605 2505
worm.co.uk

SCULPTORS AND SCULPTURE, AND ORNAMENT SUPPLIERS

After the Antique
01366 327210
aftertheantique.com

Contemporary Chandelier Company
01939 232 652

Chilstone
01892 740 866
chilstone.com

Martin Cook Studio
01494 880 724
martincookstudio.co.uk

Rachel Dein
07986 821 559
racheldein.com

Ian Gill Sculpture
01279 851 113
iangillsculpture.com

David Harber
01235 859 300
davidharber.co.uk

The Garden Gallery
01794 301 144
gardengallery.uk.com

Haddonstone
01604 770 711
haddonstone.co.uk

Matt Maddocks
07717 623 429
maddocks.uk.com

Suzie Marsh
01840 213 468
suziemarshsculpture.co.uk

John O'Connor
07979 522 495
johnoconnorsculptor.co.uk

Hannah Peschar
01306 627269
hannahpescharsculpture.com

Les Botta
00 33 562 085 497
lesbotta.com

Paul Margetts
01562 730 003
forging-ahead.co.uk

Patio & Terrace
07970 906 224
patioandterrace.co.uk

Red Dust Ceramics
01434 344 923
reddustceramics.co.uk

Mark Reed
01760 441 555
markreedsculpture.com

Lily Sawtell
01934 713 380
lilysawtell.com

Michael Speller
07930 480 347
michaelspeller.com

Surrey Sculpture Society
surreysculpture.org.uk

Neil Wilkin
01570 493 061
neilwilkin.com

Johnny Woodford
07770 758 393
johnnywoodford.co.uk

TRELLIS AND WOODEN STRUCTURES

Handspring Design
0114 221 7785
handspringdesign.co.uk

Stuart Garden Architecture
01984 667 458
stuartgarden.com

WATER GARDENING

Dorset Water Lily Company
01935 891 668
dorsetwaterlily.co.uk

Paul Dyer
0800 919 833
waterfeatures.co.uk

Fairwater
01903 892 228
fairwater.co.uk

Lilies Water Gardens
01306 631 064
lilieswatergardens.co.uk

Penlan Perennials
01239 842 260
penlanperennials.co.uk

Wasserpflanzenkulturen
Eberhard Schuster
00 49 386 322 2705
seerosenforum.de

World of Water
01580 243333
worldofwater.com

USEFUL CONTACTS

Association of Professional Landscapers (APL)
0118 930 3132
www.landscaper.org.uk

British Association of Landscape Industries (BALI)
0247 669 0333
www.bali.org.uk

Institution of Lighting Engineers
01788 576492
www.ile.org.uk

The Landscape Institute (LI)
020 7299 4500
www.landscapeinstitute.org

Royal Institute of Chartered Surveyors (RICS)
0870 3331600
www.rics.org

Society of Garden Designers (SGD)
01989 566695
www.sgd.org.uk

Designers' details

The publisher would like to thank the following garden designers for their contributions:

Acres Wild
01403 891 084
acreswild.co.uk

Marcus Barnett
020 7736 9761
marcusbarnett.com

Jinny Blom
020 7253 2100
jinnyblom.com

Declan Buckley
020 7359 9076
buckleydesignassociates.com

Maurice Butcher
01428 712 362
burlingtongardendesign.com

George Carter
01362 668 130
georgecartergardens.co.uk

Tommaso del Buono and Paul Gazerwitz
020 7613 1122
delbuono-gazerwitz.co.uk

Nicholas Dexter
07947 600 4394
ndg.de.com

Vladimir Djurovic
00 96 1486 2444
vladimirdjurovic.com

Prof. Nigel Dunnett
n.dunnett@sheffield.ac.uk
nigeldunnett.com

Andrew Fisher Tomlin and Dan Bowyer
020 8542 0683
andrewfishertomlin.com

Adam Frost
01780 740 531
adamfrost.co.uk

Annie Guilfoyle
01730 812 943
annieguilfoyle.com

Bunny Guinness
01780 782 518
bunnyguinness.com

Stephen Hall (Giles Landscapes)
01354 610 453
gileslandscapes.co.uk

Paul Hervey-Brookes
0121 629 7797
paulherveybrookes.com

Tony Heywood and Alison Condie
020 7723 0543
heywoodandcondie.com

Kazuyuki Ishihara
00 81 036 690 8787
kaza-hana.jp

Sam Joyce (The Galium Garden)
01291 621 767
thegaliumgarden.co.uk

Maggie Judycki (Green Themes, Inc.)
00 1 703 323 1046
greenthemes.com

Raymond Jungles
00 1 305 858 6777
raymondjungles.com

Arabella Lennox-Boyd
020 7931 9995
arabellalennoxboyd.com

Catherine MacDonald (Landform Consultants)
01276 856 145
landformconsultants.co.uk

Paul Martin
paulmartindesigns.com

Steve Martino
00 1 602 957 6150
stevemartino.net

Claire Mee
020 7385 8614
clairemee.co.uk

Ian Kitson
07742 301 799
iankitson.com

Philip Nixon
01451 828 282
philipnixondesign.com

Piet Oudolf
00 31 314 381 120
oudolf.com

Gabriella Pape and Isabelle van Groeningen
00 49 308 322 0900
koenigliche-gartenakademie.de

Christine Parsons (Hallam Garden Design)
0114 230 2540
hallamgardendesign.co.uk

Pip Probert (Outer Spaces Landscape and Garden Design)
0151 346 2224
outerspaces.org.uk

Sara Jane Rothwell (London Garden Designer)
07976 155 282
londongardendesigner.com

Charlotte Rowe
020 7602 0660
charlotterowe.com

Martin Royer
023 8025 1595
martinroyer.co.uk

Studio Lasso/Haruko Seki
studiolasso.co.uk

Andy Sturgeon
01273 672 575
andysturgeon.com

Jo Thompson
020 7127 8438
jothompson-garden-design.co.uk

Renata Tilli
00 55 115 095 3300

Bernard Trainor
00 1 831 655 1414
bernardtrainor.com

Cleve West
020 8977 3522
clevewest.com

Nick Williams-Ellis
01386 700 883
nickwilliamsellis.co.uk

Ruth Wilmott
020 8742 0849
ruthwillmott.com

Andrew Wilson
020 3002 6601
wmstudio.co.uk

Acknowledgements

The publisher would like to thank the following for their kind permission to reproduce their photographs:

(Key: a-above; b-below/bottom; c-centre; f-far; l-left; r-right; t-top)

2 DK Images: Peter Anderson/ Design: Cleve West, RHS Chelsea Flower Show 2011.

4 The RHS Images Collection: RHS/Neil Hepworth, design Pip Probert.

4–5 DK Images: Brian North/ Design: Catherine MacDonald (t); **The RHS Images Collection:** RHS/Sarah Cuttle, design John Warland (c); **DK Images:** Peter Anderson/Design: Heather Culpan and Nicola Reed (b).

6 The RHS Images Collection: RHS/Neil Hepworth, design: Adam Frost, RHS Chelsea Flower Show 2015.

8 The Garden Collection: Jonathan Buckley/Design: Judy Pearce (bl); **The RHS Images Collection:** RHS/Neil Hepworth (br).

9 The Garden Collection: Derek Harris (bl); Torie Chugg/RHS Chelsea 2008 (br); **Harpur Garden Library:** Jerry Harpur/ Design: Amir Schlezinger (cb).

10–11 GAP Photos: Andrea Jones/Design: Joe Swift and The Plant Room (b); Tim Gainey (t).

12 The Garden Collection: Andrew Lawson/Design: Jinny Blom (tl); **MMGI:** Marianne Majerus/Design: Sara Jane Rothwell (tr); **Photolibrary:** David Cavagnaro (bl).

13 Harpur Garden Library: Jerry Harpur/Design: Shunmyo Masuno (tl); **MMGI:** Marianne Majerus/Palazzo Cappello, Venice (bl); **Photolibrary:** Michael Howes (br); **Richard Felber:** Design: Raymond Jungles Landscape Architect (tr).

14 Charles Hawes: "Artificial Paradise". Design: Catherine Baas & Jean-Francis Delhay (France), Chaumont International Gardens Festival 2003 (tl); **MMGI:** Marianne Majerus/Claire Mee Designs (br); Marianne Majerus/Design: Andy Sturgeon, RHS Chelsea 2006 (tr); Marianne Majerus/ Design: Charlotte Rowe (bl).

15 The Garden Collection: Liz Eddison (tr); **DK Images:** Peter Anderson/RHS Chelsea Flower Show 2009 (tl); **Photolibrary:** Michael Howes/Design: Dean Herald, Flemings Nurseries, RHS Chelsea 2006 (br).

16 The Interior Archive: Simon Upton (tr); **MMGI:** Bennet Smith/Design: Mary Nuttall (tl); Marianne Majerus/Henstead Exotic Garden/Andrew Brogan, Jason Payne (tc); **Photolibrary:** John Ferro Sims (br); **Richard Felber:** Design: Raymond Jungles Landscape Architect (bc).

17 Helen Fickling: Design: Williams, Asselin, Ackaqui et Associés/International Flora, Montreal (br); **Charles Hawes:** Design: Laureline Salisch & Seun-Young Song, Ecole Supérieure d'Art et de Design (ESAD) Reims, Chaumont International Festival 2007 (tr); **MMGI:** Marianne Majerus/ Design: Arabella Lennox-Boyd, RHS Chelsea 2008 (tl); Marianne Majerus/Design: Charlotte Rowe (tc); **Clive Nichols:** Data Nature Associates (bl); Design: Stephen Woodhams (bc).

18 MMGI: Marianne Majerus/ Design: Will Giles, The Exotic Garden, Norwich (tr); **Photolibrary:** Linda Burgess (tl).

19 MMGI: Bennet Smith/Design: Denise Preston, Leeds City Council, RHS Chelsea 2008 (tl); **Undine Prohl:** Dry Design (tr).

20 GAP Photos: Nicola Stocken/ Design: Andy Sturgeon.

21 GAP Photos: Jerry Harpur/ Design: Scenic Blue, RHS Chelsea 2007 (t).

22 The Garden Collection: Nicola Stocken Tomkins (l).

MMGI: Marianne Majerus, Design: Sara Jane Rothwell (t).

24 Alamy Images: CW Images (tl); **DK Images:** Alex Robinson (tr); **GAP Photos:** John Glover (cl); **DK Images:** Peter Anderson/Design: Kati Crome and Maggie Hughes, RHS Chelsea Flower Show 2013 (cfr); **DK Images:** Jon Spaull (bl); **MMGI:** Marianne Majerus/ Kingstone Cottages (br).

25 Jason Liske: www.redwood design.com/Design: Bernard Trainor (tr); **GAP Photos:** Elke Borkowski/Design: Adam Woolcott (cr); Clive Nichols (cl); **MMGI:** Marianne Majerus/Claire Mee Designs (fbr); Marianne Majerus/Design: Bunny Guinness (b).

26–27 The Garden Collection: Jonathan Buckley/Design: Diarmuid Gavin.

27 Design: Amanda Yorwerth.

28 The Garden Collection: Derek St Romaine/Design: Phil Nash (r); **MMGI:** Marianne Majerus/Design: Laara Copley-Smith (c); Marianne Majerus/ Palazzo Cappello, Malipiero, Barnabo, Venice (l).

29 DK Images: Design: Sarah Eberle, RHS Chelsea 2007 (tl); **MMGI:** Marianne Majerus/ Design: Lynne Marcus (cl).

30–31 GAP Photos: Andrea Jones/Design: Adam Frost, Sponsor: Homebase

32–33 Case-study: Design: Fran Coulter, Owners: Jo & Paul Kelly.

33 The Garden Collection: Liz Eddison/Design: Kay Yamada, RHS Chelsea 2003 (br); **Harpur Garden Library:** Marcus Harpur/Design: Justin Greer (fbr); **MMGI:** Marianne Majerus/ Design: Jessica Duncan (cr); Marianne Majerus/Design: Wendy Booth, Leslie Howell (ftr).

34 MMGI: Marianne Majerus/ Claire Mee Design (t); Marianne Majerus/Design: Lynne Marcus, John Hall (b).

34–35 Marion Brenner: Design: Andrea Cochran Landscape Architecture.

35 Jason Liske: www. redwooddesign.com/Design: Bernard Trainor (tr).

36 Nicola Browne: Design: Jinny Blom (br); **DK Images:** Design: Graduates of the Pickard School of Garden Design (cl).

36–37 Harpur Garden Library: Jerry Harpur/Architect: Piet Boon, Planting Design: Piet Oudolf.

37 DK Images: Dwesign: Paul Williams (bl); **The Garden Collection:** Gary Rogers/ Chatsworth House (br); **Charles Hawes:** Designed & created by Tony Ridler, The Ridler Garden, Swansea, Ammonite sculpture by Darren Yeadon (ca).

38 MMGI: Bennet Smith/Design: Ian Dexter, RHS Chelsea 2008 (c); Marianne Majerus/Design: Anthony Tuite (b).

38–39 The Garden Collection: Nicola Stocken Tomkins.

39 DK Images: Design: Paul Hensey, RHS Tatton Park 2008 (b); **MMGI:** Marianne Majerus/ Design: Paul Southern (c).

40 Garden Exposures Photo

Library: Andrea Jones/Design: Dan Pearson & Steve Bradley (cl); The Garden Collection: Liz Eddison/Design: Alan Sargent, RHS Chelsea 1999 (bl).

40–41 The Garden Collection: Jonathan Buckley/Design: Joe Swift & Sam Joyce for the Plant Room.

41 Roger Foley: Scott Brinitzer Design Associates (br); MMGI: Marianne Majerus/Design: Paul Cooper (bc).

42 MMGI: Marianne Majerus/Design: Sara Jane Rothwell.

43 Bord Bia: Jacqueline Leenders/Design: Paul Martin (bl); GAP Photos: Lynn Keddie (ca); MMGI: Marianne Majerus/Design: Charlotte Rowe (tl); Marianne Majerus/ Design: Nicola Gammon, www.shootgardening.co.uk (tr); Marianne Majerus/Design: Fiona Lawrenson & Chris Moss (fbr); Derek St Romaine: Design: Koji Ninomiya, RHS Chelsea 2008 (br).

44 DK Images: Peter Anderson/RHS Hampton Court Flower Show 2014 (tr); MMGI: Marianne Majerus/ Fiveways Cottage (cla); Marianne Majerus/Design: Paul Dracott (bl); B & P Perdereau: Design: Yves Gosse de Gorre (br).

45 The Garden Collection: Jonathan Buckley/Design: Diarmuid Gavin (bl); MMGI: Marianne Majerus/ Design: Lynne Marcus (tl); Marianne Majerus/Design: Arabella Lennox-Boyd, RHS Chelsea 2008 (tr); Marianne Majerus/ Design: Chris Perry, Claire Stuckey, Jill Crooks & Roger Price, RHS Chelsea 2005 (br).

46 Harpur Garden Library: Jerry Harpur/Design: Made Wijaya & Priti Paul (bc); Photolibrary: Peter Anderson/Design: Martha Schwartz (br).

47 DK Images: Design: Marcus Barnett & Philip Nixon, RHS Chelsea 2007 (t); The Garden

Collection: Derek Harris (c); MMGI: Marianne Majerus/ Leonards Lee Gardens, West Sussex (b).

48 GAP Photos: Richard Bloom (cr); MMGI: Marianne Majerus/ Design: Ali Ward (bc); Photolibrary: David Dixon (bl).

49 Peter Anderson: (t); GAP Photos: Clive Nichols/Chenies Manor, Bucks (cl); MMGI: Andrew Lawson/Sticky Wicket, Dorset (bc); Marianne Majerus (bl) (br).

50–51 DK Images: Brian North/ Design: Catherine MacDonald, RHS Hampton Court Flower Show 2012

52 Helen Fickling: International Flora, Montreal (tr); Harpur Garden Library: Jerry Harpur/ Design: Jimi Blake, Hunting Brook Gardens (c); MMGI: Marianne Majerus/Design: Julie Toll (bl).

53 GAP Photos: J S Sira/Chenies Manor, Bucks (bc); MMGI: Andrew Lawson/Design: Philip Nash, RHS Chelsea 2008 (fbr); Bennet Smith/Paul Hensey with Knoll Garden, RHS Chelsea 2008 (tl); Marianne Majerus/ Design: Piet Oudolf (ca); Marianne Majerus/Les Métiers du Paysage dans toute leur Excellence, Jardins, Jardins aux Tuileries 2008. Christian Fournet (bl); Clive Nichols: Design: Wendy Smith & Fern Alder, RHS Hampton Court 2004 (cr); Photolibrary: Mark Bolton (tc).

54 (left to right): DK Images; Clive Nichols: Design: Fiona Lawrenson; The Garden Collection: Jonathan Buckley; Forest Garden Ltd: tel: 0844 248 9801 www.forestgarden.co.uk; The Garden Collection: Jonathan Buckley; Photolibrary. Roger Foley: Design: Raymond Jungles Landscape Architect (bc); The Garden Collection: Derek St Romaine/Design: Philip Nash (br); Photolibrary: Marie O'Hara/Design: Andrew Duff (bl).

55 GAP Photos: Rob Whitworth/Design: Mandy Buckland (Greencube Garden and Landscape Design), RHS Hampton Court Palace Flower Show 2010.

56 DK Images: Peter Anderson/ Design: Joe Swift, RHS Chelsea Flower Show 2012.

57 DK Images: Design: Heidi Harvey & Fern Adler, RHS Hampton Court 2007 (t); GAP Photos: J S Sira/Kent Design (b).

58 Alamy Images: Mark Summerfield (bl); DK Images: Design: Phillippa Probert, RHS Tatton Park 2008 (br); Harpur Garden Library: Jerry Harpur/ Design: University College Falmouth Students, RHS Chelsea 2007 (t); Jerry Harpur/East Ruston Old Vicarage, Norfolk (bc).

59 Harpur Garden Library: Jerry Harpur/Design: Julian & Isabel Bannerman (cl); Marcus Harpur/ Design: Kate Gould, RHS Chelsea 2007 (cr); MMGI: Marianne Majerus (bl); Marianne Majerus/Design: Lynne Marcus & John Hall (bc); Marianne Majerus/Design: Michele Osborne (ca); Photolibrary: John Glover (tc); Stephen Wooster (cb).

60 Marion Brenner: Design: Shirley Watts, Alameda CA www.sawattsdesign.com (br); GAP Photos: Michael King/ Ashwood Nurseries (bl); MMGI: Marianne Majerus/Design: Jonathan Baillie (bc); Clive Nichols: Wingwell Nursery, Rutland (tr); Undine Prohl: Design: Ron Wigginton (cr); DK Images: Design: Adam Frost, RHS Chelsea 2007 (c).

61 The Garden Collection: Jonathan Buckley/Design: Diarmuid Gavin (bc); MMGI: Marianne Majerus/Gardens of Gothenburg, Sweden 2008 (tr); Photolibrary: Botanica (br); Howard Rice (bl).

62–63 The RHS Images

Collection: RHS/Neil Hepworth, design Pip Probert, RHS Show Tatton Park 2016.

64 DK Images: Design: Bob Latham, RHS Chelsea 2008 (bl); Design: Del Buono Gazerwitz, RHS Chelsea 2008 (br); Peter Anderson/Design: Harry and David Rich, RHS Chelsea Flower Show 2013 (tl); Harpur Garden Library: Jerry Harpur/Design: Sam Martin, London (ca).

65 GAP Photos: Rob Whitworth/Design: Angela Potter & Ann Robinson (bc); Harpur Garden Library: Jerry Harpur/Design: Philip Nixon (tl); Marcus Harpur/Design: Growing Ambitions, RHS Chelsea 2008 (tr); MMGI: Marianne Majerus/Design: Jilayne Rickards (bl); Marianne Majerus/The Lyde Garden, The Manor House, Bledlow, Bucks (br).

66 DK Images: Design: Paul Dyer, RHS Tatton Park 2008 (br); MMGI: Marianne Majerus/ Design:Peter Chan & Brenda Sacoor (c).

68 DK Images: Design: Helen Derrrin, RHS Hampton Court 2008 (t); www.indian-ocean.co.uk (c); www.outer-eden.co.uk (b).

68–69 The RHS Images Collection: RHS/Neil Hepworth, design Charlie Albone, RHS Chelsea Flower Show 2016.

69 Nicola Browne: Design: Craig Bergman (tc); GAP Photos: Elke Borkowski (cr); MMGI: Marianne Majerus/Design: Diana Yakeley (br); www.wmstudio.co.uk (cl).

70 DK Images: Design: Francesca Cleary & Ian Lawrence, RHS Hampton Court 2007 (tr); Design: Noel Duffy, RHS Hampton Court 2008 (bl); James Merrell (tl); GAP Photos: John Glover/Design:Dan Pearson, RHS Chelsea 1996 (br).

71 DK Images: Brian North/ Design: The Naturally

Fashionable Garden designer NDG+, RHS Chelsea Flower Show 2010 (bl); Design: Philip Nash, RHS Chelsea 2008 (tc); **The Garden Collection:** Torie Chugg/Design: Sue Tymon, RHS Hampton Court 2005 (c); **The Interior Archive:** Fritz von der Schulenburg (tr); **Red Cover:** Karyn Millet (tl); www.dylon.co.uk (br).

72 Nicola Browne: Design: Piet Oudolf (tr); **DK Images:** Design: Sadie May Stowell, RHS Hampton Court 2008 (tl); Design: Sim Flemons & John Warland, RHS Hampton Court 2008 (br); **The Garden Collection:** Nicola Stocken Tomkins/Design: M Hall, Blowzone. RHS Hampton Court 2003 (bl).

73 The RHS Images Collection: RHS/Neil Hepworth, design Chris Beardshaw, RHS Chelsea Flower Show 2016 (t); **Helen Fickling:** Design: May & Watts, Loire Valley Wines, RHS Hampton Court 2003 (c); **MMGI:** Marianne Majerus/Design: Lynne Marcus (bl).

74–75 The RHS Images Collection: RHS/Sarah Cuttle, design Ruth Willmott, RHS Chelsea Flower Show 2015.

76 The Garden Collection: Marie O'Hara (br); Nicola Stocken Tomkins (bc); Steven Wooster/Design: Anthony Paul (tl); **MMGI:** Marianne Majerus/Design: Charlotte Rowe (bl); Marianne Majerus/Design: Lucy Sommers (tr); **Clive Nichols:** Design: Mark Laurence (tc).

77 Nicola Browne: Design: Kristof Swinnen (tl); **The Garden Collection:** Liz Eddison/Design: David MacQueen, Orangebleu, RHS Chelsea 2005 (bc); **Harpur Garden Library:** Marcus Harpur/Design: Charlotte Rowe (br); **Clive Nichols:** Spidergarden.com/RHS Chelsea 2000 (c); **Red Cover:** Kim Sayer (bl); Mike Daines (cra).

78 www.janinepattison.com.

79 (left to right): **Clive Nichols:** Design: Charlotte Rowe; **Helen Fickling:** Claire Mee Designs; **Clive Nichols:** Garden & Security Lighting; **GAP Photos:** Graham Strong. **Photolibrary:** Botanica (bl); **Red Cover:** Ken Hayden (bc); **Shutterstock** (br).

80 DK Images: Peter Anderson/Design: Adele Ford and Susan Willmott, RHS Hampton Court Palace Flower Show 2013.

81 GAP Photos: John Glover (b).

82 GAP Photos: Jerry Harpur (tl); **MMGI:** Marianne Majerus (tc).

83 Brian North: (br); **Photolibrary:** Howard Rice/ Cambridge Botanic Garden (cr).

84 GAP Photos: Elke Borkowski (bc); Jerry Harpur/Design: Julian & Isabel Bannerman (tr); **The Garden Collection:** Derek Harris (tc); **MMGI:** Marianne Majerus/Design: Bunny Guinness (cl).

85 Marion Brenner: Design: Mosaic Gardens, Eugene Oregon.

86 The Garden Collection: Andrew Lawson (tc); Nicola Stocken Tomkins (tr); **MMGI:** Marianne Majerus/Design: Susan Collier (bl); Marianne Majerus/ RHS Wisley/Piet Oudolf (br).

87 The Garden Collection: Andrew Lawson (b); Derek St Romaine/Glen Chantry, Essex (cl); **MMGI:** Marianne Majerus/ Woodpeckers, Warks (tr).

88 GAP Photos: Clive Nichols/ Design: Duncan Heather (br); **MMGI:** Marianne Majerus (bc); Marianne Majerus/Design: Jill Billington & Barbara Hunt. "Flow" Garden, Weir House, Hants (bl).

89 DK Images: Steven Wooster. "Flow Glow" Garden for RHS Chelsea 2002 by Rebecca Phillips, Maria Ornberg & Rebecca Heard (r); **GAP Photos:** Elke Borkowski (l).

90–91 DK Images: Peter Anderson/RHS Chelsea Flower Show 2009.

92 GAP Photos: Elke Borkowski (bl); John Glover (r).

93 DK Images: Design: Tom Stuart-Smith, RHS Chelsea 2008 (tr); **GAP Photos:** Elke Borkowski (br) (tl); J S Sira (cl); S & O (bc).

94 GAP Photos: Geoff du Feu (bl); Jerry Harpur/Design: Isabelle Van Groeningen & Gabriella Pape. RHS Chelsea 2007 (tc); **Clive Nichols:** RHS Wisley (tr).

94–95 GAP Photos: Mark Bolton.

95 GAP Photos: Elke Borkowski (tc) (cr); **Harpur Garden Library:** Jerry Harpur/Design: Beth Chatto (tr); Marcus Harpur/ Writtle College (br).

96 GAP Photos: Jonathan Buckley/Design: John Massey, Ashwood Nurseries (c); **MMGI:** Marianne Majerus/Mere House, Kent (tr); Marianne Majerus/ Ashlie, Suffolk (bl).

97 GAP Photos: Clive Nichols (cl); Elke Borkowski (tl); Jonathan Buckley/Design: Wol & Sue Staines (panel right). **The Garden Collection:** Jonathan Buckley (bc).

99 MMGI: Marianne Majerus/ Design: Declan Buckley (tl); Marianne Majerus/Design: Philip Nash, RHS Chelsea 2008 (tc); Marianne Majerus/Tanglefoot (bl); **Photolibrary:** Howard Rice (tr).

100 Charles Mann.

101 MMGI: Marianne Majerus/ Design: Sally Hull (b).

104 MMGI: Marianne Majerus/ Design: Julie Toll (bl).

105 DK Images: Design: Kate Frey, RHS Chelsea 2007 (t); **MMGI:** Marianne Majerus/ Design: Wendy Booth & Leslie Howell (b).

106–107 The RHS Images Collection: RHS/Sarah Cuttle, design Martin Royer, RHS Hampton Court Palace Flower Show 2016.

108 MMGI: Marianne Majerus/ Design: James Lee (l); Marianne Majerus/P & M Hargreaves, Grafton Cottage, Staffs (c).

109 DK Images: Design: Jason Lock & Chris Deakin, RHS Chelsea 2008 (fbl); **GAP Photos:** Jerry Harpur/Design: Roberto Silva (cla); **The Garden Collection:** Derek St Romaine/ Glen Chantry, Essex (fbr); Nicola Stocken Tomkins (tr); **MMGI:** Marianne Majerus (cb); Marianne Majerus/Design: Charlotte Rowe (clb); **Photolibrary:** Ron Evans (crb).

110 The Garden Collection: Derek Harris/Design: Lindsey Knight (cl); Nicola Stocken Tomkins (br); **Ian Smith:** Design: Acres Wild (bl).

111 Nicola Browne: Design: Jinny Blom (c); **Jason Liske:** www.redwooddesign.com/ Design: Bernard Trainor (bc); **Photolibrary:** Jerry Pavia (t).

112 GAP Photos: Clive Nichols/ Design: Tony Heywood Conceptual Gardens.

113 The Garden Collection: Nicola Stocken Tomkins (t).

118 MMGI: Marianne Majerus/ Design: Charlotte Rowe (l) (c) (r).

121 www.sketchup.com: (br) (bc).

122 DK Images: Design: Heidi Harvey & Fern Adler, RHS Hampton Court 2007 (bc); **MMGI:** Marianne Majerus/ Leonardslee Gardens, West Sussex (br).

123 GAP Photos: Elke Borkowski (c); **MMGI:** Marianne Majerus/Coworth Garden Design (br).

124–125 DK Images: Peter Anderson/Design: Robert

Myers, RHS Chelsea Flower Show 2011.

126 DK Images: Design: Robert Myers, RHS Chelsea 2008 (tr); **The Garden Collection:** Nicola Stocken Tomkins (b); **Charles Mann:** Sally Shoemaker, Phoenix AZ (cr); **MMGI:** Marianne Majerus/Scampston Hall,Yorks/Design: Piet Oudolf (tc); Marianne Majerus/Rectory Farm House, Orwell/Peter Reynolds (c).

127 DK Images: Design: Cleve West, RHS Chelsea 2008 (l).

128 DK Images: Design: Fran Coulter, Owners: Bob & Pat Ring (br); **GAP Photos:** Dave Zubraski (7); Sarah Cuttle (2); **Clive Nichols:** (4).

129 DK Images: Design: Paul Williams (t); Design: Adam Frost (b); **GAP Photos:** Adrian Bloom (1/t); Richard Bloom (5/t) (5/b).

130–131 Garden Exposures Photo Library: Andrea Jones/Design: Jack Merlo, Flemings Nurseries, RHS Chelsea 2005 (b).

132 Alamy Images: Holmes Garden Photos (bl); **The Garden Collection:** Derek St Romaine/Design: Woodford West, RHS Chelsea 2001 (br); **MMGI:** Marianne Majerus/Gainsborough Road, Alastair Howe Architects (bc).

133 Roger Foley: (br); **Harpur Garden Library:** Jerry Harpur/Design: Philip Nixon, RHS Chelsea 2008 (bl); **MMGI:** Marianne Majerus/Design: Jonathan Baille (bc).

134 MMGI: Bennet Smith/Design: Mary Nuttall (bl); Marianne Majerus/ Design: Charlotte Rowe (br).

135 GAP Photos: Lynne Keddie (bl); **Steve Gunther:** Design: Steve Martino (bc); **MMGI:** Marianne Majerus/Gunnebo House, Gardens of Gothenburg Festival, Sweden 2008, Joakim Seiler (br).

136 MMGI: Marianne Majerus/Design: Tom Stuart-Smith, RHS Chelsea 2000.

137 GAP Photos: Brian North (r).

138–139 The RHS Images Collection: RHS/Neil Hepworth, design Charlie Albone, RHS Chelsea Flower Show 2016.

139 The Garden Collection: Design: Tom Stuart-Smith, RHS Chelsea 2005 (4); **Harpur Garden Library:** Jerry Harpur (tl); **Clive Nichols:** Design: Dominique Lafourcade, Provence (l); www.stonemarket.co.uk (5).

140 GAP Photos: Jerry Harpur/Design: L Giubbilei (clb); Jo Whitworth (cla); **MMGI:** Marianne Majerus/Design: Del Buono Gazerwitz (tr); **Photolibrary:** Marijke Heuff (br).

141 Andrew Lawson: Design: Christopher Bradley-Hole (b); **Charles Mann:** Sally Shoemaker, Phoenix AZ (tl); **B & P Perdereau:** Design: Yves Gosse de Gorre (c).

142–143 The RHS Images Collection: RHS/Neil Hepworth, design Tommaso del Buono and Paul Gazerwitz, RHS Chelsea Flower Show 2014.

144 MMGI: Marianne Majerus/Design: Charlotte Rowe (br) (l).

145 The Garden Collection: Andrew Lawson (2/c); **MMGI:** Marianne Majerus (1/b), (2/b), (4/b); Marianne Majerus/Design: George Carter (cb); Marianne Majerus/Port Lympne, Kent (t).

146 MMGI: Marianne Majerus/Mannington Hall, Norfolk.

147 GAP Photos: FhF Greenmedia (r).

148–149 The Garden Collection: Nicola Stocken Tomkins.

149 The Garden Collection: Nicola Stocken Tomkins (3);

Harpur Garden Library: Marcus Harpur/Design: Gertrude Jekyll, Owners: Sir Robert and Lady Clark, Munstead Wood, Surrey (b); **MMGI:** Marianne Majerus/Bryan's Ground, Herefordshire (2);

150 GAP Photos: John Glover/Five Oaks, Sussex (c); John Glover/Design: Rosemary Verey (bl); **Photolibrary:** Juliette Wade (tl).

150–151 Harpur Garden Library: Jerry Harpur.

151 Roger Foley: Design: Oehme van Sweden (tr); **GAP Photos:** Elke Borkowski (cr); John Glover/Design: Fiona Lawrenson (c).

152–153 The RHS Images Collection: RHS/Neil Hepworth, design Jo Thompson, RHS Chelsea Flower Show 2015.

154 The Garden Collection: Liz Eddison/Design: Gabriella Pape & Isabelle Van Groeningen, RHS Chelsea 2007 (br); **Clive Nichols:** (4); **Photolibrary:** Kit Young (l); Tracey Rich (6).

155 Nicola Browne: Design: Jinny Blom (t).

156 Marion Brenner: Design: Roger Warner, Calistoga, California.

157 Alamy Images: LOOK Die Bildagentur der Fotografen GmbH (b); **Marion Brenner:** Design: Bernard Trainor, Monterey, California (t).

158–159 Jason Liske: www.redwooddesign.com/ Design: Bernard Trainor

159 DK Images: Design: Robert Myers, RHS Chelsea 2008 (3); **GAP Photos:** Jerry Harpur/Design: Roja Dove (l); **B & P Perdereau:** Design: Michel Semini (tl); **Photolibrary:** Robert Harding (6).

160 Alamy Images: Roger Cracknell (bl); **Marion Brenner:** Design: Isabelle Greene & Associates, Santa Barbara California (br); **The Garden Collection:** Steven Wooster/Design: Anthony Paul (c); **B & P Perdereau:** Design: Jean Mus (t).

161 Corbis: Pieter Estersohn/Beateworks (cl); **The Garden Collection:** Liz Eddison/Design: Andrew Walker, RHS Tatton Park 2007 (tl); **Jason Liske:** www.redwooddesign.com/ Design: Bernard Trainor (bl) (cr).

162–163 DK Images: Peter Anderson/Design: Cleve West, RHS Chelsea Flower Show 2011.

164 GAP Photos: Janet Johnson (6); Jerry Harpur/Design: Karla Newell (br) (7); Neil Holmes (2); **Photolibrary:** Mark Bolton (5).

165 Ian Smith: Design: Acres Wild (cb); **B & P Perdereau:** Design: Michel Semini (t) (5).

166 The RHS Images Collection: RHS/Neil Hepworth, design Luciano Giubbilei, RHS Chelsea Flower Show 2014.

167 DK Images: Design: Robert Myers, RHS Chelsea 2008; **Harpur Garden Library:** Jerry Harpur/Design: Vladimir Djurovic (b).

168–169 James Silverman: www.jamessilverman.co.uk/Architect: Marcio Kogan, Brazil.

169 Alamy Images: Andrea Jones/Design: Buro Landrast, Floriade (4); Matthew Noble Horticultural/Design: Lizzie Taylor & Dawn Isaac, RHS Chelsea 2005 (2); **DK Images:** Design: Marcus Barnett & Philip Nixon, RHS Chelsea 2007 (1); Design: Denise Preston, RHS Chelsea 2008 (3); Design: Philip Nixon, RHS Chelsea 2008 (5); **Peter Anderson:** (tl).

170 Henk Dijkman: www.puurgroen.nl (tl); **MMGI:** Marianne Majerus/Design: Sara Jane Rothwell (bl).

170–171 Marion Brenner: Design: Joseph Bellomo Architects, Palo Alto CA.

171 Henk Dijkman: www.puurgroen.nl (bc); **Harpur Garden Library:** Jerry Harpur/

Collection: Jane Sebire/Sheffield Botanic Gardens (b).

222–223 GAP Photos: Clive Nichols/Design: Nigel Dunnett and The Landscape Agency.

224 GAP Photos: Jo Whitworth (6); The Garden Collection: Jane Sebire/Design: Nigel Dunnett (br) (4).

225 The Garden Collection: Gary Rogers/Design: Rendel & Dr James Bartons (t) (6); MMGI: Marianne Majerus (l).

226 Clive Nichols: Design: Stephen Woodhams.

227 GAP Photos: Brian North/Design: Jo Penn, RHS Chelsea 2006 (b); MMGI: Marianne Majerus/Design: Ali Ward (t).

228–229 Harpur Garden Library: Jerry Harpur/Design: Philip Nixon.

229 GAP Photos: Clive Nichols/Design: Amir Schlezinger My Landscapes (3); Jerry Harpur/Design: Fiona Lawrenson & Chris Moss (4); Jerry Harpur/Design: Luciano Giubbilei (l); MMGI: Marianne Majerus www.finnstone.com (2); Marianne Majerus/Design: Lucy Sommers (5).

230 Henk Dijkman: www.puurgroen.nl (tr); DK Images: Design: Mark Gregory, RHS Chelsea 2008 (tl); Loupe Images: Ryland, Peters & Small Ltd (bl).

230–231 MMGI: Marianne Majerus/Design: Charlotte Rowe.

231 Harpur Garden Library: Jerry Harpur/Design: Christoph Swinnen, Sint Niklaas, Belgium (b); MMGI: Marianne Majerus/Design: Sara Jane Rothwell (r).

232–233 GAP Photos: J S Sira/Design: Paul Hervey-Brookes, built by Big Fish Landscapes, Sponsor: BrandAlley.

234 GAP Photos: Clive Nichols (l); Harpur Garden Library: Jerry Harpur/Design: Andy Sturgeon, London (br) (2) (4);

Photolibrary: John Glover (3).

235 DK Images: Design: Sam Joyce, Owner: Jacqui Hobson.

236 Andrew Lawson: Design: Arabella Lennox-Boyd.

237 MMGI: Marianne Majerus/Design: Anthony Paul Landscape Design (b).

238 GAP Photos: Jerry Harpur (t).

238–239 Helen Fickling: Design: Andy Sturgeon.

239 DK Images: Steven Wooster (2) (4); GAP Photos: Jerry Harpur/Pashley Manor (3); S & O (6).

240 GAP Photos: John Glover/Design: Penelope Hobhouse (tr); Jerry Harpur/Design: Britte Schoenaic (br); Harpur Garden Library: Jerry Harpur/Design: Christopher Lloyd, Great Dixter (bl); B & P Perdereau: Design: Piet Blankaert (tl).

240–241 Andrew Lawson: Design: Arabella Lennox-Boyd.

241 The Garden Collection: Andrew Lawson/Design: Oehme van Sweden (tr); Harpur Garden Library: Jerry Harpur/Design: Piet Oudolf (r).

242–243 The RHS Images Collection: RHS/Neil Hepworth, design Jo Thompson, RHS Chelsea Flower Show 2016.

244 GAP Photos: Clive Nichols (2); Fiona McLeod (7); Leigh Clapp (6); Richard Bloom (3); MMGI: Marianne Majerus/Design: Piet Oudolf (br).

245 Photolibrary: John Glover (t).

246 The RHS Images Collection: RHS/Sarah Cuttle, design John Warland, RHS Chelsea Flower Show 2016.

247 GAP Photos: Richard Bloom (t); MMGI: Andrew Lawson/Design: Philip Nash, RHS Chelsea 2008 (b).

248 Michael Schultz Landscape Design: (br).

248–249 Harpur Garden Library: Jerry Harpur/Design: Steve Martino.

249 DK Images: Design: Matthew Rideout, RHS Hampton Court 2008 (l); Design: Paul Cooper, RHS Chelsea 2008 (3); GAP Photos: Fiona McLeod/Design: Cleve West, RHS Chelsea 2006 (5); The Garden Collection: Liz Eddison/Design: Reaseheath College, RHS Tatton Park 2007 (6); Harpur Garden Library: Jerry Harpur/Design: Sonny Garcia (4); .

250 Helen Fickling: Design: Marie-Andrée Fortier, Art & Jardins, International Flora, Montreal, Canada (b); Harpur Garden Library: Jerry Harpur/Design: Vladimir Sitta (c).

250–251 Helen Fickling: Architect: Claude Cormier, International Flora, Montreal, Canada (t).

251 Marion Brenner: Design: Andrea Cochran Landscape Architect, San Francisco (c); Harpur Garden Library: Jerry Harpur/Design: Steve Martino (cr); Steve Gunther: Architect: Ricardo Legorreta/Landscape Architect: Mia Lehrer & Associates, LA (br); Harpur Garden Library: Jerry Harpur/Design: Peter Latz & Associates, Chaumont Festival, France (bl).

252–253 The RHS Images Collection: RHS/Neil Hepworth, design Andy Sturgeon, RHS Chelsea Flower Show 2016.

254 MMGI: Marianne Majerus/Design: Paul Cooper (br) (2) (6).

255 Clive Nichols: Tony Heywood Conceptual Gardens (t). DK Images: Peter Anderson/Design: Darren Hawkes, RHS Chelsea Flower Show 2013 (2).

256–257 GAP Photos: Tim Gainey (t).

258 DK Images: Design: Sam Joyce (bc); The Garden Collection: Gary Rogers (br).

260 DK Images: Design: Helen Williams, RHS Hampton Court 2008.

261 GAP Photos: Jerry Harpur (b); Photolibrary: Michele Lamontagne (t).

265 MMGI: Marianne Majerus/Design: Ian Kitson & Julie Toll (br); www.stonemarket.co.uk (bl).

267 GAP Photos: Clive Nichols/Design: Sarah Layton (br).

271 DK Images: Mark Winwood/Courtesy of Capel Manor, Design: Irma Ansell (bl); GAP Photos: Fiona Lea (br); MMGI: Marianne Majerus/Design: Jill Billington & Barbara Hunt. "Flow" Garden, Weir House, Hants (cr).

279 GAP Photos: Clive Nichols/The Parsonage, Worcs. (b).

281 MMGI: Marianne Majerus (br).

282–283 Clive Nichols: Design: Helen Dillon.

285 GAP Photos: Neil Holmes (tr).

287 DK Images: Design: Xa Tollemache.

288 Photoshot: Photos Horticultural (br).

290–291 GAP Photos: Tim Gainey (t).

292 GAP Photos: Rob Whitworth (bl).

293 DK Images: Peter Andreson/ Design: Cleve West, RHS Chelsea Flower Show 2012 (br).

295 Garden World Images: Paul Lane (tl).

296 Garden World Images: Carolyn Jenkins (cl).

297 The Garden Collection: Torie Chugg (c).

299 MMGI: Marianne Majerus/Design:Tom Stuart-Smith (bl).

300 Garden World Images: Nicholas Appleby (bc).

302 MMGI: Marianne Majerus/ Saling Hall, Essex (bl).

304 GAP Photos: Nicola Stocken (bl).

307 Garden World Images: Gilles Delacroix (bl).

308 GAP Photos: Fiona McLeod (bl).

316 www.davidaustinroses.com (c).

318 Garden World Images: Martin Hughes-Jones (cl).

323 DK Images: Roger Smith (tl); The Garden Collection: Nicola Stocken Tomkins (tc).

328 GAP Photos: Clive Nichols (tl); Photolibrary: Kate Gadsby (c).

331 GAP Photos: Neil Holmes (cr); The Garden Collection: Andrew Lawson (c).

332 MMGI: Marianne Majerus (cl).

333 The Garden Collection: Derek Harris (tr).

335 GAP Photos: Visions (ca); Photolibrary: Joan Dear (bl); Sunniva Harte (cr).

337 Alamy Images: Martin Hughes-Jones (ca) (cl).

338 Photolibrary: Mark Bolton (c).

340 GAP Photos: Howard Rice (bc); Photolibrary: Mayer/Le Scanff (br).

342 The Garden Collection: Andrew Lawson (bc).

343 GAP Photos: J S Sira (c).

346 Garden World Images: (bl).

348 GAP Photos: Paul Debois (tl).

352 GAP Photos: Elke Borkowski (bl); Jerry Harpur (br); www.stonemarket.co.uk (tr) (cr).

353 www.stonemarket.co.uk (top row) (bl); www.bradstone.com/garden (c) (cr); www.organicstone.com (bc).

354 DK Images: Design: Martin Thornhill, RHS Tatton Park 2008 (cr); www.stonemarket.co.uk (tl) (tc); Forest Garden Ltd, tel: 0844 248 9801 www.forest garden.co.uk (cl); Images supplied courtesy of Marshalls www.marshalls.co.uk/transform (bc); www.jcgardens.com (br).

355 DK Images: Design: Jane Hudson & Erik de Maejer, RHS Chelsea 2004 (tc); Design: Jon Tilley, RHS Tatton Park 2008 (bl); Design: Martin Thornhill, RHS Tatton Park 2008 (br); GAP Photos: J S Sira (cl); Howard Rice (bc); www.specialistaggregates.com (cr).

356 DK Images: Steven Wooster/Design: Claire Whitehouse, RHS Chelsea 2005 (c); Design: Geoff Whitten (br); GAP Photos: Elke Borkowski (bl); www.bradstone.com/garden (bc); Images supplied courtesy of Marshalls www.marshalls.co.uk/transform (tc).

357 DK Images: Design: Paul Hensey with Knoll Gardens, RHS Chelsea 2008 (c); Design: Toby & Stephanie Hickish, RHS Tatton Park 2008 (bc); Design: Niki Ludlow-Monk, RHS Hampton Court 2008 (br); Design: Ruth Holmes, RHS Hampton Court 2008 (cr); GAP Photos: Leigh Clapp/Design: David Baptiste (bl).

358 DK Images: Design: Helen Williams, RHS Hampton Court 2008 (cr); www.grangefencing.co.uk (tl); www.jacksons-fencing.co.uk (tr); Forest Garden Ltd, tel: 0844 248 9801 www.forestgarden.co.uk (cl) (c); www.kdm.co.uk (bc).

359 GAP Photos: Leigh Clapp (bc); MMGI: Marianne Majerus/Design: Hans Carlier (tr); Forest Garden Ltd, tel: 0844 248 9801 www.forestgarden.co.uk (tc) (bl); www.stonemarket.co.uk (br).

360 DK Images: Brian North/ RHS Hampton Court Palace Flower Show 2010 (tl); Design: Mark Sparrow & Mark Hargreaves, RHS Tatton Park 2008 (bl); Alamy Images: Francisco Martinez (tc); GAP Photos: Jerry Harpur (tr); Photolibrary: John Glover/ Design: Jonathan Baillie (c); www.breezehouse.co.uk (cl); www.cuprinol.co.uk (bc).

361 DK Images: Design: Jackie Knight Landscapes, RHS Tatton Park 2008 (tc); Design: Mark Gregory, RHS Chelsea 2008 (bc); www.garpa.co.uk (br); MMGI: Marianne Majerus/ Design: Earl Hyde, Susan Bennett (cl); Marianne Majerus/ Elton Hall, Herefordshire (c); www.jcgardens.com (cr); www.cuprinol.co.uk (tl) (bl).

362 DK Images: Design: David Gibson, RHS Tatton Park 2008 (cl); Design: Cleve West, RHS Chelsea 2008 (bl); GAP Photos: Elke Borkowski (cr); Jo Whitworth/Design: Tom Stuart-Smith, RHS Chelsea 2006 (br).

363 DK Images: Design: Tim Sharples, RHS Hampton Court 2008; GAP Photos: Tim Gainey (bl); The Garden Collection: Nicola Stocken Tomkins (tr); www.hayesgardenworld.co.uk (cr)

364–365 The RHS Images Collection: RHS/Neil Hepworth, design Charlie Albone, RHS Chelsea Flower Show 2016 (b).

All other images: © Dorling Kindersley

For further information see: www.dkimages.com

Thanks to the following people for allowing us to photograph and feature their gardens:

Zelda and Peter Blackadder, Jacqui Hobson, Jo and Paul Kelly, Bob and Pat Ring, Amanda Yorwerth.

Thanks to the following companies for their help on this project:

Blue Wave 00 45 7322 1414 bluewave.dk

Brandon Hire 0870 514 3391 brandontoolhire.co.uk

Garpa 01273 486 400 garpa.co.uk

Marshalls 0370 120 7474 marshalls.co.uk

Organicstone 01452 411 991 organicstone.com

Ormiston Wire 020 8569 7287 ormiston-wire.co.uk

Stonemarket 0345 302 0603 stonemarket.co.uk

Thanks to Marie Lorimer for indexing.

Thanks to the following DK staff for their work on the original edition of the book:

Senior Editor Zia Allaway
Senior Art Editor Joanne Doran
Airedale Publishing Ruth Prentice, David Murphy, Murdo Culver
Photographers Peter Anderson, Brian North
Illustrators Peter Bull Associates, Richard Lee, | Peter Thomas
Plan Visualizers Joanne Doran, Vicky Read
Managing Editor Anna Kruger
Managing Art Editor Alison Donovan
Publisher Jonathan Metcalf
Associate Publisher Liz Wheeler
Art Director Bryn Walls

Index

About the contributors

Editor-in-Chief

Chris Young is Head of Editorial for the Royal Horticultural Society and Editor of its members' magazine, *The Garden*. He studied landscape architecture at the University of Gloucestershire, England, and was Editor of *Garden Design Journal* (UK), the magazine for members of the Society of Garden Designers, for five years. He has won two Garden Media Guild awards for his writing, and is also author of *Take Chelsea Home* (Mitchell Beazley). Chris enjoys all aspects of gardening and garden making, and is currently working on his new garden on the Northamptonshire/Rutland borders.

Authors

Andi Clevely has worked in gardening for over 50 years and is the best-selling author of *The Allotment Book*, as well as over 20 other titles. He also writes for magazines and has twice been awarded Practical Journalist of the Year by the Garden Media Guild. He lives in mid-Wales, where he tends a wild garden and allotment on a rocky hillside.

Jenny Hendy has a degree in botany and is an author, garden designer, teacher, and presenter. She has written books on a wide range of subjects, including design, planting techniques, and topiary, and writes for the gardening press. She is a regular contributor to BBC local radio and runs gardening workshops for adults and children near her home in North Wales.

Richard Sneesby is a landscape architect, garden designer, and lecturer, based in Cornwall, with over 25 years' experience in the design of private and public landscapes and gardens. He has presented a number of television series, writes regularly for the garden press, and runs workshops for garden and landscape designers.

Paul Williams has spent a lifetime in horticulture, working and designing with plants. Trained at Pershore College of Horticulture, he has used his passion for plants and gardens to build a thriving horticultural consultancy and design practice. He has written several books on plants and gardening, and lectures in the UK and Japan on gardening.

Andrew Wilson is a multi-award-winning garden designer, Director of Garden Design Studies at the London College of Garden Design, co-director of design practice Wilson McWilliam Studio, and a lecturer and respected author. Together with his design partner, Gavin McWilliam, he has won a string of awards for his show gardens, both in the UK and internationally. He is also a Fellow and former Chairman of the Society of Garden Designers.

REVISED EDITION

DK UK
Senior Editor Alastair Laing
Art Editor Anne Fisher
Editor Zia Allaway
Design Assistance Philippa Nash
Picture Research Martin Copeland, Myriam Mégharbi
Cover Design Nicola Powling
Pre-Production Producer Robert Dunn
Producer Luca Bazzoli
Managing Editor Stephanie Farrow
Managing Art Editor Christine Keilty
Art Director Maxine Pedliham
Publishing Director Mary-Clare Jerram

DK INDIA
Project Editor Janashree Singha
Editor Nishtha Kapil
Assistant Editor Devangana Ojha
Managing Editor Soma B. Chowdhury
Managing Art Editor Arunesh Talapatra
Pre-Production Manager Sunil Sharma
Senior DTP Designer Tarun Sharma
DTP Designers Manish Upreti, Umesh Singh Rawat

ROYAL HORTICULTURAL SOCIETY
Image Library Ewan Guilder
Publisher Rae Spencer-Jones
Head of Editorial Chris Young

This edition published in Great Britain in 2017
in association with The Royal Horticultural Society by
Dorling Kindersley Limited, One Embassy Gardens,
8 Viaduct Gardens, London SW11 7BW

The authorised representative in the EEA is Dorling Kindersley
Verlag GmbH, Arnulfstr. 124, 80636 Munich, Germany.

First edition published in Great Britain in 2009
Copyright © 2009, 2013, 2017, Dorling Kindersley Limited
Text copyright © 2009, 2013, 2017, Royal Horticultural Society
and Dorling Kindersley Limited

A Penguin Random House company
10 9 8
018–299174–Sep/2017

A CIP catalogue record for this book
is available from the British Library
ISBN 978-0-2412-8613-5

Printed and bound in Slovakia

A WORLD OF IDEAS:
SEE ALL THERE IS TO KNOW

www.dk.com